ENVIRONMENTAL SCIENCE, ENGINEERING AND TECHNOLOGY SERIES

BUILDINGS AND THE ENVIRONMENT

Environmental Science, Engineering and Technology Series

Nitrous Oxide Emissions Research Progress
Adam I. Sheldon and Edward P. Barnhart (Editors)
2009. ISBN: 978-1-60692-267-5

Fundamentals and Applications of Biosorption Isotherms, Kinetics and Thermodynamics
Yu Liu and Jianlong Wang (Editors)
2009. ISBN: 978-1-60741-169-7

Environmental Effects of Off-Highway Vehicles
Douglas S. Ouren, Christopher Haas, Cynthia P. Melcher, Susan C. Stewart, Phadrea D. Ponds, Natalie R. Sexton, Lucy Burris, Tammy Fancher and Zachary H. Bowen
2009. ISBN: 978-1-60692-936-0

Agricultural Runoff, Coastal Engineering and Flooding
Christopher A. Hudspeth and Timothy E. Reeve (Editors)
2009. ISBN: 978-1-60741-097-3

Agricultural Runoff, Coastal Engineering and Flooding
Christopher A. Hudspeth and Timothy E. Reeve (Editors)
2009. ISBN: 978-1-60876-608-6
(Online book)

Conservation of Natural Resources
Nikolas J. Kudrow (Editor)
2009. ISBN: 978-1-60741-178-9

Conservation of Natural Resources
Nikolas J. Kudrow (Editor)
2009. ISBN: 978-1-60876-642-6
(Online book)

Directory of Conservation Funding Sources for Developing Countries: Conservation Biology, Education and Training, Fellowships and Scholarships
Alfred O. Owino and Joseph O. Oyugi
2009. ISBN: 978-1-60741-367-7

Forest Canopies: Forest Production, Ecosystem Health and Climate Conditions
Jason D. Creighton and Paul J. Roney (Editors)
2009. ISBN: 978-1-60741-457-5

Soil Fertility
Derek P. Lucero and Joseph E. Boggs (Editors)
2009. ISBN: 978-1-60741-466-7

The Amazon Gold Rush and Environmental Mercury Contamination
Daniel Marcos Bonotto and Ene Glória da Silveira
2009. ISBN: 978-1-60741-609-8

Process Engineering in Plant-Based Products
Hongzhang Chen
2009. ISBN: 978-1-60741-962-4

Buildings and the Environment
Jonas Nemecek and Patrik Schulz (Editors)
2009. ISBN: 978-1-60876-128-9

ENVIRONMENTAL SCIENCE, ENGINEERING AND TECHNOLOGY SERIES

BUILDINGS AND THE ENVIRONMENT

JONAS NEMECEK
AND
PATRIK SCHULZ
EDITORS

Nova Science Publishers, Inc.
New York

LIBRARY OF CONGRESS CATALOGING-IN-PUBLICATION DATA

Buildings and the environment / [edited by] Jonas Nemecek and Patrik Schulz.
p. cm.
Includes index.
ISBN 978-1-60876-128-9 (hardcover)
1. Buildings--Environmental engineering. 2. Indoor air pollution--Prevention. 3. Buildings--Energy conservation. 4. Buildings--Environmental aspects. I. Nemecek, Jonas. II. Schulz, Patrik. III. Title.
TH6028.B85 2009
696--dc22

2009028419

Published by Nova Science Publishers, Inc. ✢ New York

CONTENTS

PREFACE

The rise of living standards and the deterioration of thermal conditions in the urban environment are mainly responsible for the increase in the energy consumption in buildings. Low energy design of urban environment and buildings in densely populated areas requires consideration of a wide range of factors, including urban setting, transport planning, energy system design and architectural and engineering details. Furthermore, sustainability has acquired great importance due to the negative impact of various developments on the environment. This book focuses on ways to reduce building energy consumption by designing buildings which are more economical in their use of energy for heating, lighting, cooling, ventilation and hot water supply. Furthermore, people living in developed countries are exposed to many physical and mental stresses caused by the urban environment, including air, water and ground pollution. Various trails concerning the domiciliary environment are discussed to support and promote human health in indoor conditions, such as those involving the use of fewer chemicals in buildings. This book also details the establishment of negatively-charged indoor air conditions and investigate their biological effects. Other chapters in this book include a study of methods on how to reduce the energy consumption in different kinds of of school buildings, findings on a new generation of decision-making tools for estimating the environmental effects on residential property value and methods on how to reduce airborne contamination in hospital buildings.

Chapter 1 - Modelling of fires has contributed significantly to the modern development of fire safety science and the blossoming of the discipline of fire engineering. Suitable numerical simulation tools have effectively taken centre stage for practicing fire engineers to exploit the freedoms offered under the performance-based fire engineering approach. In this article, the authors review the state-of-the-art deterministic fire models based on the approach of field modelling utilizing the Computational Fluid Dynamics (CFD) techniques. Special attention is also directed towards the necessity to adopt relevant models and the historical development in handling the combustion and soot chemistry and radiation heat transfer in fire simulations. The structure of this article consists of three main parts: (i) Historical development of fire models; (ii) Implementation of relevant combustion, radiation and soot sub-models and (iii) Possible aspects of the future development in fire modelling. In the first part, discussion will centre on the revelation of the complex chemo-physical behaviour of building fires and the challenges being faced in numerical modelling. A comprehensive review of the historical development of fire models will then follow by the unveiling of the basic framework of field modelling technique. Based on the framework of field modelling, theoretical concepts and mathematical formulation of combustion, radiation and soot sub-models that have been proposed within literature will be summarized in second part. As a more in-depth demonstration of the necessity and effects of these sub-models, numerical studies devoted to assess the capability of some selected sub-models will be presented in which predicted results have been rigorously validated against full-scale experimental data. These studies show that good agreement can be achieved through the present model with regards to the spatial temperature and velocity distributions. Finally, for the last part, limitations and assumptions taken by the models will be addressed and some suggestions of possible direction for further fire model development are discussed.

Chapter 2 - The use of renewable energy sources is a fundamental factor for a possible energy policy in the future. Taking into account the sustainable character of the majority of renewable energy technologies, they are able to preserve resources and to provide security, diversity of energy supply and services, virtually without environmental impact. Sustainability has acquired great importance due to the negative impact of various developments on environment. The rapid growth during the last decade has been accompanied by active construction, which in some instances neglected the impact on the environment and human activities. Policies to promote the rational use of electric energy and to preserve natural non-renewable resources are of paramount importance. Low energy design of urban environment and buildings in densely populated areas requires consideration of wide range of factors, including urban setting, transport planning, energy system design and architectural and engineering details. The focus of the world's attention on environmental issues in recent years has stimulated response in many countries, which have led to a closer examination of energy conservation strategies for conventional fossil fuels. One way of reducing building energy consumption is to design buildings, which are more economical in their use of energy for heating, lighting, cooling, ventilation and hot water supply. Passive measures, particularly natural or hybrid ventilation rather than air-conditioning, can dramatically reduce primary energy consumption. However, exploitation of renewable energy in buildings and agricultural greenhouses can, also, significantly contribute towards reducing dependency on fossil fuels. Therefore, promoting innovative renewable applications and reinforcing the renewable energy market will contribute to preservation of the ecosystem by reducing emissions at local and global levels. This will also contribute to the amelioration of environmental conditions by replacing conventional fuels with renewable energies that produce no air pollution or greenhouse gases. This article presents review of energy sources, environment and sustainable development. This includes all the renewable energy technologies, energy savings, energy efficiency systems and measures necessary to reduce climate change.

Chapter 3 - Processes related to planning, redesign and services assignment inside buildings, squares, parks, sidewalks or more generally indoor and open environments are usually realized without the aid of a decision support system able to consider pedestrian mobility. Usually engineers, architects and planners handle pedestrian mobility in these environments through design patterns, experience and rough estimations. A decision support system able to manage pedestrian mobility, in all its aspects and consequences, is a precious instrument for these professional Figures, that will rapidly become more and more important in the next years. This importance is due to the ability pedestrian mobility simulation gives to these professionals: the power to predict consequences of their choices. It can be expected that pedestrian mobility simulation will help in the evaluation of key parameters of areas including: level of usage ("How many persons pass through this alley?", "How long people stays in front of this showcase?"), kind of population ("What are the kinds of people staying in this waiting room?"), security ("Are these streets sufficiently lighted?") or safety ("Is this alley sufficiently wide for the fluxes passing in?") etcetera. The importance of these kinds of questions (and obviously of their answers) will reasonably become more important due to the evolution of economic and environmental world conditions:

- private cars will reduce their number (due to costs and pollution);
- collective solutions will augment their numbers (busses, undergrounds, trains, airplanes).

These two simple assertions will cause an augment of pedestrian mobility impact in many environments (parks, roads, buildings, stations etcetera). Holding a suitable instrument for managing pedestrian mobility before the definitive point-break (i.e. the number of existing private car starts to decrease) is important key in order to avoid having in the future additional problems caused by raised density of pedestrians.

This chapter describes, after discussing the evolution of pedestrian mobility modeling, several potentialities of this research field through the Distrimobs simulator. A highlight of the most prominent components of the Distrimobs software architecture is given. It includes contents about environment representation, interactions with other existing software and data sources, behaviors of simulated pedestrians (artificial intelligence) and handling of high computational costs (parallelization).

Finally, some applications of the simulator are presented, including the historic Palazzo Poggi building, the pedestrianization of the medieval center of Bologna, the changes brought by Calatrava's Bridge in Venice and the requalification of the Scalo Farini district in Milano.

Chapter 4 - Soil, water, and air pollution is an important issue for human society. In recent years, water pollution from industrial and household sources is becoming more and more serious and the world is facing clean water crisis. Wastewater treatment and recycling will provide a solution. For water treatment, adsorption and advanced oxidation process are the two important physical and chemical technologies. In this paper, the authors will report an investigation of using these two methods in treatment of dye containing wastewater.

In adsorption, zeolites synthesized from fly ash (FA) have been tested for methylene blue adsorption. One-step hydrothermal and two-step fusion and hydrothermal syntheses were investigated for zeolite synthesis. The effect of aging process during the two-step synthesis was studied. It was found that zeolite P would be formed during the hydrothermal reaction; however, the conversion efficiency depended on the preparation method. One-step hydrothermal treatment could not convert fly ash to zeolite efficiently while the two-step fusion and hydrothermal treatment with aging process would achieve a complete conversion of fly ash to zeolite. Those synthesized zeolites show higher dye adsorption capacity than fly ash itself and the fusion-hydrothermal synthesized zeolite exhibits the highest adsorption capacity. Fly ash will exhibit adsorption capacity of 5×10^{-6} mol/g, while the synthesized zeolite (FA-3) will show adsorption capacity of 9×10^{-5} mol/g.

For advanced oxidation process, a nanosized magnetite (Fe_3O_4) has been employed as a heterogeneous Fenton catalyst for decolorization of methylene blue in aqueous solution. It is found that the nanosized magnetite is quite effective for dye decolorization at low catalyst loading and H_2O_2 concentration. A 100% decolorization of methylene blue can be achieved within 30 min at the conditions of 2 g/L magnetite and 300 μM H_2O_2. The catalyst exhibits remarkable stability of performance and low metal loss. The maximum concentration of Fe in solution after 3 h oxidation is less than 1.9 mg/L. Life cycle tests also indicate that no decrease in decolorization efficiency was observed after 6 round tests.

Chapter 5 - Natural ventilation is an efficient energy-saving method, used in architecture for the design and construction of low energy-consuming buildings. The present chapter investigates the air flow in and around a naturally ventilated building, which represents a student dormitory, using computational fluid dynamics (CFD) techniques. Turbulence is simulated applying the standard k-ε and the RNG k-ε models, both modified to account for wind and buoyancy forces. The study focuses on single-sided natural ventilation in a typical room located on the ground floor of the student dormitory. Two cases are investigated: rooms with a) two openings, and b) one opening, on the windward side of the room. Internal furnishings are taken into account and the room includes three thermal sources (occupant, TV, PC). Two different mechanisms of natural ventilation are examined: a) displacement ventilation (two openings) and b) mixing ventilation (one opening). The numerical results are compared with those obtained by two well-known empirical models, related to the effective velocity of incoming air and to the height of the neutral level in the case of one opening. It is concluded that the numerical results are in acceptable agreement with those obtained by the empirical models, especially when the standard k-ε model is used. Finally, the mathematical model described is used to evaluate thermal comfort inside the room, and it is found that the case with two openings on the windward side of the room is the best design.

Chapter 6 - Indoor air condition may be involved to the health status□Among various factors in indoor air conditions, few reports were published concerning negatively charged-particles. In this review, the authors detaile the biological effects of negatively-charged indoor air conditions, particularly on the psycho-neuro-endocrino-immune network. The short-term (2.5 hours) and mmedium-term (2 weeks) abidance of negatively-charged air particle dominant room resulted activation of natural killer cell activity induced by recurrent transient activation of interleukin 2. These results indicate that negatively-charged indoor air condition is better for immune status and may improve daily lives.

Chapter 7 - A Coruña, located in northwestern Spain, has a mild climate with a high relative humidity as a consequence of winds. Public buildings, like Spanish school buildings, present the highest energy consumption in air conditioning during the winter season, but during the spring the heating system is employed only if indoor conditions are below certain temperature and relative humidity values. Some energy saving methods could be employed to reduce or, in some cases, replace the heating systems and, as a consequence, reduce energy consumption. Some of these methods are centred in adequate design and application of an HVAC system while others are focused on the study of heat and mass transfer through building envelopes. A clear example of these methods was shown when permeable coverings [1, 2] were employed to reduce relative humidity peaks in office buildings, but there are other parameters, such as thermal inertia, that could help us to compare indoor ambience with respect to energy savings. In this sense, actual software, such as HAM tools, could be employed to simulate indoor conditions and phenomena of material and energy transfer through building envelopes and their effect on indoor conditions. This chapter describes a study of methods to reduce the energy consumption in different kinds of schools buildings by means real sampled data and HAM tool simulations. Results showed that parameters such as thermal inertia could be influenced principally by solar heat gains, changing the buildings' time constant. Other parameters like air changes per hour or permeable coverings present a clear enhancement of indoor ambiences.

Chapter 8 - Since the dawn of civilisation, human beings have used underground spaces for living and storing food. There are numerous examples of traditional underground constructions throughout the world, such as homes in North Africa, China, Tunisia and the United States. In Spain, caves were used as houses up to the last century and underground cellars were used for preparing, maturing and conserving wine.

Wine is a "living" product that requires specific hygrothermal conditions for its preparation and aging which affect its final quality. For this reason, in places with fierce climates, wine was traditionally matured in underground cellars since the heat capacity and density of the ground dampened the extreme temperatures of the outside, providing suitable conditions for maturing and conserving wine. However, since the last century, wine production has moved to buildings above ground that require a large energy consumption to maintain suitable conditions of temperature and relative humidity.

With the energy crisis, it is necessary to develop bioclimatic strategies that reduce energy consumption. Traditional underground cellars are a good example of techniques adapted to the environment since they provide optimum conditions for wine aging without energy costs. The viability of applying this type of architecture today can be found in Spain where various of its best and most prestigious wines continue to be produced in underground cellars.

Traditional underground constructions have certain advantages in hygrothermal control for maturing wine compared to modern commercial (above ground) wineries. Previous work [1] has shown that there are large differences in the thermal behaviour of underground cellars compared to modern wineries. The conditions inside traditional cellars meet the optimal climatic conditions for maturing wine while this can only be achieved in commercial wineries with air conditioning. The traditional constructions are more stable with a small annual temperature variation and a higher relative humidity level even when commercial wineries are only partially above ground. The thermal behaviour inside a totally above ground commercial winery will be even more unfavourable for the wine aging process, assuming that the energy exchange between a below-grade building wall and the surrounding soil is smaller than that between above-grade walls and the surrounding ambient conditions.

For this reason, traditional underground cellars have certain economical and environmental advantages. Traditional underground cellars provide suitable conditions to achieve high-quality wines by natural energy, so the operational cost of these cellars is lower than that of commercial wineries, because of both the cooling energy savings and the reduced wine losses.

Chapter 9 - Ongoing research to develop a new generation of decision-making tools for estimating the environmental effects on residential property value has significantly increased the demand for land surface data, information on the state of living environment, and the corresponding need for the advanced use of geographic information systems (GIS). In order to provide the foundation for price estimates, all the existing data focused on building and environment are integrated in the framework of a GIS spatial database. Traditional methods mostly emphasize the relationship between the effects of accessibility to central locations, and ignore location-specific attributes of housing. However, little has been done on high-rise, densely populated residential areas. Thus, the paper aims to investigate the neighboring and environmental characteristics of the selected site. A more advanced approach based on spatial analysis and modeling in the GIS environment is used to manage spatio-temporal data, to process aerial images and satellite images, to import measurements from GPS, to create digital terrain models, to analyze topography together with environmental data, and to visualize the results. The partial results based on the traveling time to the closest services and the origin-destination cost matrixes are derived from the network analysis. The landscape characteristics can be demonstrated on animated sequences showing the flight over a landscape model that is based on the digital terrain model with draped images over it to show the area. The living environment contains a few compartments: air pollution, water pollution, waste management, noise assessment, and monitoring of environmental impacts on population health. Air pollution is estimated by continuous map surfaces, predicting the values of pollutants concentration for the selected site in dependence on the sample points at the air quality monitoring stations. Water pollution covers surface water pollution, drinking water supply and quality, waste water, accidental contaminant spills, and optionally, flood control measures. The waste compartment is focused on the system of municipal waste management with sorting of reusable components of municipal waste, and hazardous chemicals information. Noise assessment covers road traffic noise and air traffic noise. The environmental impacts on population health come out from reports of the national institute of public health. In addition to the main compartments, other data can complement the partial inputs to the system (energy prices, public transport schema, locations of neighbor natural reservation sites and historical sites). In the framework of spatio-temporal analysis, the spatial weighting matrixes are used for prediction of the final rating. The final results are represented by the thematic map layers in the GIS project based on ArcGIS and its extensions. The attached case study shows a scenario in dependence on setting the weight parameters. Unexpected findings are caused by rapid changes in the dense living environment and slow conversion of the reality market in the selected site in Prague, the Czech Republic. For all that, the case study brings better understanding of how the residential site rating depends on various environmental attributes.

Chapter 10 - Molds are spore-forming filamentous microfungi that are saprophytic in the environment. Spores inhaled by humans result in life-threatening infections in immunocompromised patients. This chapter is focused on hospital buildings, where a large concentration of immunocompromised patients is present. The major environmental risk factor occurs during hospital construction. Indeed, demolition renovation or construction work has an impact on the environmental fungal load, and there is a well-recognised relationship between outbreaks of invasive filamentous fungal infections in immunocompromised patients and construction work.

The main environmental risk is due to airborne contamination that is dependent both on outdoor and indoor flora in the hospital. In our experience, most indoor contamination comes from micromycetes, which have a particular ability to adapt to the indoor environment. Thus, *Penicillium* spp. and *Aspergillus* spp. are the most frequent molds indoors. The authors have also shown that the basidiomycete *Bjerkandera adusta* can find a suitable ecological niche within hospitals, confirming that hospitals have their own flora. Factors such as moisture, ventilation, temperature, or organic matter present in building materials probably participate in the selection of indoor-adapted fungi. Finally, the significance of fungi, such as *Fusarium* spp., occasionally described in hospital water networks, is unclear.

Seasonal variations in the distribution of species have also been described in our study and others. Indoors, *Aspergillus* was found to be particularly abundant in winter, *Alternaria* and *Penicillium* in autumn, and *Cladosporium* in spring or summer. Outdoors, *Penicillium* spp. and *Aspergillus* spp. were more dominant in winter.

Reducing airborne contamination in hospital buildings is thus an important health concern. High efficiency particulate air filtration with or without laminar air flow ventilation has been shown to reduce airborne fungal contamination. Recently, cheaper alternative devices like mobile filtration have been studied, and the authors have shown that Plasmair™ units reduce indoor airborne contamination both during periods with and without construction work. The authors and others also found that airborne contamination was not increased if construction work was performed after implementing protective measures such as keeping windows and doors closed, and reinforcing biocleaning.

Finally, although there is no agreement on a threshold of spore concentration, the authors believe that environmental surveillance must be established both in clinical units and in mycology laboratories in order to avoid false positive results.

In conclusion, in each hospital a multidisciplinary approach involving biologists, clinicians, hygienists, epidemiologists and environmental engineers is necessary to implement a prevention strategy adapted to local conditions.

Chapter 11 - This chapter reviews selected monitoring and modelling studies co-conducted by the authors in the past decade. These studies were either awarded by influential professional organizations (URISA, CPGIS, HKSTS) or granted by national foundation (NSFC) etc. As the study area is the Macao Peninsula, a highly compact and the wealthiest urban area in Asia, the study approaches and results can be used as a base for further discussions of the building form and environment in compact urban areas. The findings will also be very relevant to urban planners as they prepare for higher population densities in cities throughout the world.

Chapter 12 - For more relevant informational retrieval and matching of user request with metadata about informational recourses it is necessary to formalize the user knowledge about subject domain of search and handle the semantics of user's request as well as the semantics or informational recourses. The authors propose to base on Semantic Web technologies and use the ontologies and associated with them thesauri of the appropriate subject domains for representation of domain knowledge. The algorithms of formation and normalization of the multilinguistic thesauri, and also methods of their comparison are given in this work.

Chapter 13 - One of the significant steps in ontological analysis is the development of relations between the ontology's concepts. All mistakes of this phase result to the great problems on the subsequent steps and in ontology use. Though the authors propose to use some elements of mereology for more precise definition of relation "part-of" subclasses. This approach is applicable for development of user thesauri that used in different distributed systems (such as informational retrieval systems) as a special case of domain ontology.

In: Buildings and the Environment
Editors: Jonas Nemecek and Patrik Schulz

ISBN: 978-1-60876-128-9

Chapter 1

NUMERICAL APPROACHES FOR FIRE MODELLING WITH TURBULENT, COMBUSTION, SOOT AND RADIATION CONSIDERATIONS

Sherman C.P.Cheung[1], G.H.Yeoh[2] and J.Y. Tu[1*]
[1]School of Aerospace, Mechanical and Manufacturing Engineering, RMIT University, Victoria 3083, Australia
[2]Australian Nuclear Science and Technology Organisation (ANSTO), PMB 1, Menai, NSW 2234, Australia

ABSTRACT

Modelling of fires has contributed significantly to the modern development of fire safety science and the blossoming of the discipline of fire engineering. Suitable numerical simulation tools have effectively taken centre stage for practicing fire engineers to exploit the freedoms offered under the performance-based fire engineering approach. In this article, we review the state-of-the-art deterministic fire models based on the approach of field modelling utilizing the Computational Fluid Dynamics (CFD) techniques. Special attention is also directed towards the necessity to adopt relevant models and the historical development in handling the combustion and soot chemistry and radiation heat transfer in fire simulations. The structure of this article consists of three main parts: (i) Historical development of fire models; (ii) Implementation of relevant combustion, radiation and soot sub-models and (iii) Possible aspects of the future development in fire modelling. In the first part, discussion will centre on the revelation of the complex chemo-physical behaviour of building fires and the challenges being faced in numerical modelling. A comprehensive review of the historical development of fire models will then follow by the unveiling of the basic framework of field modelling technique. Based on the framework of field modelling, theoretical concepts and mathematical formulation of combustion, radiation and soot sub-models that have been proposed within literature will be summarized in second part. As a more in-depth demonstration of the necessity and effects of these sub-models, numerical

* Corresponding Author : SAMME, RMIT University, Bundoora, Melbourne, Victoria 3083, Australia Email: jiyuan.tu@rmit.edu.au Phone no.: +61-3-9925 6191Fax no.: + 61-3-9925 6108

studies devoted to assess the capability of some selected sub-models will be presented in which predicted results have been rigorously validated against full-scale experimental data. These studies show that good agreement can be achieved through the present model with regards to the spatial temperature and velocity distributions. Finally, for the last part, limitations and assumptions taken by the models will be addressed and some suggestions of possible direction for further fire model development are discussed.

Keywords: Fire modelling; Combustion; Computational Fluid Dynamics

1. INTRODUCTION

Long before the beginning of human history, fire had already been widely used by mankind and made a veritable revolution in the human history. In our time, fire has become a "tool" being applied for various purposes. For example: to be used for cooking foods; to be a source of power or heat for industrial processes; and even warming up our own house for enjoyment. Owning to fact that fire is now a part of our daily life, the word "Fire" is often interpreted as a natural combustion progress which is well controlled by people, or the term so-called "wanted fire".

Unfortunately, right behind the wonderful side, fire shows a dangerous and destructive downside claiming thousands and thousands of lives and costing millions of property damages every year. These uncontrollable fires often referred to as "fire disasters" or "unwanted fires". In essence, in parallel exploring the benefits of fire, our ancestor also learnt the importance of preventing the "unwanted fire" simultaneously. Until now, all types of buildings and constructions are equipped with fire prevention and extinguishing systems. Every country has their own minimal requirement of such prevention and extinguishing system clearly stated in their law book or "fire code". Even after exhaustive efforts, "unwanted fire" still occurs, almost everyday.

Amongst the many incidents of uncontrollable fires, *unwanted fires* in enclosures are most frequently encountered. Significant examples of some major fire disasters recorded in history are the Kings Cross Fire in the London Underground occurred on 18th November 1987 and the collapse of the World Trade Center Towers in New York on 11th September 2001. The hazard that these fires represents is usually associated with the uncontrolled nature of the exothermic chemical reactions especially between organic or combustible materials and air and its interaction with the structural components. What follows from the analysis of this fire hazard is that it cannot, in general, be totally eliminated but can be reduced to an acceptably low level via appropriate design considerations and procedures.

Physically speaking, fire dynamics embraces numerous complicated physical and chemical interactions, which include fluid dynamics, thermodynamics, combustion, radiation or even multi-phase effect. During the early investigations of fire research, a great deal of attention has been focused in better understanding the fire behaviors using experimental techniques and theoretical approaches. Experiments provide useful observations and measurements of the flaming process while theoretical models employ a mathematical description of the physical phenomena through input of experimental data. There are, however, limitations in fully applying experimental techniques and theoretical approaches to a range of fire problems. Conducting full-scale experiments can be rather expensive due to the high costs of construction of a fire facility and the instrumentation and hardware required for data collection. On the other hand, in spite of the low computational costs associated with the use of theoretical approaches, these models are still highly dependent on the experimental data that are correlated from and the specific geometrical configuration where the fire experiments are carried out.

With the advent of digital computers, the use of numerical methodologies in fire modelling offers nonetheless fire modellers the flexibility of aptly simulating the fire behaviours in different enclosure configurations hence overcoming the constraints in experimental techniques and theoretical approaches. Suitable numerical simulation tools have effectively taken centre stage for practicing fire engineers to exploit the freedoms offered under the performance-based, fire safety engineering approach. The core of all fire modelling activities remains essentially on the proper treatment of the gas phenomena of the fire itself for any subsequent assessment of impact on structures, people, or environment. Unfortunately, this effectively requires the fire modellers to gain a fundamental knowledge of fire dynamics and the theoretical formulation of fire models which is certainly a time-consuming and pain-taking task. The main aim of this paper therefore focuses on reviewing the state-of-the-art of deterministic fire models based on the approach of field modelling; briefly discusses the theoretical development and necessity of adopting relevant sub-models; and projects the possible future research directions of fire modelling technique. Readers interested in more details of the model development and its derivation are directed to the text book by Yeoh and Yuen (2009). Before starting off our discussion, it is essential to introduce some characteristics of fire dynamics in the below section.

1.1. Characteristic of Fire Dynamics

Most fires involve combustible gases. According to Drysdale (1999), the term fuel, required for combustion, can be defined as whatever the state of matter that may be in the form of gases, liquids or solids burning in atmosphere. Consider the burning of a solid fuel such as an item of furniture within a room for the purpose of illustration. For almost all types of solids, chemical decomposition (pyrolysis) is responsible for yielding products of sufficiently low molecular weight that can be volatilize from the surface and enter the flame as fuel. Clearly, the visible flame is a gas phase phenomenon and flaming combustion of a solid fuel, as described by the pyrolysis process, or a liquid fuel must involve the conversion of fuel to gaseous form. For a burning liquid, the process is generally simpler where volatiles are released due to evaporative boiling at the surface.

As the combustible volatiles fuel the burning fire as shown in Figure 1, they react with the *oxygen* in the air which is being entrained into the combustion zone under appropriate conditions in the gas phase. The air entrainment may be driven by force or free *convection* depending on the environmental conditions surrounding the burning fire. Assuming a carbon-based fuel, *combustion products* such as carbon dioxide and water vapour are generated for a complete chemical reaction. In reality, such a chemical reaction is an idealization of the actual gas phase combustion whereby the complexity of the overall process is hidden. The true picture usually consists of a series of elementary reaction steps taking part in the course of combustion. Heat is released following the chemical reaction – an exothermic process. With very few exceptions, particulate smoke is usually produced in all fires. Depending on the nature of different fuels, smoke can contain high concentrations of finely-dispersed particles, commonly known as soot, or narcotic gases such as carbon monoxide in addition to the production of combustion products. Radiation from these hot combustion gases and non-luminous flames also plays an important consideration in fires. For an unabated fire, radiation heat transfer causes the heat supplied by the visible flame onto the solid surface to sustain a continuous supply of volatiles into the combustion zone for flaming combustion. The example of the burning of a solid fuel in an open space as described in Figure 1 can usually be categorized as a free-standing fire. As it will become more transparent in the later sections, due to instability of the air entrainment, the free standing fire preserves a unique periodic oscillatory motion close to its fuel bed which has often been referred to as the "puffing" behaviour (Bryan and Nelson, 1970; Portscht, 1975; McCaffrey, 1979).

Another important classification, enclosure fire, exhibit some distinguishing features and characteristics that are usually different from a free-standing fire. Here, the confining rigid boundaries surrounding the enclosure greatly influence the fire dynamics.

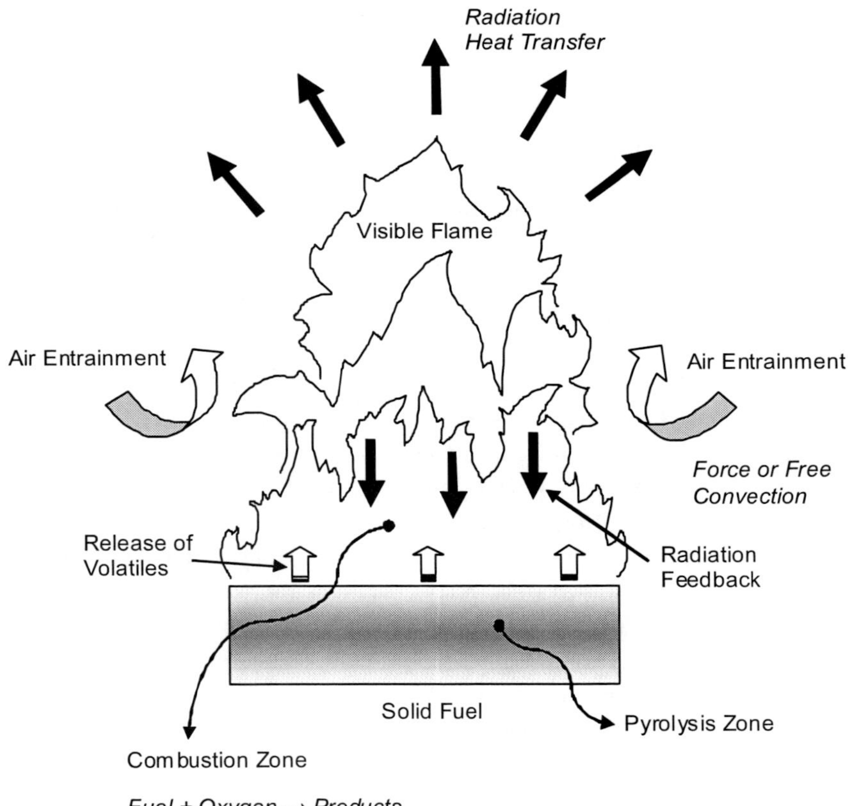

Figure 1. Schematic representation of a burning solid fuel in air showing the respective physical processes involved

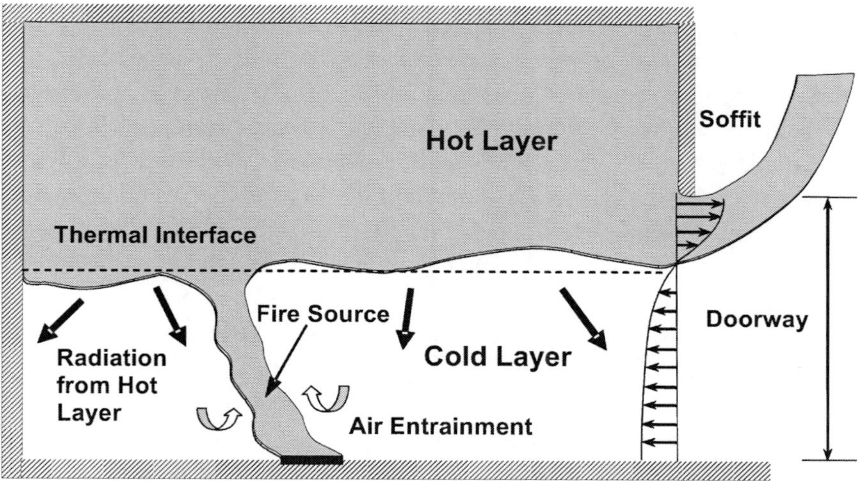

Figure 2. Schematic representation of a fire source in a single compartment enclosure

As shown in Figure 2, for an *enclosure fire*, a fire source that could be a gas, liquid or solid fuel ignites and burns at the centre of the enclosure. *Combustion* takes place resulting in a heat release due to the exothermal reaction; this fire source can now be considered as a heat source discharging energy onto the surrounding fluid or wall boundaries by three modes of heat transfer: *conduction*, *convection* and *radiation*. For natural fires, the hot fluid heated by the fire source is driven by buoyancy; the development of an upward flow, normally *turbulent* in nature, transfers the hot fluid into the upper region of the room. At this early stage of the fire growth, the pyrolysis rate and energy release rate are affected only by the burning of the fuel itself and not by the presence of the boundaries of the compartment. In time, the presence of the wall boundaries and the soffit of the room create a "reservoir" for the accumulation of the hot gases or *smoke* below the ceiling, which is commonly referred as the "Hot Layer". Whilst hot gases accumulate in the upper region, fresh air from the outside surrounding is entrained into the room through the doorway to supply oxygen to the fire in the lower level. This represents the "Cold Layer" region of the compartment. In fire engineering, the two-layer flow pattern is a prevalent feature identifying compartment fires. The height of this interface is often considered as an important datum of measure in fire experiment; the term "Thermal Interface" is usually introduced to characterize the averaged height of the interface. As this interface continues to descend due to increasing concentrations of hot gases especially down to the level below the depth of the soffit, the amount of hot gases eventually exceeds the volume of the reservoir and some of these hot gases will spill into the surrounding below the soffit as shown in Figure 2. These two counter flows have a tendency of creating a rapid change of velocity at the doorway. As a result, *positive* pressure is sustained at the top to exhaust the hot gases whilst *negative* pressure entrains the fresh air at the bottom of the doorway.

Free-standing and enclosure fires as described are just some of the many representative fires that have been studied rather extensively to gain invaluable insights into the fire phenomena. Based on the above description, one should be noticed that all the aforementioned physical processes are closely coupled and interact with each other. Obviously, from the standpoint of modelling consideration, isolating any one of the processes in the calculation would impose numerical errors to the final predictive results. With the same reason, there has been a vast of research works conducted to advance the capability of attaining solutions through conservations equations of fluid dynamics and heat transfer coupled with suitable models capturing the complex physical processes associated with combustion, radiation, smoke movement and production and solid pyrolysis. These significant advancements were mostly developed on the basis of field modelling approach – also known as Computational Fluid Dynamics (CFD) technique. With greater speeds expected in digital computers (arriving sooner than later), field modelling will undoubtedly play an even greater role and cement its place as the preferred approach in the areas of fire science and fire engineering The next section introduces the concept of field modelling and some commonly adopted physical sub-models. Special attention is also directed toward the theoretical considerations and assumptions behind models formulation revealing the advantages and disadvantages of each models.

2. Framework of Fire Modelling

2.1. Equations of Fluid Motion

Under the framework of Computational Fluid Dynamics (CFD), to model fluid motion within the physical domain, it requires the sub-division of the domain into a number of smaller, non-overlapping sub-domains. This subsequently results in the generation of a *mesh* (or grid) of *cells* (elements or control volumes) covering the whole domain. These cells (or control volume) can be considered as an infinitesimal fluid element of the size large enough to encapsulate molecules at macroscopic length scales. These fluid elements can be then taken as continuous medium which follow the neighbouring fluid motion. Figure 3 illustrates the consideration of infinitesimal fluid element within the free-standing and single compartment fire problems. The behaviour of the fluid motion can thus be described in terms of macroscopic properties such as the velocity, pressure, density and temperature as well as their space and time derivatives. These macroscopic properties are not influenced by individual molecules and this leads immediately to the development of fundamental equations in partial differential form.

In essence, the fundamental equations of fluid motions are derived from the conservation laws of physics namely the *conservation of mass*, *Newton's second law for the conservation of momentum* and *first law of thermodynamics for the conservation of energy*. For simplicity, detailed derivation of governing equation is excluded in this paper. Interested readers are directed to other text books (Versteeg and Malalasekera, 1995; Tu et al., 2008; Yeoh and Yuen, 2009).

Due to the fact that all these fundamental equations share some commonalities in its formulations, it is a common practice to express the governing equations with a general differential form; namely "General Transport Equation":

$$\frac{\partial \rho\phi}{\partial t} + \nabla\cdot\left(\rho\vec{U}\phi\right) = \nabla\cdot\left(\Gamma_\phi \nabla\phi\right) + S_\phi \tag{1}$$

where ϕ is a general variable which includes all the field dependent variables: u, v and w velocities in x, y and z Cartesian coordinates, enthalpy, species concentrations etc; Γ_ϕ and S_ϕ denote the diffusion coefficient and the source term of its dependent variable ϕ. The details of all the terms are summarized in then Table 1.

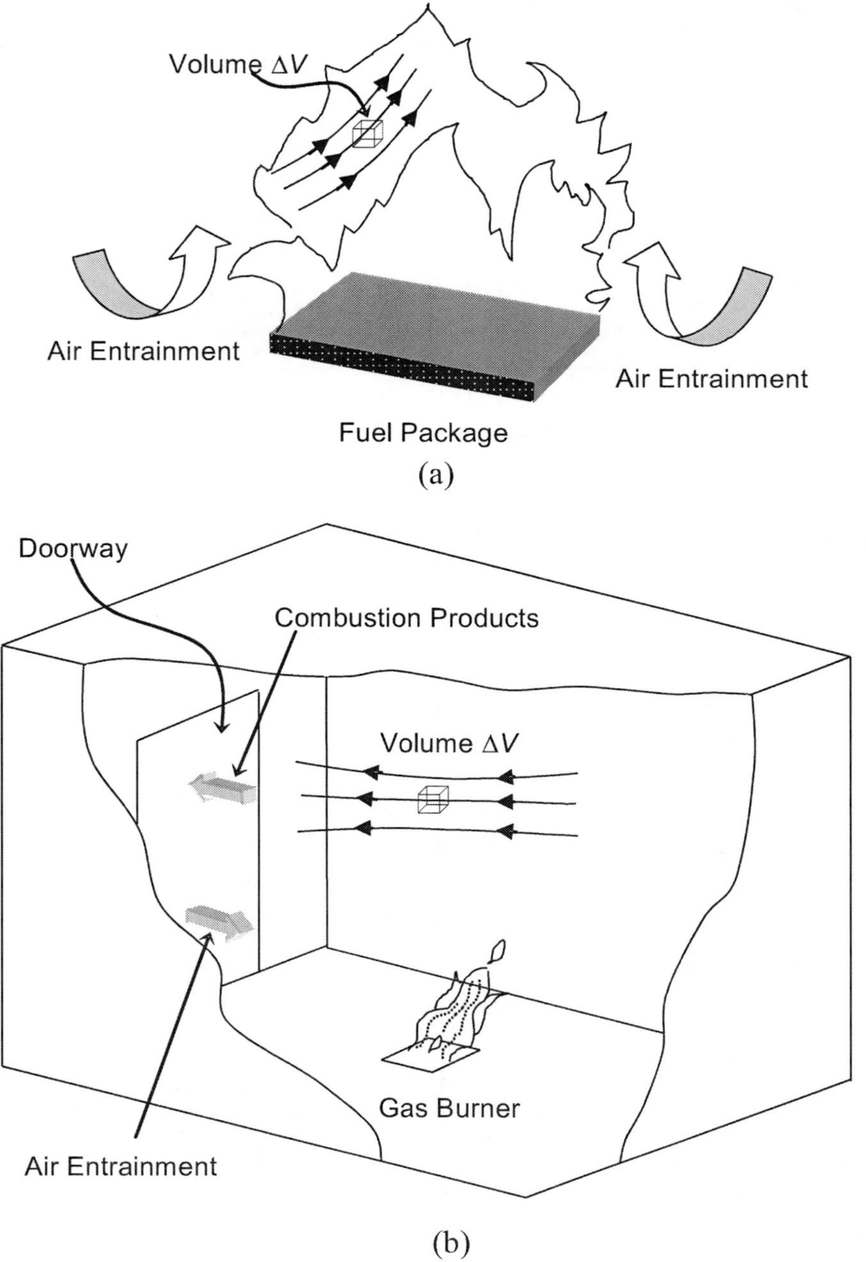

Figure 3. Infinitesimal fluid element approach. Representation models of a fluid flow in: (a) Free-standing fire and (b) Single compartment fire

Table 1. Transport Equation for field dependent variable ϕ

ϕ	Γ_ϕ	S_ϕ
1	0	0
u	μ	$-\frac{1}{\rho}\frac{\partial p}{\partial x}+\frac{\partial}{\partial x}\left[\mu\frac{\partial u}{\partial x}\right]+\frac{\partial}{\partial y}\left[\mu\frac{\partial v}{\partial x}\right]+\frac{\partial}{\partial z}\left[\mu\frac{\partial w}{\partial x}\right]+\frac{\partial}{\partial x}\left[\lambda\left(\frac{\partial u}{\partial x}+\frac{\partial v}{\partial y}+\frac{\partial w}{\partial z}\right)\right]+\sum F_x^{body\ forces}$
v	μ	$-\frac{1}{\rho}\frac{\partial p}{\partial y}+\frac{\partial}{\partial x}\left[\mu\frac{\partial u}{\partial y}\right]+\frac{\partial}{\partial y}\left[\mu\frac{\partial v}{\partial y}\right]+\frac{\partial}{\partial z}\left[\mu\frac{\partial w}{\partial y}\right]+\frac{\partial}{\partial y}\left[\lambda\left(\frac{\partial u}{\partial x}+\frac{\partial v}{\partial y}+\frac{\partial w}{\partial z}\right)\right]+\sum F_y^{body\ forces}$
w	μ	$-\frac{1}{\rho}\frac{\partial p}{\partial z}+\frac{\partial}{\partial x}\left[\mu\frac{\partial u}{\partial z}\right]+\frac{\partial}{\partial y}\left[\mu\frac{\partial v}{\partial z}\right]+\frac{\partial}{\partial z}\left[\mu\frac{\partial w}{\partial z}\right]+\frac{\partial}{\partial z}\left[\lambda\left(\frac{\partial u}{\partial x}+\frac{\partial v}{\partial y}+\frac{\partial w}{\partial z}\right)\right]+\sum F_z^{body\ forces}$
h	$\frac{k}{C_p}$	$\dot{Q}_s$
φ	ρD	$\dot{R}_s$

Eq. **Error! Reference source not found.** illustrates the various physical transport processes occurring in the fluid flow: the rate of change of ϕ which is the local acceleration term accompanied by the advection terms on the left hand side is respectively equivalent to the diffusion term (Γ_ϕ is designated as the diffusion coefficient) and the source term (S_ϕ) on the right hand side. It is noted that the additional body force terms in the momentum equations $F_x^{body\ forces}$, $F_y^{body\ forces}$ and $F_z^{body\ forces}$ consists of the pressure and non-pressure gradient terms and other possible sources such as gravity that influence the fluid motion. By setting the transport property ϕ equal to 1, u, v, w, h, φ and selecting appropriate values for the diffusion coefficient Γ_ϕ and source terms S_ϕ presented in Table 1, one can easily obtain the expressions for each of the partial differential equations for the conservation of mass, momentum, energy and scalar property.

For the energy equation, one may notice that enthalpy is selected as a transport variable. To be consistent, it is common practice to transform the energy equation by replacing the heat flux according to the local enthalpy gradient instead of the temperature gradient. The diffusion term is thus expressed in terms of thermal conductivity and specific heat of constant pressure. Furthermore, as most of the fire problem involve combustion processes in low Mach number and weakly-compressible situation, the kinetic energy $\frac{1}{2}\left(u^2+v^2+w^2\right)$ in the definition of enthalpy, the pressure work term $\partial p/\partial t$ and the dissipation function that represents the source of energy due to work done deforming the fluid element are usually ignored in most practical applications.

2.2. Turbulence Closure

The governing equations mentioned above at best only provide a description of the fire dynamics associated with a laminar combusting flow system. To adequately resolve practical fires, additional models to better capture all the physical processes involving turbulence, combustion, radiation, smoke (soot particles) movement and production are generally needed to predict the fire characteristics occurring in real fires. These models are incorporated within field modelling before computing the approximated solutions of the velocity, pressure, temperature, etc. In the context of fire engineering, most practical fires are turbulent in nature. Some fundamental concepts of turbulence modelling will be expounded in this section.

2.2.1. Time averaging turbulence modelling approach

In many engineering applications, the knowledge of mean values of flow properties is usually of greater significance rather than their fluctuating components. . It is thus more preferable that there be some means of practically resolving the random transient distribution of the property ϕ with time by decomposing the instantaneous property ϕ as a steady mean motion $\overline{\phi}$ and a fluctuating, or eddy, motion ϕ' as:

$$\phi=\overline{\phi}+\phi' \text{ and } \overline{\phi}=\frac{1}{t_o}\int_0^{t_o}\phi dt \tag{2}$$

where t_o is a large enough time interval exceeding the time scales of the slowest variations (due to largest eddies). This approach presents an attractive way to characterize a turbulent flow by the mean values of flow properties ($\overline{u}$, $\overline{v}$, $\overline{w}$, $\overline{p}$ etc.) with its corresponding statistical fluctuating property (u', v', w', p' etc.) where its time-averaged quantities, by definition, are zero.

Through the Eq. (2), instantaneous density, velocities, enthalpy, scalar property, etc. can be expressed in terms of their mean and fluctuating quantities. Substituting into the governing equations and taking the time average, a system of equations commonly known as the *Reynolds-Averaged Navier-Stokes* (RANS) equations can be derived. The main core of turbulence closure in the RANS model is based on the two equations turbulence model proposed by Launder and Spalding (1972, 1974). In the two equations k-ε turbulence closure, based on the eddy viscosity hypothesis, flow values in the Navier-Stokes equations are averaged with respect to time.

Moreover, as fire simulation often involves large variation of density, most of field models use Favre-Averaging (mass weighted) technique (Favre, 1965) to describe the Navier-Stokes equations. The Favre-averaged approach is described for the derivation of the conservation equations. If we define a mass-weighted mean property ϕ as:

$$\tilde{\phi} = \frac{\overline{\rho\phi}}{\bar{\rho}} \tag{3}$$

the instantaneous property ϕ may now be written according to:

$$\phi = \tilde{\phi} + \phi'' \tag{4}$$

where ϕ'' is the superimposed velocity fluctuation. Multiplying Eq. (4) by density ρ, we obtain:

$$\rho\phi = \rho\left(\tilde{\phi} + \phi''\right) = \rho\tilde{\phi} + \rho\phi'' \tag{5}$$

By time-averaging the above equation:

$$\overline{\rho\phi} = \bar{\rho}\tilde{\phi} + \overline{\rho\phi''} \tag{6}$$

From the definition of Eq. (3), it follows that

$$\overline{\rho\phi''} = 0 \tag{7}$$

Substituting the instantaneous density, velocities, enthalpy, scalar property, etc. expressed in terms of their mass-weighted mean and fluctuating quantities in the form of Eq. (4) into the governing equations and taking the time average, the system of equations known as the *Favre-Averaged Navier-Stokes* equations can be alternatively expressed in compact form as shown in Table 2.

The equations through Favre-averaging are much simpler by the mere presence of only the Reynolds and scalar stress terms: $\overline{\rho u_i'' u_j''}$, $\overline{\rho u_i'' h''}$ and $\overline{\rho u_i'' \varphi''}$. Since most experimental sampling probes measure values that approximate mass-weighted concentrations rather than time-averaged concentrations, Favre-averaged equations are very amenable to obtain practical solutions in many numerically related fire investigations. To solve the system of equations, the Reynolds and scalar stresses must be related to the mean quantities of the flow field.

In the past decades, the k-ε model enjoyed broad popularity and was widely adopted in the modelling of fires. However, as the model was not developed for fire simulations, parametric and modification studies have been performed to adjust the model constants or source terms. Nam and Bill (1993) employed a field model (i.e. PHONICES) to study the k-ε model numerically with the pool fire and heptane spray fire plume. By using the original model constants, they found that the original model under-predicted the plume widths. Based on the above observation, they altered the constant C_{μ} by doubling its original value. The results obtained with the modified constant were then significantly improved.

Table 2. Favre-Averaged Navier-Stokes equations in Cartesian coordinates

Favre-Averaged Mass	
$\frac{\partial \bar{\rho}}{\partial t}+\frac{\partial}{\partial x_j}\left(\bar{\rho}\tilde{u}_j\right)=0$	j = 1,2,3
Favre-Averaged Momentum	
$\frac{\partial}{\partial t}\left(\bar{\rho}\tilde{u}_i\right)+\frac{\partial}{\partial x_j}\left(\bar{\rho}\tilde{u}_i\tilde{u}_j+\overline{\rho u_i'' u_j''}\right)=-\frac{\partial \bar{\sigma}_{ij}}{\partial x_j}+\bar{S}_{u_i}$ where $\bar{\sigma}_{ij}=\bar{p}\delta_{ij}-\mu\left(\frac{\partial \tilde{u}_i}{\partial x_j}+\frac{\partial \tilde{u}_j}{\partial x_i}\right)+\frac{2}{3}\mu\frac{\partial \tilde{u}_i}{\partial x_j}\delta_{ij}$	j = 1,2,3
Favre-Averaged Enthalpy	
$\frac{\partial}{\partial t}\left(\bar{\rho}\tilde{h}\right)+\frac{\partial}{\partial x_j}\left(\bar{\rho}\bar{u}_j\tilde{h}+\overline{\rho u_j'' h''}\right)=\frac{\partial}{\partial x_j}\left[\frac{k}{C_p}\frac{\partial \tilde{h}}{\partial x_j}\right]+\bar{S}_h$	j = 1,2,3
Favre-Averaged Scalar Property	
$\frac{\partial}{\partial t}\left(\bar{\rho}\tilde{\varphi}\right)+\frac{\partial}{\partial x_j}\left(\bar{\rho}\bar{u}_j\tilde{\varphi}+\overline{\rho u_j''\varphi''}\right)=\frac{\partial}{\partial x_j}\left[\bar{\rho}D\frac{\partial \tilde{\varphi}}{\partial x_j}\right]+\frac{\partial}{\partial x_j}\left[\overline{D\rho\frac{\partial \varphi''}{\partial x_j}}\right]+\bar{S}_{\varphi}$	j = 1,2,3

Besides fire plume study, Fletcher et al. (1994) employed the *k*-ε model with additional buoyancy terms by Goussebaile and Viollet (1982) to simulate tunnel fire situation. They reported that, without buoyancy consideration, the vertical temperature profiles were poorly predicted. Moreover, when buoyancy is neglected, the stratification almost totally vanished. Therefore, they concluded that the buoyancy terms in the *k*-ε model play a critical role in correctly reproducing stratified flow.

Similar observations were also reported by Rodi and his co-workers (Hossain and Rodi, 1982; Rodi, 1985). To include the buoyancy effect, they attempted to re-derive the two equations turbulence model from Reynolds Stresses correlation. From their analysis, they concluded that turbulence is generated by the temperature fluctuations or by indirect density changes. They further proposed to replace the original model C1 by a function of a buoyancy parameter which has taken the horizontal and vertical buoyant flow into account. Predictions from the proposed model agreed well with the full range measured data between inertia-dominated flows and buoyancy-dominated flows.

Another version of buoyancy-modified *k*-ε model was proposed by Woodburn and Britter (1996a, 1996b). Different from the model by Rodi, they believed that the value of the Prandtl number varies significantly in strongly stratified flows. Since Prandtl number is assumed as constant in the standard model, they proposed dynamic function to make the Prandtl number to be self-adaptive to the flow situation. Although prediction results by-in-large compared well with measured data, the proposed function is relatively complicated. Thus, it is rather cumbersome to implement. Moreover, it was also noticed that this model was tailored for stratified flows. Numerical studies by Wang and co-workers (Wang et al., 1995; Wang and Joulain, 1996) proved the model was inadequate when applied to wall fires.

The aforementioned model and numerical studies were developed based on the two equations k-ε model. In fact, to best of the author's knowledge, most of the fire modelling studies employed the same turbulence model. Other means turbulence closures in fire modelling are limited. For example: Than and Savilonis (1993) used the Algebraic Stress model as turbulent closure. A similar study can also be found by examining Motevalli's (1994) work.

Liu and Wen (2002) modified the second moment closure model proposed by Hanjalic and Jakirlic (1993) to simulate the buoyant diffusion flame measured by McCaffrey (1979). Comparing the predicted results with the Low Reynolds Number (LRN) k-ε model by Ince and Launder (1989) and the standard k-ε model, they found that both the standard and LRN version k-ε model under-predicted the buoyancy-produced turbulence and the centreline velocities. However, in their study, buoyancy terms are not included in the two versions of k-ε model.

Although the Reynolds Stress model and the second moment closure model have their advantages, they are complicated and are known to be unstable in character. For example: Woodburn and Drysdale (1998) reported convergence difficulty in their numerical study of flame development in trenches. Due to the high computational cost, these models have not been widely employed in building fire studies.

2.2.2. Large eddy simulation (LES)

Recently, beside the RANS model, the concept of Large Eddy Simulation in turbulence modelling emerged in computational fluid dynamics. Certainly, researchers in fire modelling also started to employ the concept in their field models. Since the concept is still relatively new, numerical study on the concept and its performance in building fire simulation is limited. However, recent studies have been performed, which presented considerable and encouraging results. In the review article by Novozhilov (2001), he concluded that the LES concept is very likely to be the next focus of the fire field models development.

The basic idea behind large eddy simulation (LES) is that the turbulent eddies that account for most of the mixing or large scale motion are large enough to be calculated with sufficient accuracy from the equations of fluid mechanics. The hope is that the small scale eddy motion is approximated by some appropriate models, which must be ultimately justified by appeal to experiments. The establishment of the LES method has its roots in the prediction of atmospheric flows since the 1960s and recently in fire engineering, the development of the fire dynamics simulator (FDS) computer code by NIST which is increasingly being adopted for practical engineering investigations of fires.

In LES, the governing equations are formally derived by applying a filtering operation, which proceeds according to:

$$\bar{\phi}(x_i',t) = \int_{\Delta} \phi(x_i',t) G(|x_i - x_i'|) dx_i' \tag{8}$$

where G is a filter function. In practice, LES is usually conducted so that a finite computational mesh with the truncation error from the numerical discretisation of the flow equations is considered as the filtering operator. The advantage of this approach is that no explicit filtering operation is required. Nonetheless, the danger is that the truncation error at the smallest resolved scales, i.e. at the highest wave numbers, can be substantially large. Although explicit filters can be employed to remove these errors, *mesh refinement* has shown to significantly improve the numerical results at a much faster rate. In essence, a denser grid without imposing any explicit filtering produces better results albeit the smallest scales are influenced by the numerical error. This error is however removed to the high wave numbers, whose contribution to the results is apparently small. Within the finite volume method, it is rather sensible to consider the filter width to be of the same order as the grid size. In three-dimensional computations with grid cells of different grid sizes along the Cartesian coordinate directions, the filter width is often taken to be the cube root of the grid volume:

$$\Delta = \sqrt[3]{\Delta x \Delta y \Delta z} \tag{9}$$

In a rough sense, the flow eddies larger than the filter width are considered to be *large eddies* while eddies smaller than the filter width are *small eddies* which require modelling.

Sub-grid scale (SGS) model

In LES, the small dissipative scales are not solved accurately. The prime objective of the subgrid scale (SGS) models are to represent the kinetic energy losses due to the viscous forces and do not attempt to produce the SGS stresses accurately but rather to only account for their effect in a statistical sense. Most models are prescribed through the eddy-viscosity concept; it therefore shares many similarities to that used in FANS modelling. Smagrorinsky (1963) suggested that the Boussinesq hypothesis can be invoked to provide a good description of the unresolved eddies of the resolved flow since the smallest turbulent eddies are almost isotropic. For the unresolved SGS turbulent stresses τ_{ij}, they are modelled accordingly as:

$$\tau_{ij} = \bar{\rho}\widetilde{u_i u_j} - \bar{\rho}\tilde{u}_i\tilde{u}_j = -2\mu_T^{SGS}\tilde{S}_{ij} + \frac{1}{3}\tau_{kk}\delta_{ij} \tag{10}$$

$$\tilde{S}_{ij} = \frac{1}{2}\left(\frac{\partial \tilde{u}_i}{\partial x_j} + \frac{\partial \tilde{u}_j}{\partial x_i}\right) - \frac{1}{3}\frac{\partial \tilde{u}_k}{\partial x_k}\delta_{ij}$$

where μ_T^{SGS} is the SGS eddy viscosity and $\tilde{S}_{ij}$ is the strain rate of the large scale or resolved field. The Smagorinsky-Lilly model assumes that the SGS eddy viscosity can be described in terms of a length and a velocity scale. Taking the length scale to be the filter width Δ, the velocity scale can be expressed as the product of the length scale and the average strain rate of the resolved flow; the SGS eddy viscosity takes the following dependency:

$$\mu_T^{SGS} = \bar{\rho} C_1 \Delta^2 \left| \tilde{S} \right| \tag{11}$$

where C_1 is an empirical constant and $\left| \tilde{S} \right| = \sqrt{2\tilde{S}_{ij}\tilde{S}_{ij}}$. The stress tensor τ_{kk} in Eq. (10) can be similarly modelled as:

$$\tau_{kk} = 2\bar{\rho} C_k \Delta^2 \left| \tilde{S} \right| \tag{12}$$

Erlebacher et al. (1992) have found that τ_{kk} may be ignored for practical calculations since $C_k \ll C_1$.

The Smagroinsky constant, $C_S = \sqrt{C_1}$, generally varies between a range between 0.065 and 0.3 depending on the particular fluid flow problem. Difference in C_S is attributed to the effect of the mean flow strain or shear. This gives an indication whereby the behavior of the small eddies is not as universal as has been surmised in the beginning. On a theoretical analysis of the decay rates of isotropic turbulent eddies in the inertial subrange of the energy spectrum, Lilly (1966, 1967) has, for example, obtained values of C_S between 0.17 and 0.21. After reviewing other works, Rogallo and Moin (1984), suggested values between 0.19 and 0.24 across a range of grids and filter functions. Zhou et al. (2001) have indicated that a little larger C_s is more applicable for thermal flows of which they have employed a value of 0.23 in their LES study. For open buoyant fires with fully developed turbulence, the current authors have employed a value of C_S equivalent to 0.2 with much success. In most internal flow calculations, C_S = 0.1 – 0.13 is nonetheless commonly adopted in practice as suggested by Piomelli et al. (1988) and Scotti et al. (1993). Note that there is a difference in the way of how the turbulent viscosity is evaluated between the LES and FANS approaches.

In addition to the Smagorinsky-Lilly model, other basic subgrid-viscosity models such as the Structure Function model by Métais and Lesieur (1992) and the Mixed Scale Model by Sagaut (1996) have also been proposed. In the Structure Function model, the subgrid eddy viscosity is alternatively evaluated according to:

$$\mu_T^{SGS} = \bar{\rho} C_2 \Delta \sqrt{\bar{F}_2(\Delta)} \tag{13}$$

where $\bar{F}_2$ is the second-order structure function constructed with the filtered velocity field:

$$\begin{aligned} \bar{F}_2(\Delta) &= \int_{|x'|=\Delta} \left[\tilde{u}(x) - \tilde{u}(x+x')\right]^2 d^3x' \\ &\approx \frac{1}{6}\sum_{i=1}^{3}\left\langle \left[\tilde{u}(x)-\tilde{u}(x+\Delta x_i)\right]^2 + \left[\tilde{u}(x)-\tilde{u}(x-\Delta x_i)\right]^2 \right\rangle \left(\frac{\Delta}{\Delta x_i}\right)^{2/3} \end{aligned} \tag{1}$$

In Eq. (1), the structure function $\overline{F}_2$ has been approximated based on a local statistical average of the square (filtered) velocity differences with the six immediately adjacent cells. A constant value of $C_2 = 0.063$ is prescribed as suggested by Métais and Lesieur (1992). The Mixed Scale Model (MSM), as proposed by Sagaut (1996), accounts for the contribution of the resolved field gradients, the kinetic energy of the highest resolved modes and the cut-off length scale Δ. The viscosity is defined as:

$$\mu_T^{SGS} = \overline{\rho} C_3 \left|\tilde{\omega}\right|^{1/2} \Delta^{3/2} \left(q_c^2\right)^{1/4} \qquad (15)$$

where $\tilde{\omega}$ is the vorticity of the resolved scales defined by $\tilde{\omega} = \nabla \times \tilde{u} \equiv \mathrm{Curl}(\tilde{u})$ and the quantity $q_c^2 = \frac{1}{2} u_i' u_i'$ is the kinetic energy of the test field $u' = \tilde{u} - \hat{\tilde{u}}$, which is extracted from the resolved velocity field through the application of a test filter associated with the cut-off length scale $\hat{\Delta} > \Delta$, usually taken as $\hat{\Delta} = 2\Delta$. More discussions on the evaluation of the test filtered velocity $\hat{\tilde{u}}$ will be expounded below. The value of the constant C_3 according to Saguat (1996) is 0.1.

All the above models have been designed assuming that the simulated flow is turbulent, fully-developed and isotropic, and therefore do not incorporate any information related to an eventual departure of the simulated flow from these assumptions. In order to obtain an automatic adaptation of the models for inhomogeneous flows, simulations of engineering flows are more likely to be based on the dynamic formulations of these models. Nonetheless, application of the various dynamics SGS models (Ghosal *et al.*, 1995; Germano *et al.* 1991; Meneveau *et al.*, 1996) in fire simulations is still subject to further investigation which is not included in the present review.

The Large Eddy Simulation (LES) approach has recently become the central focus of fire modelling community (Xin et al., 2002,2005; McGrattan and Forney, 2004; Kang and Wen, 2004). McGrattan *et al.* (1994) first developed a two-dimensional LES fire model using finite difference and stream function formulation. Based on these two dimensional LES codes, they gradually extended the complexity of the model to a famous three-dimensional explicit LES fire model (Baum *et al.* (1994); McGrattan *et al.*, (1996, 1998)). Beginning from the year 2000, the computer code was publicly released as a freeware in the webpage of the National Institute of Standard and Technology (NIST) and named as – Fire Dynamic Simulator (FDS) (McGrattan and Forney, 2004).

Based on the spatial filtering technique, LES can provide information to match macroscopic observables (scales that are resolved on computational mesh) while the microscopic (unresolved) information are indirectly reflected by the formulation of subgrid-scale (SGS) turbulence model. As the dynamic behaviours of large eddies are resolved directly, the temporal vortical structures are expected to be better captured by the LES as opposed to the RANS approach (Tieszen et al., 1996; Baum et al., 1994).

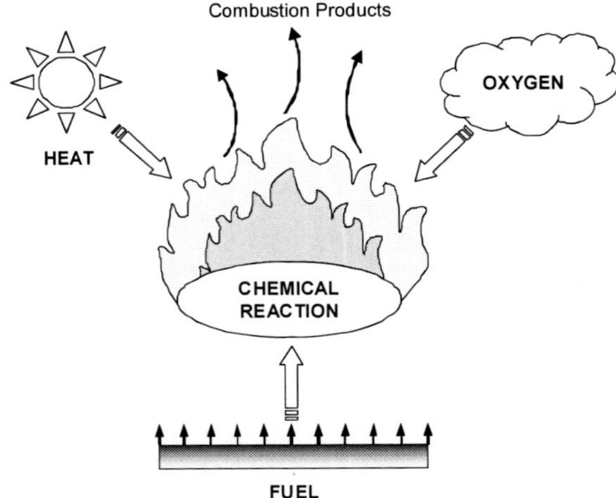

Figure 4. A schematic representation of flaming fire condition (Fire Triangle)

2.3. Combustion Modelling

2.3.1. Premixed and Diffusion Flame

Before taking a close look with various combustion models, let us first consider the three essential elements of combustion processes as illustrated in Figure 4 (also known as "Fire Triangle"). A *flaming fire* is categorically a rapid oxidation process. Chemical reaction which resides at the centre can be only realized when adequate supplies of fuel and oxygen react in an environment with sufficient heat. To create a self-sustaining combustion process, it is the chemical reaction that actually feeds the fire more heat and allows it to continue. A burning fire can therefore be regarded as a manifestation of the chemical reaction. It is nonetheless recognized that the mode of burning may depend somewhat on the physical condition and distribution of the fuel, and its surrounding environment, than on its chemical nature.

Assuming there has sufficient heat to initiate the combustion; the mode of burning then depends on the availability or mixing of fuel and oxygen (or generally oxidants). Based on the mixing nature of fuel and oxidants, model of burning can be broadly categorized in two types: Premixed and Diffusion Flame. Consider a naturally flaming process as described by a burning candle in Figure 5. This common flame is a classical example of a diffusion flame. The principal characteristic of the diffusion flame is that the fuel and oxidizer (oxygen) are initially separated and combustion occurs in the zone where the gases are mixed. A candle flame operates through the evaporation of the fuel rises in a laminar flow of hot gas which reacts with the oxygen diffusing into the flame from the surrounding air. The identifiable luminous yellow zone of the candle flame represents the soot being produced that becomes incandescent from the heat of the flame. In the candle flame, a zone of blue tip exists because of the occurrence of some premixing of the fuel and oxidizer at the bottom edge of the wick where the flame is quenched, which incidentally has a similar appearance to a premixed flame. In a premixed flame, the structure of the flame is rather different. A fully aerated Bunsen burner is the simplest example in developing such a flame of which the oxygen supply is premixed with the fuel prior to combustion. A blue-coloured flame naturally emerges with less black body radiating soot being produced. The slow diffusion process of a burning candle represents a typical laminar flame. Most practical fires occur however at high flame speeds. The combustion process is now characterized by turbulent flows. Turbulent combustion is strongly governed by the *eddy mixing* process between the fuel and oxidizer. The focus of this section is therefore concentrated on the modelling technique of turbulent diffusion flame.

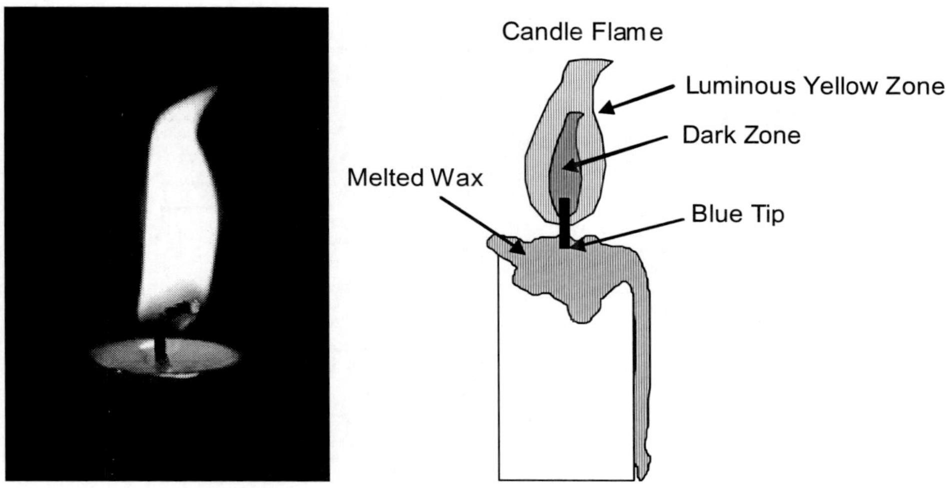

Figure 5. An example of a naturally flaming process: a burning candle

In a state of combustion, a flame is the product of a highly exothermic reaction and it can be regarded as a body of gaseous material consisting of reacting gases and finely dispersed carbonaceous particles (soot) that emit at specific wavelength bands depending on the combustion chemistry on the fuel involved. Consider a naturally flaming process as described by a burning candle in Figure 3.2. This common flame is a classical example of a diffusion flame. Another example is a Bunsen burner with a closed oxygen valve. The principal characteristic of the diffusion flame is that the fuel and oxidizer (oxygen) are initially separated and combustion occurs in the zone where the gases are mixed. A candle flame operates through the evaporation of the fuel rises in a laminar flow of hot gas which reacts with the oxygen diffusing into the flame from the surrounding air. The identifiable luminous yellow zone of the candle flame represents the soot being produced that becomes incandescent from the heat of the flame. In the candle flame, a zone of blue tip exists because of the occurrence of some premixing of the fuel and oxidizer at the bottom edge of the wick where the flame is quenched, which incidentally has a similar appearance to a premixed flame. In a premixed flame, the structure of the flame is rather different. A fully aerated Bunsen burner is the simplest example in developing such a flame of which the oxygen supply is premixed with the fuel prior to combustion. A blue-coloured flame naturally emerges with less black body radiating soot being produced.

2.3.2. Definition of mixture fraction

Under a certain set of simplifying assumptions, the basis of the non-premixed modeling approach is that the instantaneous thermochemical state of the fluid is related to a conserved scalar quantity known as the mixture fraction f. The concept of mixture fraction is essentially a numerical construct used in analysis of non-premixed combustion to describe the degree of scalar mixing between the fuel and oxidant. It is a local quantity within the flow field that varies both spatially and temporally.

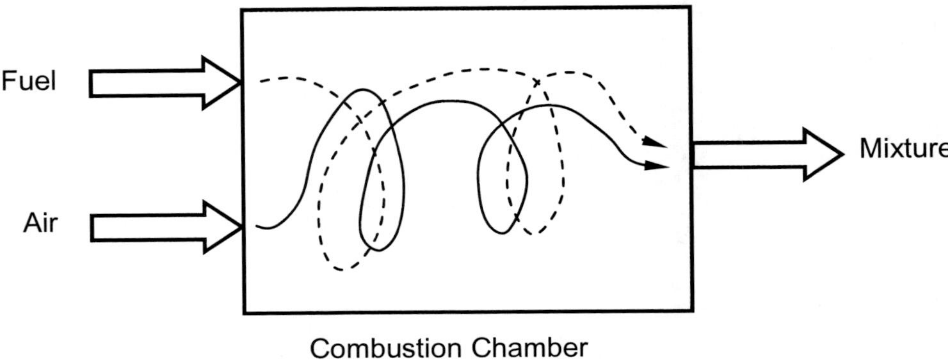

Figure 6. Mixing of steady fuel and air streams in a combustion chamber

The mixture fraction is best understood by visualizing the mixing process in a generic combustion chamber as illustrated in Figure 6. Supposing that a resultant mixture (M) leaves at one end after the fuel (F) and air (A) enter the system via two feeding inlets and thereafter thoroughly mix in the chamber, the resulting mixture from this two-stream mixing process can be written as:

$$f\beta_F + (1-f)\beta_A = \beta_M \tag{16}$$

The above equation subsequently leads to

$$f \equiv \frac{\beta_M - \beta_A}{\beta_F - \beta_A} \tag{17}$$

This property β of the mixture which is free from sources and sinks obeys the relation of a conserved property. In a non-reacting mixture of pure fuel and air, the mass fraction of fuel is equivalent to the value of f, and the mass fractions of oxygen and nitrogen are equivalent to $1-f$. If $Y_{fu,F}$ is taken to be the mass fraction of fuel in the fuel stream (unity for pure fuel) and $Y_{ox,A}$ and $Y_{in,A}$ be the mass fractions of oxygen and nitrogen in the air stream, the mass fractions of fuel, oxygen and nitrogen in a completely *unburnt* mixture are respectively:

$$Y_{fu} \equiv fY_{fu.F}$$

$$Y_{ox} \equiv (1-f)Y_{ox,A}$$
$$Y_{in} \equiv (1-f)Y_{in,A} \tag{18}$$

Within the combustion chamber, a special value of the mixture fraction, f_{st}, exists which divides the flame into two distinct regions. For a single-step chemical reaction, with no oxidant in the fuel feeding stream and no fuel in the oxidant stream, f_{st} corresponds to the stoichiometric condition:

$$f_{st} \equiv \frac{Y_{ox,A}}{rY_{fu,F} + Y_{ox,A}} \tag{19}$$

2.3.3. Mixed-is-burnt model

Chemically speaking, most chemical reactions of practical fire problem have high rates. The chemical reaction time in such case is therefore negligibly short in comparison to the mixing time due to turbulence. Based on the above phenomenon, the *fast chemistry assumption* is commonly adopted for the turbulent flame/fire modelling which drastically simplifies the problems of chemistry-turbulence interactions since the molecular-species concentrations are directly related to a conserved scalar, and the statistics of all thermodynamics variables are obtainable from the knowledge of statistics of that scalar (Bilger, 1980). The *Mixed-is-Burnt* model is one of the combustion models developed specifically for the simplest reaction scheme with the basis of fast chemistry assumption. This approach assumes that the chemistry is infinitely fast and irreversible, with fuel and oxidant species never co-exist in space and proceeds into complete combustion resulting in a one-step conversion to final products. Such a chemical reaction allows the species mass fractions to be determined directly from the given reaction stoichiometry without any required knowledge for the reaction rate or chemical equilibrium information.

Consider the fast chemical reaction for a stoichiometric combustion of one mole of fuel in an oxidant stream consisting only of nitrogen and oxygen:

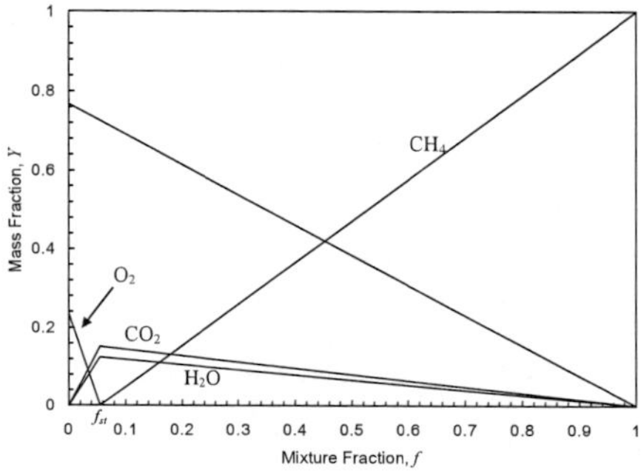

Figure 7. Complete combustion state relationships for methane

$$\mathrm{F}+\nu_{O_2}\left(\mathrm{O}_2+\frac{0.79}{0.21}\mathrm{N}_2\right)\rightarrow \nu_{CO_2}\mathrm{CO}_2+\nu_{H_2O}\mathrm{H}_2\mathrm{O}+\nu_{O_2}\frac{0.79}{0.21}\mathrm{N}_2 \tag{20}$$

For the combustion methane, the stoichimetric coefficients of ν_{O_2}, ν_{CO_2} and ν_{H_2O} as described in Eq. (20) are 2 moles, 1 mole and 2 moles respectively. In the fuel stream, if pure fuel exists, the mass fraction of fuel $Y_{fu,F}$ is equivalent to unity at $f = 1$. Since oxygen and nitrogen co-exist together in the air stream, the evaluation of the mass fraction of oxygen $Y_{ox,A}$ can be ascertained by the number of moles in the reactants' side (left hand side) of the single-step reaction Eq. (20).

The linear relationship between the instantaneous mass fractions of fuel, oxygen, carbon dioxide, water vapour and nitrogen and the mixture fraction f for a non-premixed combustion system in the limit of fast chemistry is shown in Figure 7. Relationships for other fuels besides methane could also be similarly derived following the similar procedures.

Eq. (20) could also be further generalized by introducing the equivalence ratio ϕ, which is expressed (including excess fuel or oxygen) as:

$$\begin{gathered}\mathrm{F}+\frac{1}{\phi}\nu_{O_2}\left(\mathrm{O}_2+\frac{0.79}{0.21}\mathrm{N}_2\right)\rightarrow\\ \left(1-\frac{1}{\phi}\right)\mathrm{F}+\left(\frac{1}{\phi}-1\right)\nu_{O_2}\mathrm{O}_2+\frac{1}{\phi}\nu_{CO_2}\mathrm{CO}_2+\frac{1}{\phi}\nu_{H_2O}\mathrm{H}_2\mathrm{O}+\frac{1}{\phi}\nu_{O_2}\frac{0.79}{0.21}\mathrm{N}_2\end{gathered} \tag{21}$$

The equivalence ratio may be defined as the ratio of actual fuel-oxidant ratio (= mass of fuel / mass of oxidant) to the ratio of fuel-oxidant for a stoichiometric process. It can be related to the mixture fraction as:

$$\phi=\frac{f}{1-f}\frac{1-f_{st}}{f_{st}} \tag{2}$$

For fuel-lean condition, the system lies in between $0 < \phi < 1$. At stoichiometric, $\phi = 1$. For fuel-rich condition, we have $1 < \phi < \infty$.

On the basis where equal diffusivities are assumed, the species transport equations can be reduced to a single equation for the mixture fraction f. The reaction source terms for each chemical reactant can be eliminated and represented by the mixture fraction f. The Favre-averaged transport equation for the mixture fraction with equal diffusivities is:

$$\frac{\partial}{\partial t}\left(\bar{\rho}\tilde{f}\right)+\frac{\partial}{\partial x_j}\left(\bar{\rho}\bar{u}_j\tilde{f}\right)=\frac{\partial}{\partial x_j}\left[\left(\frac{\mu}{Sc}+\frac{\mu_T}{Sc_T}\right)\frac{\partial\tilde{f}}{\partial x_j}\right] \tag{23}$$

Although the assumption of equal diffusivities appears to be problematic for laminar flows, it is generally acceptable for turbulent flows to adopt a global mass diffusion coefficient since turbulent convection generally overwhelms the molecular diffusion; the specification of detailed laminar diffusion properties is therefore unwarranted.

Early attempts to analyze turbulent diffusion flames using the *mixed-is-burnt model* have entailed the consideration whereby the chemical reaction possessed the features of being single-step, irreversible and infinitely fast. It should be noted that such an approach despite its inherent simplicity is still prevalently used in field modeling and will continue to be adopted so long as the detailed chemistry particularly for the combustion of most solid fuels is absent. As a further extension to the *mixed-is-burnt model*, the *chemical equilibrium model* could be adopted to express all the intermediate species within a hydrocarbon fuel as a function of the conserved scalar (or mixture fraction). This particular model assumes that the chemistry is rapid enough for chemical equilibrium to always exist at the molecular level. Details of the *chemical equilibrium model* can be found in (Sivathanu and Faeth, 1990; Yeoh and Yuen, 2009) and the references therein.

2.3.4. Laminar flamelet model with probability density function (PDF) approach

Another strategy to better describe the combustion process of turbulent non-premixed combustion is the assumption whereby the microscopic element in the model, describing the local mixture state and burning, can be taken to have the structure of an undisturbed laminar diffusion flame. This flame sheet according to Williams (1975) and Peters (1984) can be treated as an ensemble of counterflow diffusion flames called flamelets. This approach (commonly known as *laminar flamelet model*) differs from that of *mixed-is-burnt model* in the use of laminar flamelet relationships in place of equilibrium.

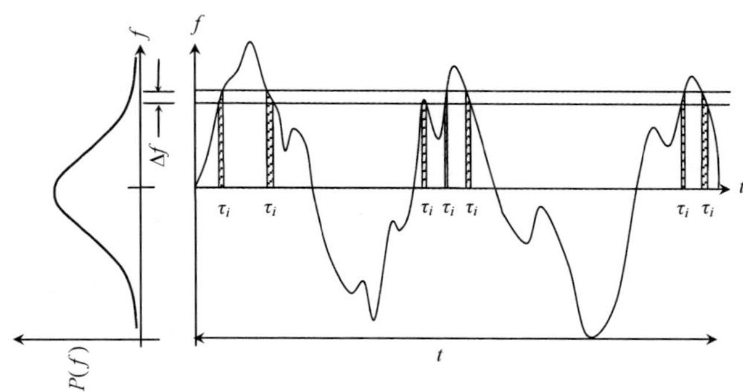

Figure 8. Probability density function of the mixture fraction

Nonetheless, in turbulent reacting flow, the fuel and oxygen concentration fluctuations generally have a measurable value that varies from one location to another in the flame zone. Owing to these fluctuations as observed by Bilger (1972) and El Ghobashi (1974), the fuel and the oxidant exist at the same location but at different times. To account the effects of these fluctuations, closure between laminar flamelet relationships and turbulent fluctuation has to be developed. The probability density function (PDF) approach is one of commonly adopted approach.

The concept of the probability density function is illustrated in Figure 8. From a physical viewpoint, the fluctuating value of the mixture fraction f spends some fraction of time in the range denoted as Δf (see right hand side of Figure). Denoting the probability density function as $P(f)$, it should take on values such that the area under its curve in the band denoted by Δf is equal to the fraction of time that spends in this range. From a mathematical viewpoint, it takes the form:

$$P(f)\Delta f = \lim_{t\to\infty} \frac{1}{t} \sum_i \tau_i \tag{24}$$

where t is the time scale and τ_i is the amount of time that f spends in the Δf band. From Eq. (24), it also implies that the sum of the values of $P(f)\Delta f$ over a long period of time must be equal to unity, i.e.

$$\int_{-\infty}^{\infty} P(f)\,df = 1 \tag{25}$$

Hence, the Favre-averaged value of $\tilde{f}$ can be expressed as:

$$\tilde{f} = \int_{-\infty}^{\infty} f P(f)\,df \tag{26}$$

which represents the first moment of $P(f)$ about the origin $f = 0$. The effect of turbulent fluctuations on the local flow properties can be introduced by including second and higher-order correlations of the concentration. The concentration fluctuation g can be defined as:

$$g = \widetilde{f''^2} = \int_{-\infty}^{\infty} \left(f - \tilde{f}\right)^2 P(f)\,df \tag{27}$$

which the above equation constitutes a second-order closure. In addition to solving the Favre-averaged mixture fraction, which is described by equation (3.4.67), the local values of the concentration fluctuation g can also be obtained from the corresponding conservation equation expressed in the convection-diffusion form as:

$$\frac{\partial}{\partial t}(\bar{\rho} g) + \frac{\partial}{\partial x_j}\left(\bar{\rho}\bar{u}_j g\right) = \frac{\partial}{\partial x_j}\left[\left(\frac{\mu}{Sc} + \frac{\mu_T}{Sc_T}\right)\frac{\partial g}{\partial x_j}\right] + C_{g_1}\left(\frac{\mu}{Sc} + \frac{\mu_T}{Sc_T}\right)\frac{\partial \tilde{f}}{\partial x_j}\frac{\partial \tilde{f}}{\partial x_j} - C_{g_2}\bar{\rho}\frac{\varepsilon}{k} g \tag{28}$$

The last two terms of Eq. (28) which represents the source terms of the governing equation denote the production of concentration fluctuation due to non-uniformity of mixture fraction and dissipation of the fluctuations due to the molecular diffusion.

The mean properties based upon Favre averaging can be obtained from the knowledge of the distribution of $P(f)$, which in turn can be ascertained via different PDF shapes of varying complexity. Assuming the shape of PDF is known, at any location in space, the density-weighted average of any scalar property ϕ, which can be expressed as a function of the instantaneous mixture fraction f, may be obtained from:

$$\tilde{\phi} = \int_0^1 \phi(f) P(f) df \tag{29}$$

And the mean mass fraction of ith species is given by:

$$\tilde{Y}_i = \int_0^1 Y_i(f) P(f) df \tag{30}$$

Obviously, the shape of the function $P(f)$ is strongly influenced by the nature of the turbulent fluctuations in f. In practice, $P(f)$ is expressed as a mathematical function that approximates the PDF shapes that have been observed experimentally. Three mathematical PDF shape are usually adopted. There are:

- Double delta functions (El Ghobashi, 1975; Khalil, 1977) at $f = 0$ and $f = 1$, i.e. square wave distribution of f with time
- A Gaussian distribution (Lockwood and Naguib, 1975; Kent and Bilger, 1976) between $f = 0$ and $f = 1$ together with two delta functions at $f = 0$ and $f = 1$
- A beta function (Jones, 1980) that automatically bounds between $f = 0$ and $f = 1$

Details of each presumed function can be found from the above references and various text books (Kuo, 1986; Yeoh and Yuen, 2009). Moreover, it has to be emphasized that the above presumed PDF approach is not limited for laminar flamelet approach describe below. Depends on the desired levels of complexity for the chemistry of the non-premixed combustion, other state relationships obtained form chemical equilibrium, generalized empirical relationships or experimentally determined profiles can also be adopted.

Coming back to the laminar flamelet model, the model views a turbulent flame as an ensemble of laminar diffusion flamelets. Similar to the fast chemistry assumption where the notion is taken whereby the chemical time scale is very much smaller than the convective and diffusive time scales, the fuel and oxidant are considered to react in narrow regions in the vicinity of the stoichiometic flame surfaces. In turbulent flows, where the local diffusion time scales can vary considerably, the fast chemistry assumption may not be locally valid. In such situation, the non-equilibrium effects of chemical reaction are indispensable and should be incorporated in modeling. The laminar flamelet is essentially an extension of the conserved scalar formulation to include non-equilibrium effects. The main advantage of the model is the feasibility of incorporating detailed chemistry with relative ease and simplicity by decoupling of the combustion chemistry modeling from the calculations of the flow field.

There are two methods of generating the laminar flamelets. The first method involves solving governing equations for opposed flow diffusion flame situations in the physical space. There are a number of available computer programs for laminar flamelet calculations such as

RUN1DL (Rogg, 1993, and Rogg and Wang, 1997), OPPDIF (Lutz et al., 1997) and more recently, COSILAB (http://www.softpredict.com) of which the reader may be interested in applying these codes in field modeling. The second method which concerns transforming the governing equations of the opposed flow diffusion flames into mixture fraction space will be further elaborated below.

According to *Bilger's mixture fraction formula* (Bilger, 1988), a conserved scalar most preferably employed for combustion purposes is the mass fractions Z of the chemical elements (C, H, O) of which can be related to the mass fractions Y of species by:

$$Z_j = \sum_{i=1}^{N} \frac{a_{ij} M_j}{M_i} Y_i \tag{31}$$

where a_{ij} is the number of atoms of element j in a molecule of species i, M_i is the molecular weight of species i and M_j is the molecular weight of element j. Using these element mass fractions, the conserved scalar for the variable β for a typical reaction $\nu_C \mathrm{C} + \nu_O \mathrm{O} + \nu_H \mathrm{H} \rightarrow \mathrm{Products}$ where ν_j is the number of atoms of element j can be defined as:

$$\beta = \frac{Z_C}{\nu_C M_C} + \frac{Z_H}{\nu_H M_H} - 2\frac{Z_O}{\nu_O M_O} \tag{32}$$

the mixture fraction then becomes:

$$f = \frac{\frac{Z_C}{\nu_C M_C} + \frac{Z_H}{\nu_H M_H} - 2\frac{(Z_O - Z_{O,ox})}{\nu_O M_O}}{\frac{Z_{C,fu}}{\nu_C M_C} + \frac{Z_{H,fu}}{\nu_H M_H} - 2\frac{Z_{O,ox}}{\nu_O M_O}} \tag{33}$$

As previously indicated, the subscripts *fu* and *ox* refer to the fuel and oxidant streams.

On the basis of the developments suggested by Peters (1984, 1986), the flamelet equations for the mass fraction of species and temperature can be derived by using the co-ordinate transformation of the Crocco-type. In mixture fraction space, they are:

Species

$$\rho \frac{\partial Y_i}{\partial t} - \rho D \left(\frac{\partial f}{\partial x_j} \right)^2 \frac{\partial^2 Y_i}{\partial f^2} - \omega_i + \underbrace{R(Y_i)}_{\text{higher order terms}} = 0 \tag{34}$$

Temperature

$$\rho \frac{\partial T}{\partial t} - \rho D \left(\frac{\partial f}{\partial x_j} \right)^2 \frac{\partial^2 T}{\partial f^2} - \frac{1}{C_p} \frac{\partial p}{\partial t} + \sum_{i=1}^{N} \frac{h_i}{C_p} \omega_i + \underbrace{R(T)}_{\text{higher order terms}} = 0 \tag{35}$$

The boundary conditions are:

Oxidant stream, $f = 0$, $T = T_{ox,A}$ $Y_i = Y_{i,ox,A}$ $i = 1, . . , N$

Fuel stream, $f = 1$, $T = T_{fu,F}$ $Y_i = Y_{i,fu,F}$ $i = 1, . . , N$

The influence of the flow field is introduced into Eq. (34) and (35) by the instantaneous scalar dissipation (s^{-1}) defined by:

$$\chi = 2D\left(\frac{\partial f}{\partial x_j}\right)^2 = 2D\left[\left(\frac{\partial f}{\partial x}\right)^2 + \left(\frac{\partial f}{\partial y}\right)^2 + \left(\frac{\partial f}{\partial z}\right)^2\right] \tag{36}$$

The scalar dissipation is a parameter that controls mixing, and represents the non-uniformity of the mixture fraction, which is related to the strain. The different flamelet relationships of the temperature and mass fractions of major and minor species for different scalar dissipation rates of a C_1 skeletal reaction mechanism of methane combustion can be visulaized in Figure 9. With increasing scalar dissipation rates, the departure from chemical equilibrium is evident and at some critical scalar dissipation rate, which is just above $\chi = 10$, the flame is subsequently quenched. It should be noted that the above flamelet relationships have been obtained neglecting the higher order terms involving convection and curvature along the mixture fraction surface – $R(Y_i)$ and $R(T)$. The simpler forms of the one-dimensional equations have nevertheless been found to be sufficiently adequate for most practical purposes in fire investigations (Kang and Wen, 2004).

In principal, the two-equation models of turbulence provide the mean the scalar dissipation rate by relating it to the concentration fluctuations and the turbulent time scale ε/k. The mean scalar dissipation rate can be formulated as:

$$\tilde{\chi} = C_\chi \frac{\varepsilon}{k} g \tag{37}$$

Normally, the constant C_χ in Eq. (37) is assigned a value of $C_\chi = 2.0$. According to Bray and Peters (1994), it has nonetheless become common practice to ignore the scalar dissipation fluctuations. With this simplification, the average value of the scalar property is now given by:

$$\tilde{\phi} = \int_0^1 \phi(f, \tilde{\chi}) P(f) df \tag{38}$$

The above integration is usually not carried out during the flow field calculations. A flamelet library is generated for the mean scalar property as functions of the mean mixture fraction, concentration fluctuations and mean scalar dissipation rates. The CFD model looks up pre-integrated values from the created library.

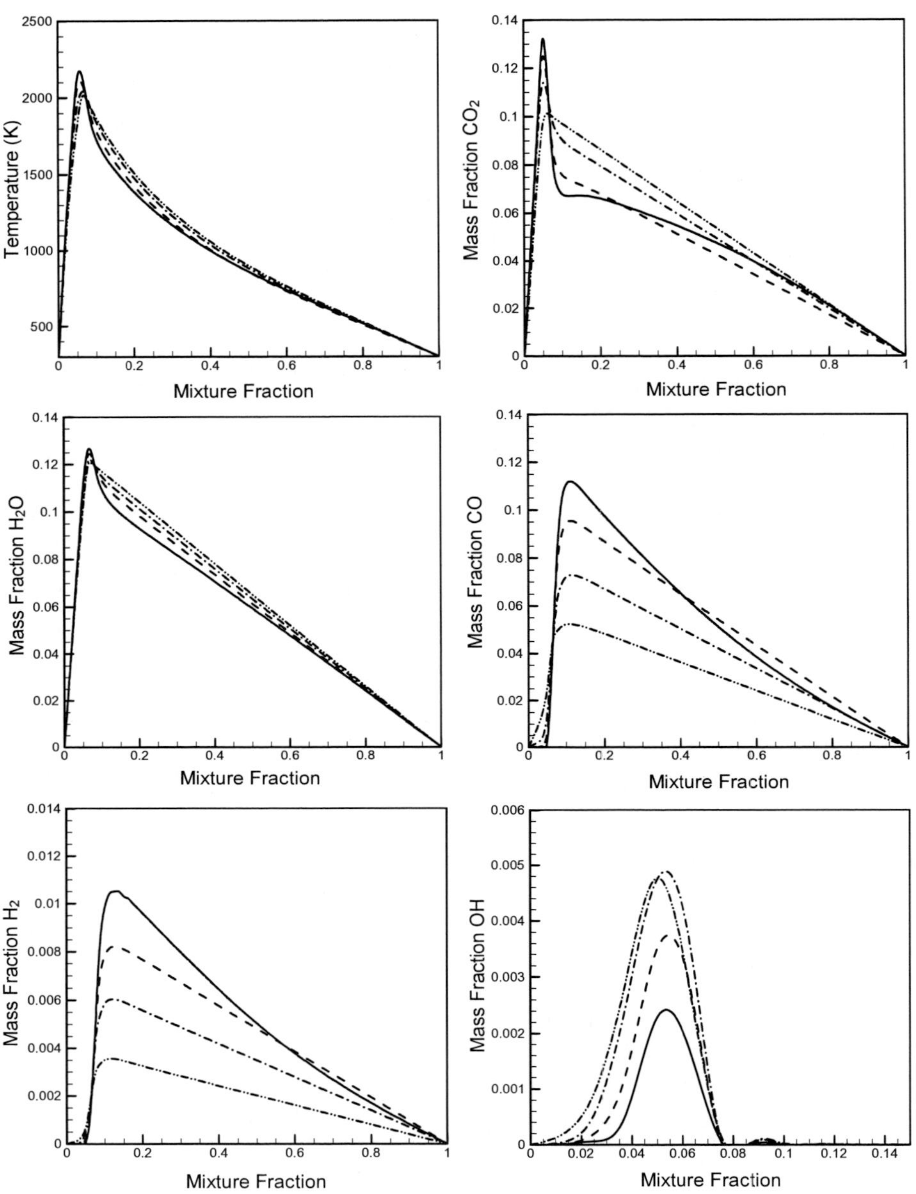

Figure 9. A sample of laminar flamelet profiles for methane combustion: χ = 0.01 ( ), χ = 0.1 ( ), χ = 1.0 ( ) and χ = 10.0 ( )

2.3.5. Eddy break-up and eddy dissipation model

In some circumstances such as fast chemistry for localized high-temperature regions or fast mixing for high turbulence intensity, the overall features of turbulent flames can be taken to be independent of the detailed chemistry. Owing to the *eddy mixing* process, the rate of combustion in this limit is assumed to be determined by the rate of intermixing on a molecular scale of fuel and oxygen eddies, which is presented by the rate of dissipation of eddies. These so-called *eddy break-up model* and *eddy-dissipation model* which is applicable for high Reynolds number represent an alternative approach to treat turbulent non-premixed flames in addition to the *mixed-is-burnt model*, and *laminar flamelet model.*

According to Spalding (1976), the eddy break-up reaction rate can be taken to be based simply on the species concentration fluctuations and the rate of break-up of eddies. For the single-step irreversible reaction for the consumption of fuel, the eddy break-up reaction can be expressed as:

$$\overline{R}_{fu} = C_R \overline{\rho} \frac{\varepsilon}{k} \left(\widetilde{Y''^2_{fu}} \right)^{1/2} \tag{39}$$

The above reaction rate requires the evaluation of the entity $\widetilde{Y''^2_{fu}}$. As reported by Bilger (1975) and Hutchinson et al. (1977), the Favre averaged conservation equation for the transport of $\widetilde{Y''^2_{fu}}$ follows the similar form of the convection-diffusion equation for its mean counterpart.

$$\frac{\partial\left(\overline{\rho}\widetilde{Y''^2_{fu}}\right)}{\partial t} + \frac{\partial\left(\overline{\rho}\tilde{u}_i\widetilde{Y''^2_{fu}}\right)}{\partial x_i} = \frac{\partial}{\partial x_i}\left[\left(\frac{\mu}{Sc} + \frac{\mu_T}{Sc_T}\right)\frac{\partial \widetilde{Y''^2_{fu}}}{\partial x_i}\right] + \overline{R}_{\widetilde{Y''^2_{fu}}} \tag{40}$$

where the mean source rate $\overline{R}_{\widetilde{Y''^2_{fu}}}$ can be modeled according to:

$$\underbrace{C_{g_1} G_{fu}}_{Term\ I} - \underbrace{C_{g_2} \overline{\rho} \frac{\varepsilon}{k} \widetilde{Y''^2_{fu}}}_{Term\ II} \underbrace{- 2A_o \frac{\overline{\rho}^2}{M_{fu} M_{ox}} \tilde{Y}_{fu} \tilde{Y}_{ox} \exp\left(-\frac{E_a}{R_u \tilde{T}}\right)\left(\frac{\widetilde{Y''_{fu}Y''_{ox}}}{\tilde{Y}_{ox}} + \frac{\widetilde{Y''^2_{fu}}}{\tilde{Y}_{fu}}\right)}_{Term\ III} \tag{41}$$

and the default constants of $C_{g_1} = 2.0$ and $C_{g_2} = 2.0$ are commonly adopted. Term I denotes the production of concentration fluctuations due to the non-uniformity of the fuel mass fraction of which G_{fu} can be formulated as:

$$G_{fu} = \left(\frac{\mu}{Sc} + \frac{\mu_T}{Sc_T}\right)\frac{\partial \tilde{Y}_{fu}}{\partial x_i}\frac{\partial \tilde{Y}_{fu}}{\partial x_i} \tag{42}$$

Term II describes the dissipation of the fluctuations due to molecular diffusion; the term ε/k that is the reciprocal of the turbulent time scale can be regarded as the rate of decay multiplier for the turbulence kinetic energy *k*. Term III appears due to the presence of the finite rate effect on the fluctuations of Y'^2_{fu}. Based on Khalil (1977), the entity $\widetilde{Y''_{fu}Y''_{ox}}$ that appears in the above equation can be obtained from the corresponding convection-diffusion conservation equation alongside with the mean source rate following Eq. (41) as:

$$\frac{\partial\left(\bar{\rho}\widetilde{Y''_{fu}Y''_{ox}}\right)}{\partial t}+\frac{\partial\left(\bar{\rho}\tilde{u}_i\widetilde{Y''_{fu}Y''_{ox}}\right)}{\partial x_i}=\frac{\partial}{\partial x_i}\left[\left(\frac{\mu}{Sc}+\frac{\mu_T}{Sc_T}\right)\frac{\partial\widetilde{Y''_{fu}Y''_{ox}}}{\partial x_i}\right]+\bar{R}_{\widetilde{Y''_{fu}Y''_{ox}}}=$$

$$\left[\underbrace{C_{g_1}\left(\frac{\mu}{Sc}+\frac{\mu_T}{Sc_T}\right)\frac{\partial\tilde{Y}_{fu}}{\partial x_i}\frac{\partial\tilde{Y}_{ox}}{\partial x_i}}_{Term\ I}-\underbrace{C_{g_2}\bar{\rho}\frac{\varepsilon}{k}\widetilde{Y''_{fu}Y''_{ox}}}_{Term\ II}\right.$$

$$\left.\underbrace{-A_o\frac{\bar{\rho}^2}{M_{fu}M_{ox}}\tilde{Y}_{fu}\tilde{Y}_{ox}\exp\left(-\frac{E_a}{R_u\tilde{T}}\right)\left(\left(\tilde{Y}_{fu}+r\tilde{Y}_{ox}\right)\frac{\widetilde{Y''_{fu}Y''_{ox}}}{\tilde{Y}_{fu}\tilde{Y}_{ox}}+\frac{\widetilde{Y''^2_{ox}}}{\tilde{Y}_{ox}}+r\frac{\widetilde{Y''^2_{fu}}}{\tilde{Y}_{fu}}\right)}_{Term\ III}\right] \quad (43)$$

The estimation of the relative contribution of the chemical kinetics and turbulent mixing to the mean source rates in the conservation equations of $\widetilde{Y''^2_{fu}}$ and $\widetilde{Y''_{fu}Y''_{ox}}$ can be best explained through the *Damköhler* number *Da*. In flame situations where the reactants are in intimate contact due to rapid mixing, characterized by large values of ε/k, the reaction is kinetically influenced, i.e. *Da* « 1. The influence of chemical kinetics on the formation of the entities $\widetilde{Y''^2_{fu}}$ and $\widetilde{Y''_{fu}Y''_{ox}}$ is hence negligible and only Terms I and II contribute to the generation/destruction of $\widetilde{Y''^2_{fu}}$ and $\widetilde{Y''_{fu}Y''_{ox}}$. Nevertheless, in the flame situations where the temperature is high enough for the reaction to proceed but the reactants are not in intimate contact, the reaction is considered to be controlled by mixing; the chemical time τ_k is very small relative to τ_s, i.e. *Da* » 1. Here again, if the combustion process is suitably represented by a single-step reaction, the respective relationships for the mean reaction rates can be obtained:

$$\tilde{R}_{fu}=\frac{1}{r}\tilde{R}_{ox}=-\frac{1}{1+r}\tilde{R}_{pr} \quad (44)$$

On the basis of Eq. (40), the eddy break-up model requires the solution to the mass fractions of the fuel, oxidant and products, in addition to three conservation equations for the transport of second order correlations: $\widetilde{Y''^2_{fu}}$, $\widetilde{Y''_{fu}Y''_{ox}}$ and $\widetilde{Y''^2_{ox}}$. The transport equation for the entity $\widetilde{Y''^2_{ox}}$ is formulated similar to Eq. (40) and (45).

Magnussen and Hjertager (1976) propose an alternative combustion model of which they have considered the reaction rate to be governed by the mean species concentrations rather than the species concentration fluctuations. Known as the eddy dissipation model, the consumption rate of the fuel in a single-step irreversible reaction in terms of mass fractions is given by the smaller (i.e. limiting value) of the two expressions below:

$$\overline{R}_{fu} = C_R \overline{\rho} \frac{\varepsilon}{k} \min\left[\tilde{Y}_{fu}, \frac{\tilde{Y}_{ox}}{r} \right] \tag{45}$$

$$\overline{R}_{fu} = C'_R \overline{\rho} \frac{\varepsilon}{k} \left(\frac{\tilde{Y}_{pr}}{1+r} \right) \tag{46}$$

Like in the eddy break-up model above, the reaction rate will proceed whenever turbulence is present (small values of ε/k) and an ignition source is not required to initiate combustion. Eq. (52) is applicable for non-premixed flames, but the inclusion of Eq. (46) according to Magnussen and Hjertager (1976) also allows the model to handle premixed flames.

As formulated from above, the eddy dissipation model is closely related to the eddy break-up model. It is nonetheless evidently clear that the eddy dissipation model differs in especially relating the dissipation of eddies to the mean concentration of intermittent quantities instead of the concentration fluctuations. Significantly, this model does not call for solution of equations for the entities: $\widetilde{Y''^2_{fu}}$, $\widetilde{Y''_{fu}Y''_{ox}}$ and $\widetilde{Y''^2_{ox}}$. Nevertheless, the simplifications of the eddy dissipation model should be viewed from the proposition whereby the mean quantities appear intermittent within the turbulent fluid.

2.4. Radiation Modelling

In practical fires, heat transfer is mainly by means of conduction and convection. However, as temperature grows up, contribution by the radiative heat transfer becomes more significant or even plays an active role in the fire development (Bishop and Drysdale, 1998). This is due to the fact that combustion products such as carbon dioxide CO_2 and water vapor H_2O, and various hydrocarbon gases absorb and radiate heat which alters the radiation intensity along its path (as shown in Figure 10).

The radiation modeling in fires concerns two key issues. Firstly, appropriate evaluation of the absorption coefficient is required to account for the complex radiative properties of the participating medium consisting of gaseous mixture of combustion products. Secondly, the availability of appropriate solution methods to best handle the radiation heat transfer, which may entail different ways of treating the angular dependence and spatial variation of intensity within the physical domain. Possible approaches with different levels of approximation of radiation properties are first presented in the following section.

2.4.1. Radiation properties approximations

Every combustion process produces combustion gases such as water vapor and carbon dioxide. These gases do not scatter radiation significantly, but they are strong selective absorbers and emitters of radiant energy. Considering the diversity of products and the probability of having some or all of these gases in any volume element of the system, the prediction of radiative properties is certainly not an easy task. In order to present a systematic methodology for the prediction of the radiative properties of combustion products, this section will first focus on the discussion of suitable relations for the properties of the combustion gases (and soot particles) and simplifications that have been made in arriving to these relations.

Gray gas model

For engineering calculations, it is always desirable to have some reliable yet simple models for predicting the radiative properties of the gases. Gray gas assumption approach is one the simplest and widely adopted model for engineering approximation where radiation properties are expressed independent to the wavelength. If scattering is considered not to be important for the combustion gases, the gray absorption/emission coefficient can be obtained from Beer-Lambert's law. For a given mean path length L, the mean absorption coefficient for the combustion gases is given as:

$$\overline{K}_{a,g} = -\frac{1}{L}\ln\left(1-\varepsilon_{gray}\right) \tag{47}$$

where the corresponding mean path length L can be evaluated based on Viskanta and Menguc (1987) for a volume (V) of a gas radiating to its entire surface area (A) can be evaluated according to:

$$L = L_m = \frac{4CV}{A} \tag{48}$$

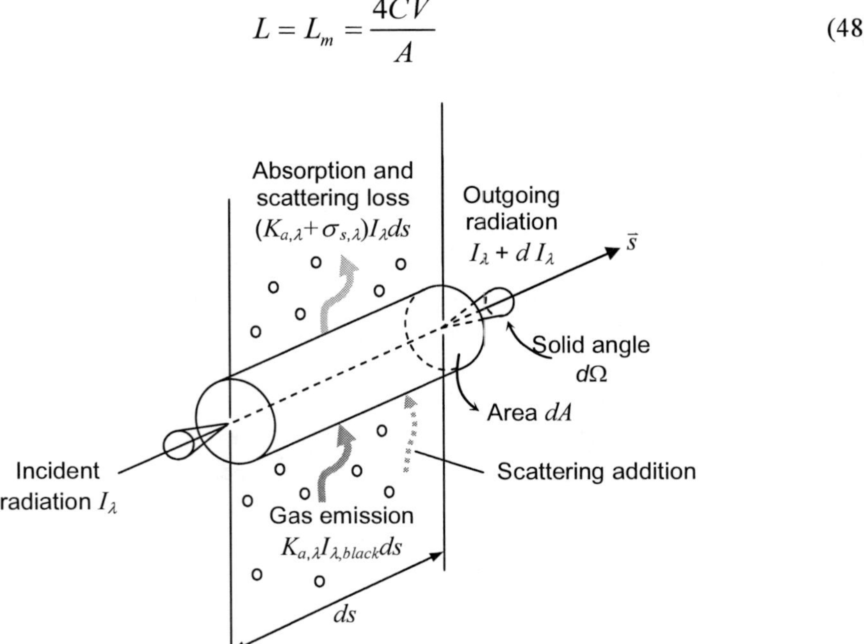

Figure 10. Change of intensity I_λ in direction $\overline{s}$ across an elemental volume

where C is the correction factor and for an arbitrary geometry its magnitude is 0.9 (Hottel and Sarofim, 1967). Considering the CO_2-H_2O mixture, the emissivity ε_{gray} in may be ascertained as:

$$\varepsilon_{gray} = C_{CO_2}\varepsilon_{CO_2} + C_{H_2O}\varepsilon_{H_2O} - \Delta\varepsilon_{CO_2-H_2O} \quad (49)$$

the term $\Delta\varepsilon_{CO_2-H_2O}$ is an additional correction to consider the overlap wavelength about 4.4 μm of CO_2 and 4.8 μm of H_2O. The gas emissivities of carbon dioxide ε_{CO_2} and water vapour ε_{H_2O} can be obtained from Hottel's charts (Hottel, 1954) of gas emittance as a function of gas temperature, gases partial pressure and mean beam path length. Alternatively, the emissivity of each individual constituent can be modelled according to Modak (1979) where Chebyshev polynomials are used instead of the Hottel's charts.

For soot particles, it is usually assumed that it is sufficient small compared to the wavelength of radiation and its complex refractive is independent of wavelength. Under these assumptions, the spectral absorption coefficient of soot that is inversely proportional to wavelength and scattering effects are negligible. The spectrally integrated absorptivity of soot of a path length L according to Felske and Tien (1974) is given by:

$$\alpha_{soot} = 1 - \frac{15}{\pi^4}\psi^{(3)}\left(1 + \frac{k_o\lambda_o T_s L}{C_2}\right) \quad (50)$$

where $k_o\lambda_o$ has been approximated by Hottel and Sarofim (1967) as a function of soot volume fraction f_v as $k_o\lambda_o \cong 7f_v$, C_2 is Planck's second constant with a value 1.4388 cm K, and $\psi^{(3)}$ is the pentagamma function (Abramowitz and Stegun, 1964).

The use of the absorption coefficients related to the mean beam length is a convenient way of scaling the radiation heat transfer in practical systems. Whereas difficulty may arise in finding an appropriate mean beam length, the mean absorption coefficient concept proposed by Hubbard and Tien (1978) can be used to calculate the mean absorption coefficient for the gas mixture. In this approach, the Planck's mean absorption coefficient which is independent of the mean beam length is determined. The Planck's mean absorption coefficient for the ith species can be expressed in discrete form as:

$$\overline{K}_{P,i} = \sum_j a_j \frac{E_{b,\lambda,j}}{\sigma T^4} \quad (51)$$

where a_j represents the integrated band intensity of each band. The overall absorption coefficient of the CO_2-H_2O and soot mixture is just the sum of the mean coefficients, which is:

$$\overline{K}_a = \overline{K}_{P,s}f_v + \left(\overline{K}_{P,CO_2}P_{CO_2} + \overline{K}_{P,H_2O}P_{H_2O}\right) \quad (52)$$

In field modelling, Wen and Huang (2000) have demonstrated the applicability of the Planck's mean absorption coefficient described in Eq. (52) to adequately represent the radiation properties of the CO_2-H_2O and soot mixture for confined jet fires in small and large compartments.

Weight sum of gray gas model

The weighted sum of gray gases model (WSGGM) first introduced by Hottel and Sarofim (1967) represents another elegant radiative gas property model which is a reasonable compromise between the oversimplified gray gas assumption and a complete model accounting for the entire spectral variations of radiation properties. The model postulates that the total emissivity and absorptivity may be represented by the sum of a gray gas emissivity weighted with a temperature dependent factor. The WSGGM entails the evaluation of the total emissivity over the distance L from the following expression:

$$\varepsilon_T = \sum_{i=0}^{I} a_{\varepsilon,i}\left(1 - e^{-k_i PL}\right) \tag{53}$$

where $a_{\varepsilon,i}$ denote the emissivity weighting factors for the ith fictitious gray gas as based on the gas temperature. The bracketed quantity in Eq. (53) is the ith gray gas emissivity with absorption coefficient k_i, and partial pressure-path length product PL. For a gas mixture, P is the sum of the partial pressures of the absorbing gases, i.e. $P = P_{CO_2} + P_{H_2O}$. Physically, the weighting factor $a_{\varepsilon,i}$ may be interpreted as the fractional amount of black body energy in the spectral regions where the gray gas coefficient k_i exists as illustrated in Figure 11. The absorption coefficient k_0 is assigned a zero value in order to account for windows in the spectrum between spectral regions of high absorption ($\sum_{i=0}^{I} a_{\varepsilon,i} < 1$), and the weighting factor for $i = 0$ is evaluated from:

$$a_{\varepsilon,0} = 1 - \sum_{i=1}^{I} a_{\varepsilon,i} \tag{54}$$

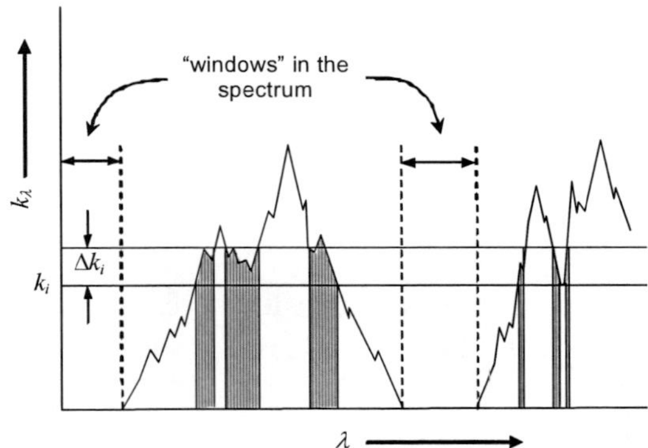

Figure 11. Interpretation of $a_{\varepsilon,i}$ in terms of the spectral energy distribution where $a_{\varepsilon,i}$ represents the fraction of black body energy in wave number region λ associated with Δk_i

A convenient representation of the temperature dependency of the weighting factors is of the polynomial form of order J – 1 given as:

$$a_{\varepsilon,i} = \sum_{j=1}^{J} b_{\varepsilon,i.j} T^{j-1} \tag{55}$$

where $b_{\varepsilon,i.j}$ are the emissivity gas temperature polynomial coefficients. Employing a statistical *narrow-band* model and experimental spectral data, Taylor and Foster (1974) have fitted the parameters for CO_2-H_2O mixture by employing a one-clear 3-gray gases model for temperatures between 1200 K and 2400 K for the ratio of partial pressure of water vapour to partial pressure of carbon dioxide: $P_{H_2O}/P_{CO_2} = 1$ or 2. It is noted that in the case of natural gas combustion (i.e. , methane-air or methane-oxygen), the ratio of water vapour to carbon dioxide partial pressures P_{H_2O}/P_{CO_2} is approximately 2. Most other hydrocarbon fuels have combustion products with a ratio of P_{H_2O}/P_{CO_2} in between 1 and 2. Smith et al. (1982) proposed a third-order polynomial fit of $b_{\varepsilon,i.j}$ for one-clear 3-gray gases model to incorporate temperatures between 600 K and 2400 K for $P_{H_2O}/P_{CO_2} = 1$ or 2. The initial emmisivity data were generated from Edwards' exponential *wide-band* model (1976). The polynomial coefficients $b_{\varepsilon,i.j}$ are presented by a third-order function of T:

$$a_{\varepsilon,i} = b_{\varepsilon,i,1} \times 10^{-1} + b_{\varepsilon,i,2} \times 10^{-4} T + b_{\varepsilon,i,3} \times 10^{-7} T^2 + b_{\varepsilon,i,3} \times 10^{-11} T^3 \tag{56}$$

In addition to CO_2 and H_2O, the model developed by Beer, Foster and Siddall (1971) further accounted the contribution of other gas species such as carbon monoxide CO and unburnt hydrocarbons (e.g. methane), which they are also significant emitters of radiation. To incorporate these species, the term $k_i P$ in Eq. (53) can be generalized as:

$$k_i P \rightarrow k_i \left(P_{CO_2} + P_{H_2O} + P_{CO} \right) + k_{HC_i} P_{HC} \tag{57}$$

For appreciably small distance L, i.e. $L \leq 10^{-4}$ m, it can be shown that the change of radiation intensity in the WSSGM is identical to the gray gas assumption. For large distance where L is much greater than 10^{-4} m, the Beer-Lambert's law should be used instead, which is given by:

$$\bar{K}_{a,g} = -\frac{1}{L} \ln\left(1 - \varepsilon_T\right) \tag{58}$$

For most practical purposes, Eq. (58) assumes that the absorptivity is taken to be equal to the emissivity. Such simplification is commonly adopted and this assumption is fully justified if the medium is not optically thin and the temperature does not differ considerably from the gas temperature.

Narrow band and wide band models

The simplifying assumption of adopting the gray gas assumption and a more sophisticated weighted sum of gray gases model is that the participating medium is usually taken to be homogeneous. For the inhomogeneous effects on radiative heat transfer in high temperature combustion gases, more complete models such as the statistical *narrow-band* or exponential *wide-band* models are necessary to accurately predict particularly the radiance emanating from non-isothermal, variable concentration carbon dioxide and water vapor mixture.

The *narrow-band* models have been developed to approximate the average behavior of the spectral absorption coefficient in a small spectra interval, which generally lead to analytical expressions of the spectral transmissivity averaged over a spectral range $\Delta\lambda$, but small enough to assume the spectral black body intensity $I_{black,\lambda}$ remains constant inside it. Hundreds of individual absorption lines are contained within this interval; statistical assumptions are made for line locations, shapes and intensities (Goody, 1952). Accurate predictions require the narrow band width to have a width of about 25 cm-1. The covered spectral range is between 150 cm-1 to 7000 cm-1 of which the total intensity calculation for CO_2-H_2O mixture requires scanning over about 300 narrow-band regions. It is not entirely surprising that the complexity of narrow-band models, and the large computational effort which they require, do not make them very attractive for engineering applications. Nevertheless, they are useful for model validation purposes. The reader may wish to obtain the computer code developed by Grosshandler (1993) from National Institute of Standards and Technology (NIST) called RADCAL which exemplifies the essential features of the narrow-band model for radiation calculations in a combustion environment.

The wide-band models significantly reduce the inefficient spectral calculations from 300 down to a total of about 30 bands or less. The radiative properties of gaseous species $\tau_{g,j}$ over the bandwidth $\Delta\lambda_j$ are determined using the exponential wide-band model of Edwards (1976). This model considers the absorption and emission of infrared radiation by a particular species is generated in between one and six or eight wide bands that are associated with vibrational modes of energy storage by the species. In contrast to those of narrow-band models, a detailed knowledge of the position and intensity of these rotational lines is considered to be unimportant in the wide-band model. The band shape is approximated by one of three simple exponential functions depending upon whether the lower limit, upper limit, or band center wave number is used to prescribe the position of the band. This model is suitable for non-isothermal gases and accounts for the effects of temperature and pressure on absorption and emission of radiation, and for the overlap of absorption bands for gaseous species such as the CO_2-H_2O mixture. More specific details on the method can be referred in Edwards (1976).

In practice, neither *narrow-band* nor *wide-band* models are actually required to accurately model the effects of radiative heat transfer in high temperature combustion gases. As a reasonable compromise, the concept of weighted sum of gray gases approach (Modest, 1991) can be applied for arbitrary solution methods in radiative transfer. In this method, the non-gray gas is replaced by a number of gray gases. The heat transfer rates are calculated independently, and the total flux is then determined by adding the fluxes of the gray gases after multiplication with certain weighting factors. It can be carried out to any desired accuracy, and since no spectral flux evaluations, followed by spectral integration, are required, computer time savings amount to factors of hundred and even thousands for comparable accuracy.

2.4.2. Radiation transfer methods

With the aforementioned models approximate the radiation properties of combustion products, discussion of modelling radiative heat transfer now leads to the solution methods of the Radiative Transfer Equation (RTE). According to Ozisik (1973) and Siegel and Howell (2002), the RTE describing the variation of a monochromatic pencil of radiation across an elemental volume taken along the path $\vec{r}$ for steady state conditions and coherent isotropic scattering is given by:

$$\frac{dI_\lambda(\vec{r},\vec{s})}{ds} = -\left(K_{a,\lambda} + \sigma_{s,\lambda}\right)I_\lambda(\vec{r},\vec{s}) + K_{a,\lambda}I_{black,\lambda}(\vec{r}) + \frac{\sigma_{s,\lambda}}{4\pi}\int_{4\pi}\Phi(\vec{s}',\vec{s})I_\lambda^-(\vec{s}')d\Omega' \quad (59)$$

where $I_{black,\lambda}$ is the Planck's intensity of black body radiation per unit wavelength at some temperature in the participating medium, $K_{a,\lambda}$ and $\sigma_{s,\lambda}$ are respectively the local spectral absorption and scattering coefficient for a wavelength λ, and Φ is the scattering phase function that specifies the fraction of incident intensity $I_\lambda^-(\vec{s}')$ from all possible $\vec{s}'$ that is scattered into direction $\vec{s}$ of which involves the integration over a unit sphere (i.e. a solid angle of 4π steradians) surrounding the point in the medium. As discussed before, it is common practice to neglect scattering the fire modelling. Without scattering, the TRE effectively simplified as:

$$\frac{dI_\lambda(\vec{r},\vec{s})}{ds} = -K_{a,\lambda}I_\lambda(\vec{r},\vec{s}) + K_{a,\lambda}I_{black,\lambda}(\vec{r}) \quad (60)$$

The radiative transfer equation in the form of Eq. (59) is an *integrodifferential* equation, a form very different from an ordinary partial differential equation, and because of this it is extremely difficult to solve exactly for multidimensional geometries. There have been many solution methods that have been developed over the years to handle radiative heat transfer. These include various analytical approximation techniques and numerical methods. This section will primarily focus on the latter algorithms that are commonly found in many general-purpose CFD applications as well as in field modeling: Monte Carlo, P-1 radiation model, discrete transfer radiative model, discrete ordinates model and finite volume method.

Monte carlo method

Monte Carlo methods for radiation heat transfer predictions are essentially purely statistical methods that yield solutions that are as accurate as exact methods. The merit of the model is its flexibility of being applied for any complex three-dimensional and non-Cartesian geometries, some known source of radiation incident on (or emitted within) the geometry, and complicated physical phenomena due to interaction of radiation with the participating medium, such as: transfer problem in furnaces (Steward and Cannon, 1971; Taniguchi *et al.*, 1984)

The model consists of simulating a finite number of photon histories (energy bundles) through the use of a random number generator (Taniguchi et al., 1988). By assigning the initial position, energy and direction of each photon, the mean free paths that the photon propagates can then be determined stochastically. With the sampled absorption and scattering coefficients, the Monte Carlo method is able to determine whether the incoming photon is absorbed or scattered by the medium. Depending if the photon is absorbed or scattered; the history of the photon is terminated or changed by assigning a new direction (Siegel and Howell, 2002; Modest, 2003).

Monte Carlo calculations provide the necessary means of attaining appropriate radiation quantities such as the surface heat flux and volumetric sources or sinks by tracking the energy absorbed or emitted within surface elements and volumes. The total energy absorbed and emitted by surfaces and volumes are usually recorded during the simulation. Additional information such as penetration distances could also be obtained and stored. In majority of applications, numerous photon histories are generated to obtain estimates that closely reflect the physical quantities in the system. As the number of photons initiated from each surface or volume element increases, this method is expected to converge to the exact solution of the problem. It should nevertheless be noted that the directions of the photons are ascertained from a random number generator; hence the method is always subjected to statistical errors and the lack of guaranteed convergence. Owning to the extensive data storage requirement, this method has not been widely adopted in engineering problems and certainly not within practical fire problem. However, it is expected that this method will become more attractive with the ever advancement of computing power. Numerical studies have been performed, which have demonstrated that Monte Carlo method can be employed in supercomputers with encourage solutions (Brown and Martin, 1984).

P-N method

The spherical harmonics *P-N* differential approximation is one of the most tedious and cumbersome of the radiative transfer methods first suggested by Jeans (1917). Nonetheless, it is also considered as the most elegant approach to handle radiative energy transfer problems because of its sound mathematical foundations. The model transforms the problem into spherical harmonics approximation Case and Zweifel (1967). The radiation intensity is the expressed by as:

$$I(x,y,z,\theta,\phi)=\sum_{n=0}^{N}\sum_{m=-n}^{n} A_n^m(x,y,z)Y_n^m(\theta,\phi) \tag{61}$$

where $Y_n^m(\theta,\phi)$ are the spherical harmonics given by

$$Y_n^m(\theta,\phi) = (-1)^{(m+|m|)/2}\left[\frac{2n+1}{4\pi}\frac{n-|m|!}{n+|m|!}\right]^{1/2} P_n^{|m|}(\cos\theta)e^{im\phi} \tag{62}$$

and P_n^m are the associated Legendre polynomials which are related to the Legrende polynomials. The upper limit N for the index n denotes the order of approximation of the method. Exact solution of the radiative transfer equation is obtained if N is taken as infinity. For most practical calculations, a finite N is assigned. If $N = 1$, the first order P-1 spherical harmonic approximation, equivalent to the Eddington approximation (1988), is attained.

The P-1 approximation is very accurate if the optical dimension of the medium is large (i.e. greater than 2). It yields however inaccurate results for thinner media particularly near the domain boundaries. Nonetheless, if the approximation of the spherical harmonics is expanded to the third order, i.e. $N = 3$, the P-3 approximation can yield accurate results for an optical dimension as small as 0.5 (Ratzel and Howell, 1983) and for anisotropic fields. Naturally, the P-3 approximation results in additional equations and they are usually more complicated than those of the P-1 approximation Menguc and Viskanta (1987). Therefore, the P-3 approximation greatly suffers at the expense of the additional computational burden and becomes impractical if extended to WSGGM; hence it has not enjoyed much success in practical applications when compared to the simpler P-1 approximation.

Discrete transfer radiative method

The discrete transfer radiative method (DTRM) proposed by Lockwood and Shah (1980) is principally built on the concept of solving representative rays in a radiating enclosure. To certain extent, it closely resembles the Monte Carlo method, but the directions of the rays are now pre-specified in advance rather than being chosen at random. The rays are solved for only along paths between two boundary walls rather than being partially reflected at walls and tracked to extinction.

Figure 12 illustrates the concept of the DTRM. Consider a centre point P on the surface element has a solid angle span over the hemisphere. Taking for an instance that the hemisphere can be divided into four equal segments, four representative rays may be visualized to impinge on point P from points Q_1, Q_2, Q_3 and Q_4 on the far walls of the enclosure. Assuming a homogeneous gas mixture and the radiation intensities arriving at P are constant over the solid angles, the fundamental RTE can then be approximated as:

$$\frac{dI}{ds} = -\overline{K}_a I + \overline{K}_a \frac{\sigma T^4}{\pi} \tag{63}$$

Integrating the above equation yields the analytical recurrence relation:

$$I^{n+1} = I^n e^{-\overline{K}_a \delta s} + \frac{\sigma T^4}{\pi}\left(1 - e^{-\overline{K}_a \delta s}\right) \tag{64}$$

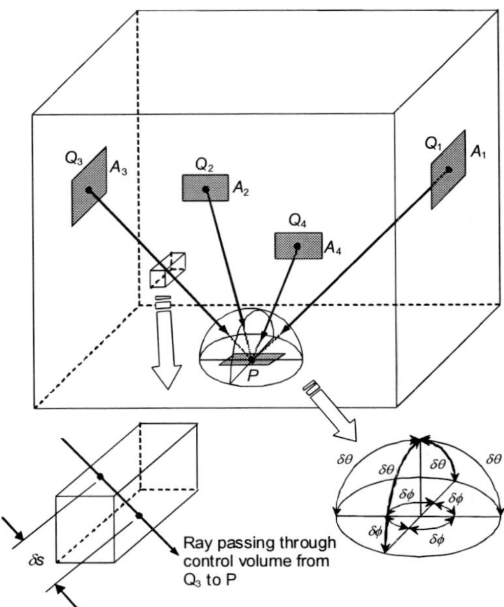

Figure 12. Illustration of the discrete transfer radiative method

where δs is the distance travelled by the ray within the control volume, and I^n and I^{n+1} are respectively the radiation intensities entering and leaving the control volume. According to Shah (1979), the hemisphere can be discretized into N_θ equal polar angles and N_ϕ azimuthal angles according to:

$$\delta\theta = \frac{\pi}{2N_\theta} \quad \delta\phi = \frac{2\pi}{N_\phi} \tag{65}$$

The total 2π hemispherical solid angle on a surface element is the sum of all finite angles such that $N_\Omega = N_\theta \times N_\phi$.. As shown in the Figure, the values of $N_\theta = 1$ and $N_\phi = 4$ gives a total of 4 rays for the angular discretization.

In practice, it is rather common to adopt a total of 8 rays or 16 rays spanning the hemisphere. It should be noted that the accuracy of the DTRM also depends on the surface discretisation (number of surface elements) in addition to the angular discretisation. With an adequately refined mesh and a sufficiently large number of rays, DTRM has shown to produce results that are rather comparable in accuracy with Monte Carlo solutions.

Numerically speaking, the DTRM method provides an "attractive" balance between computational time and solution accuracy. Consequently, the method has been widely adopted and enjoyed much success in numerous fire modelling applications (Fletcher *et al.*, 1994; Yan and Holmstedt, 1996; Novozhilov *et al.*, 1997b; Xue *et al.*, 2001; Wen *et al.*, 2001). Novozhilov *et al.* (1997b) studied the extinguishing mechanism of a burning PMMA horizontal sheet due to water sprinkler discharge. By using the DTRM to handle radiation calculations, encouraging predictions were successfully obtained. However, the method also suffers from some limitations. One of these is the requirement of exact geometrical information of the surface model which complicates the solution procedures if intricate surface configuration is involved. Moreover, as the net gain or loss of radiant energy depends on the rays, solution accuracy is found sensitive to the shapes of the control volumes and positions of the rays (Keramida *et al.*, 2001).

Discrete ordinates method

A discrete ordinate approximation to the radiative transfer equation, as the terminology suggests, is obtained by discretising the entire solid angle ($\Omega = 4\pi$) using a finite number of ordinate directions and corresponding weight factors. Originally proposed by Chandrasekhar (1960) for astrophysical problems, discrete ordinates method has enjoyed much success in applications to problems of neutron transport (Lathrop, 1976, Lewis and Miller, 1984).

The method approximates the entire solid angle using a finite number of ordinate directions and its corresponding weighting factors. The augmentation or reduction of ordinate directions along its path is then obtained by solving the RTE of which the integral term is replaced by a quadrature summed over each ordinate. Similar to the discrete transfer method, the ordinate directions and its weighting factors are specified prior to provide a desired order of accuracy. All these parameters are obtained by using Gaussian or Lobatto quadratures. These are also called S_N-approximations to symbolize the discrete ordinates approximations in which there are N discrete values of positive and negative direction cosines ξ_n, μ_n, η_n, which always satisfy the identity $\xi_n^2 + \eta_n^2 + \mu_n^2 = 1$. Based on the quadrature technique, the RTE without scattering may be written for each quadrature point n in Cartesian co-ordinates as:

$$\xi_n \frac{\partial I_n}{\partial x} + \mu_n \frac{\partial I_j}{\partial y} + \eta_n \frac{\partial I_n}{\partial z} = -(\kappa + \sigma_r) I_n + k_a \sigma T^4 + \frac{\sigma_r}{4\pi} S_m \quad (66)$$

Let us consider for the purpose of illustrating this particular method via the three dimensional rectangular enclosure containing a participating medium as depicted in Figure 13. Solution of equation requires boundary conditions at the wall as well as the temperature of the medium. By assuming the surroundings walls to be diffusely emitting-reflecting surfaces, the boundary conditions can now be expressed as:

at $x = 0$: $$I_w^n = \frac{\varepsilon_w \sigma T_w^4}{\pi} + \frac{(1-\varepsilon_w)}{\pi} \sum_{\substack{n' \\ \xi_{n'}<0}} w_{n'} |\xi_{n'}| I_w^{n'} \quad \text{for} \quad \xi_n > 0$$

at $x = L$: $$I_w^n = \frac{\varepsilon_w \sigma T_w^4}{\pi} + \frac{(1-\varepsilon_w)}{\pi} \sum_{\substack{n' \\ \xi_{n'}>0}} w_{n'} |\xi_{n'}| I_w^{n'} \quad \text{for} \quad \xi_n < 0$$

at $y = 0$: $$I_w^+ = \frac{\varepsilon_w \sigma T_w^4}{\pi} + \frac{(1-\varepsilon_w)}{\pi} \sum_{\substack{n' \\ \mu_{n'}<0}} w_{n'} |\mu_{n'}| I_w^- \quad \text{for} \quad \mu_n > 0$$

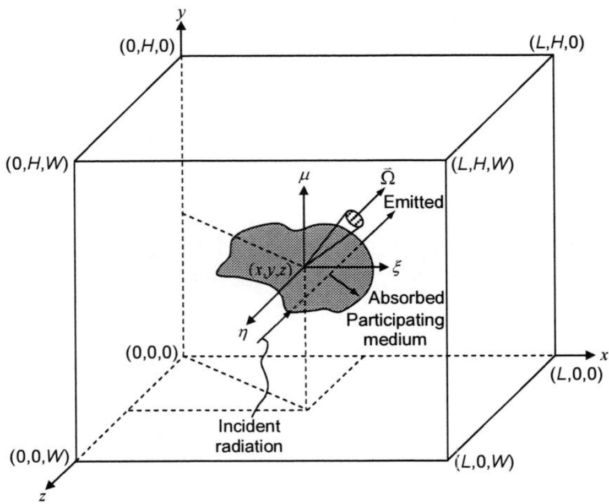

Figure 13. Illustration of the discrete ordinates method within a three-dimensional rectangular geometry

at $y = W$: $$I_w^+ = \frac{\varepsilon_w \sigma T_w^4}{\pi} + \frac{(1-\varepsilon_w)}{\pi} \sum_{\substack{n' \\ \mu_{n'}>0}} w_{n'} |\mu_{n'}| I_w^- \quad \text{for} \quad \mu_n < 0$$

at $z = 0$: $$I_w^+ = \frac{\varepsilon_w \sigma T_w^4}{\pi} + \frac{(1-\varepsilon_w)}{\pi} \sum_{\substack{n' \\ \eta_{n'}<0}} w_{n'} |\eta_{n'}| I_w^- \quad \text{for} \quad \eta_n > 0$$

at $z = H$: $$I_w^+ = \frac{\varepsilon_w \sigma T_w^4}{\pi} + \frac{(1-\varepsilon_w)}{\pi} \sum_{\substack{n' \\ \eta_{n'}>0}} w_{n'} |\eta_{n'}| I_w^- \quad \text{for} \quad \eta_n < 0$$

In the above equations, the values n and n' denote outgoing and incoming directions respectively. With the boundary conditions, spatial variation of each intensity I_n can be now calculated. The local radiative heat transfer of each particular location can thus be expressed by a simple weighted sum of angular quantities:

$$\int_{\omega=0}^{4\pi} \frac{dI_S}{dS} d\omega = \sum_{n=1}^{M} w_n I_n \tag{67}$$

The most basic discrete ordinate approximation is S_2. Only one direction is represented in an eighth of sphere. For higher order approximations such as S_4, S_6 and S_8, they consist of three, six and ten directions spanning the one-eighth of a sphere (Jamaluddin and Smith, 1988). The spatially distribution of these high-order directions is shown in Figure 14.

At the beginning of development, application of the DOM was very limited. Fiveland (1984, 1987, 1988 and 1994) preformed a series of studies and successfully extended the method to multidimensional problem. After this pioneering work by Fiveland, employing the DOM method to multidimensional problem became viable to researchers. Lewis and Miller (1984) reviewed the method and concluded that DOM can be resulted a computer algorithm which combine minimum computer memory requirements with fewer arithmetic operations per space-angle grid point. Recently, owning to its simplicity and modest data storage requirement, application of the DOM in fire simulation is becoming increasingly popular among fire researchers (Collin et al., 2005; Yeoh et al., 2002, 2003; Dembele and Wen, 2000; Wang and Joulain, 2000; Yuen et al., 2000, Wang et al., 1999; Dembele et al., 1997). Nonetheless, as the DOM calculates radiation intensity with extrapolate, it suffers from non-physical intensity (i.e. negetive intensity) problem when absorption cross sections are large or when inadequate spatial resolution is used. Special differencing schemes (i.e.. positive scheme by Kim and Lee (1988) and exponential scheme by Chai et al. (1994)) may need to deploy to ensure positive intensity.

2.5. Soot Formation and Oxidation Modelling

Particulate smoke (soot) is produced in almost all fires. As indicated by Rasbash and Drysdale (1982), most are formed in the gas phase as a result of *incomplete combustion* and *high temperature pyrolysis reactions at low oxygen* concentrations. Soot can be generated even if the original fuel is a gas or liquid in addition to those produced by the ablation of a condensed solid under high heat flux. As soot travels through the flame, it radiates away energy and cools the combustion products – the principal mechanism for radiative heat loss. At the smoke point flame height, radiative heat loss accounts for 30% of the total heat release rate (Markstein, 1986). In order to account for soot radiation in field modeling, the effective soot absorption coefficient requires the evaluation of the soot volume fraction, which is usually related to the concentration of soot particles. Essentially, the means of determining the concentration of soot particles require insights into the controlling physical and chemical mechanisms associated with the formation and oxidation of soot. A comprehensive model of the soot process must therefore include the consideration of both of these phenomena.

The main constituent of soot is mostly carbon; other elements such as hydrogen and oxygen are usually present in small amounts. In spite of the rigorous science to identify the many properties of soot, it is still not currently possible to uniquely define its chemical composition. As reviewed in Kennedy (1997), Figure 15 shows the various steps involved in the process of soot formation and oxidation. Soot production is a chemically-controlled phenomenon. During nearly all phases of soot production: inception, condensation, coagulation, surface growth, agglomeration and oxidation, chemistry plays an important role. The inception of soot starts with the formation of the aromatic species such as cyclic benzene c-C_6H_6 and phenyl c-C_6H_5 in the gas phase. These aromatic species then grow by the addition of other aromatic and smaller alkyl species into two-dimensional polyaromatic hydrocarbons (PAH). The PAH can then continue to grow by three mechanisms: condensation, coagulation and surface growth.

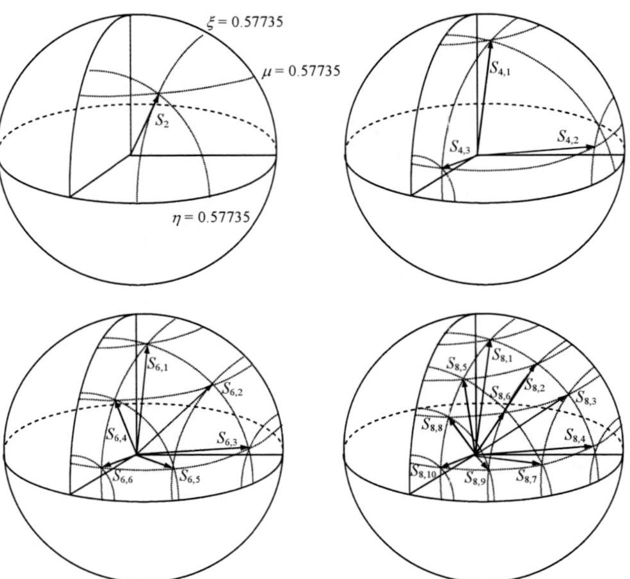

Figure 14. Illustration of the discrete ordinates method

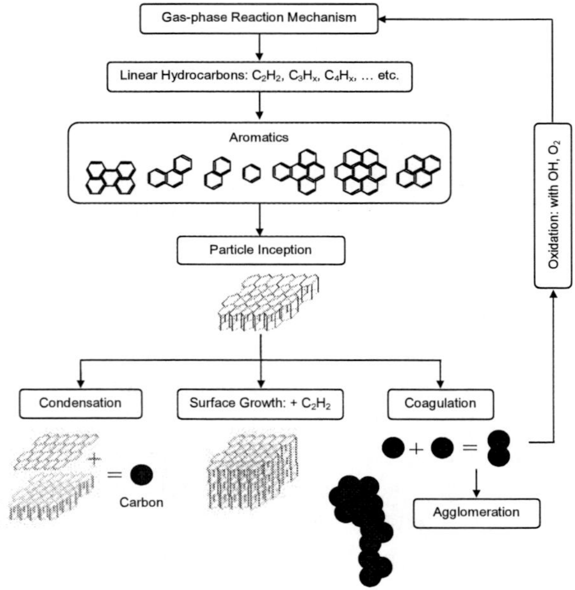

Figure 15. Road Map of soot formation and oxidation

The condensation and coagulation processes are typically physical in nature where in the former, the particles grow via condensation of a two-dimensional PAH on a three-dimensional PAH while the latter leads to the coalescence of particles leading to the formation of a larger spheroid. Surface growth of particles proceeds in conjunction with coagulation. The particles grow via chemical reactions in the gas phase. It is generally agreed that acetylene C_2H_2 is mainly responsible for the growth of soot particles. Older particles will undergo agglomeration where large clusters of particles are subsequently formed. These clusters are now the primary soot particles in the system. Finally, soot particles may burnout via oxidation process which is primarily as a result of attack by molecular O_2 and the OH radical.

Owing to the inadequate knowledge of the combustion and soot formation chemistry, as a compromise between modelling simplicity and computational accuracy, practical models such as based on empirical and semi-empirical approaches have been purposefully applied in field modelling to determine the concentration of soot particles. The main drawback of these models is the need to determine the necessary pre-exponential constants and activation energies that appear in the reaction rates through some input from experimental data. For completeness, the description of detailed models that seek to solve the rate of equations for elementary reactions leading to soot is also described.

2.5.1. Empirical models

In the early development, researchers believed that formation of soot in a premixed flame can be related to the critical or threshold equivalence ratio. Calcote and Manos (1983) proposed to use the so called "Threshold Sooting Index" (TSI) for attempting to remove all the system or burner dependence from measures of sooting tendency of fuels. Following this, Gill *et al.* (1984) employed the same definition to predict soot threshold value for fuel mixtures. In their study, individual TSIs of each component were used for calculating the overall TSI value of the mixture. Their predicted results compared quite well for a ternary mixture consisting of: isooctane, decalin and 1-methylnaphthalene. Afterwards, Olson et al. (1985) further extended the definition of TSI to laminar diffusion flames for a wide variety of fuels. Again, satisfactory results were obtained. These studies of TSI demonstrated that the TSI approach was capable for assessing the sooting tendencies of fuels and mixtures.

Beside studies on TSI, researchers attempted to relate the soot formation tendency with the critical C/H ratio as well. Takahashi and Glassman (1984) proposed to correlate the sooting tendency with the effective equivalence ratio, defined as $\psi \equiv [C + H/2]/O$. Based on their analysis of experimental result, they concluded that the sooting tendency of a fuel was a function of the C/H ration, the number of C atoms and the flame temperature instead of the structure of the fuel.

On the other hand, following the combustion modelling approach, Khan and Greeves (1974) assumed that the production of soot particles in a flame is inherently a chemically-controlled phenomenon and is governed entirely by the formation of soot particles (i.e. by the soot inception rate) which is given by:

$$\overline{R}^{+}_{soot} = C_s \overline{P}_{fu} \phi^n \exp\left(-\frac{E_a}{R_u \tilde{T}}\right) \tag{68}$$

where C_s is a constant, $\overline{P}_{fu}$ is the mean partial pressure of fuel, ϕ is the local unburnt equivalence ratio, E_a is the activation energy and R_u is the universal gas constant usually taken to be equivalent to 1.9872 cal mol^{-1} K^{-1}. Modelling parameters such as C_s, n and E_a have been ascertained through experiments performed in connection with diesel engines. Soot production is essentially zero for the equivalence ratio ϕ less than that of the incipient soot limit and for ϕ in excess of a value corresponding roughly to the upper flammability limit. Following Khan and Greeves (1974), the upper and lower limits are set to 2 and 8 respectively. The applicability of the model to a wide range of fuels greatly suggests its suitability in field modeling.

To consider soot oxidation, Magnussen and Hjertager (1976) proposed a simple method similar to the eddy dissipation concept as discussed previously. They assumed that the combustion is controlled by the rate of mixing of the particle-bearing vortices with adjacent oxygen bearing material. The consumption rate of soot is given by:

$$\overline{R}_{soot}^{-} = -C_R \overline{\rho} \tilde{Y}_s \frac{\varepsilon}{k} \tag{69}$$

In regions where the oxygen concentration is low, the oxygen becomes the limiting specie that controls the rate of consumption of soot. Being a tracer element, soot must also however compete for oxygen with the unburned fuel. This leads to

$$\overline{R}_{soot}^{-} = -C_R \overline{\rho} \left(\frac{\tilde{Y}_{ox}}{\tilde{Y}_s r_s + \tilde{Y}_{fu} r_{fu}} \right) \tilde{Y}_s \frac{\varepsilon}{k} \tag{70}$$

where C_R is a model constant assigned a value of 4 and r_s and r_{fu} are the soot and fuel stoichiometric ratios. The lower reaction rate of either equation (4.3.4) or (4.3.5) determines the local rate of soot consumption.

More recently, Lautenberger et al. (2005) proposed a simplified approach to model soot formation and oxidation in non-premixed hydrocarbon flames. The basic form of their soot model is similar to the work by Kent and Honnery (1994) of which the soot formation rate could be estimated from only the local mixture fraction and temperature. The model postulated considered only homogeneous soot formation – moderately to heavily sooty flames. Soot oxidation is treated by a global fuel-independent mechanism, which is also only a function of the mixture fraction and temperature. Here, the model assumes that the diffusion of molecular oxygen is the governing process rather than being controlled by the reaction of OH radicals impinging on the available soot surface area. For the formulation of the instantaneous soot formation and oxidation rates, they are simply determined from the product of an analytic function of mixture fraction (Z) and an analytic function of temperature (T) by:

$$R_{soot}^{+} = \dot{f}_{sf}'''(Z) g_{sf}(T)$$

$$R_{soot}^{-} = \dot{f}_{so}'''(Z) g_{so}(T) \tag{71}$$

Several analytic forms of these functions were considered by Lautenberger et al. (2005). These functions rise from a formation rate of zero at a mixture fraction of Z_L to a peak formation rate at a mixture fraction of Z_P and then fall back to zero at a mixture fraction of

Z_H. The polynomial coefficients are determined through specifying Z_L, Z_P, Z_H and $\dot{f}_{sf}'''\left(Z_p\right)$, which are then solved through the resultant set of linear equations. In order to generalize the model to a range of fuels, the values of Z_L, Z_P, Z_H for each polynomial are related to the fuel's stoichiometric mixture fraction Z_{st} by a parameter ψ of order unity. More details on the formulation of these rates as well as appropriate governing equations based on the conserved scalar approach can be found in the dissertation of Lautenberger (2002).

Although encouraging results were obtained by the aforementioned empirical models, model formulations and parameters were calibrated with the experimental data. Applicability of the models are therefore limited to the system which shares similar the characteristics of the experimental apparatus. As a result, researchers attempted to develop some more sophisticated models by using the semi-empirical approach.

2.5.2. Semi-empirical models

This next level of soot modeling aims to incorporate some considerations of the physics and chemistry of the phenomenon. In this approach, a transport equation for the particulate number density is introduced and solved in conjunction with the transport equation for the soot particles. The representation of soot properties is now characterized by two variables – the soot mass fraction or soot volume fraction, and particulate number density. Three widely used models of soot formation for turbulent combustion developed by Tesner et al. (1971a, 1971b), Moss et al. (1988) and Leung et al. (1991), are described in this section.

Tesner et al. method

Tesner et al. (1971a, 1971b) have assumed that the soot formation occurs from a gaseous parent fuel in two stages. The first stage represents the formation of radical nuclei. Radical nuclei are defined to be the active sites on particles from which the soot deposits and will eventually grow. According to Tesner et al. (1971a), the philosophy of the chain branching theory are that the increase of the rate of formation of particles is a result of a branched process, and the observed retardation is related to the acceleration of the destruction of active particles. This process is linked to the creation and rapid growth of the total surface of the soot particles on which the radical nuclei are being destroyed. The Favre-averaged form of the conservation equation for the concentration of radical nuclei $\tilde{n}_c$ can be expressed by:

$$\frac{\partial}{\partial t}\left(\bar{\rho}\tilde{n}_c\right)+\frac{\partial}{\partial x_j}\left(\bar{\rho}\tilde{u}_j\tilde{n}_c\right)=D_{nuclei}^{th}+\frac{\partial}{\partial x_j}\left[\frac{\mu_T}{Sc_T}\frac{\partial\tilde{n}_c}{\partial x_j}\right]+\bar{R}_{nuclei}^{+}+\bar{R}_{nuclei}^{-} \tag{72}$$

where D_{nuclei}^{th} is the diffusion occurs by thermophoresis, which in terms of mean quantities is given as:

$$D_{nuclei}^{th}=-0.55\frac{\partial}{\partial x_j}\left[\tilde{n}_c\frac{\mu}{\tilde{T}}\frac{\partial\tilde{T}}{\partial x_j}\right] \tag{73}$$

Defining the particle number concentrations of nuclei and soot particles according to:

$$C_n = \bar{\rho} N_o \tilde{n}_c \,;\quad C_s = \bar{\rho}\frac{\tilde{Y}_s}{m_p} \tag{74}$$

where N_o is the Avogadro's number and m_p is the mass of a soot particle, the rate of nuclei formation $\bar{R}^+_{nuclei}$ depends on a spontaneous generation and branching process described by:

$$\bar{R}^+_{nuclei} = n_o + (f - g)C_n + g_o C_n C_s \tag{75}$$

where the f and g are the linear branching and linear termination coefficients, and g_o is the coefficient of linear termination on soot particles. The spontaneous generation of radical nuclei n_o from the fuel is modelled according to the Arrhenius law as:

$$n_o = a_o \bar{\rho} \tilde{Y}_{fu} \exp\left(-\frac{E_a}{R_u \tilde{T}}\right) \tag{76}$$

The Favre-averaged conservation equation for soot particles also follows the same form of the scalar property equation as:

$$\frac{\partial}{\partial t}\left(\bar{\rho}\tilde{Y}_s\right) + \frac{\partial}{\partial x_j}\left(\bar{\rho}\tilde{u}_j\tilde{Y}_s\right) = D^{th}_{soot} + \frac{\partial}{\partial x_j}\left[\frac{\mu_T}{Sc_T}\frac{\partial \tilde{Y}_s}{\partial x_j}\right] + \bar{R}^+_{soot} + \bar{R}^-_{soot} \tag{77}$$

The rate of formation of soot particles depends on the interaction between the active particles and the original hydrocarbon molecules and on the termination process by the surface of the soot particles. The rate of soot formation $\bar{R}^+_{soot}$ which depends on the concentration of radical nuclei is modelled as:

$$\bar{R}^+_{soot} = m_p\left(a - bC_s\right)C_n \tag{78}$$

The default values for all the soot parameters are calibrated against acetylene fuel.

Magnussen and Hjertager (1976) have employed the kinetic theory of soot formation by Tesner et al. (1971a, 1971b) to a turbulent acetylene flame. On the basis of the Eddy Dissipation Concept, they have proposed to conveniently calculate the mean combustion rates of the radical nuclei $\bar{R}^-_{nuclei}$ and soot particles $\bar{R}^-_{soot}$ from the rate of combustion of fuel. For the combustion of soot particles, the rate is given by:

$$\bar{R}^-_{soot} = \min\left[-C_R\bar{\rho}\tilde{Y}_s\frac{\varepsilon}{k}, -C_R\bar{\rho}\left(\frac{\tilde{Y}_{ox}}{\tilde{Y}_s r_s + \tilde{Y}_{fu} r_{fu}}\right)\tilde{Y}_s\frac{\varepsilon}{k}\right] \tag{79}$$

The local radical nuclei can be assumed to be reduced by combustion according to

$$\bar{R}^-_{nuclei} = \frac{\tilde{n}}{\tilde{Y}_s} \bar{R}^-_{soot} \tag{80}$$

This so-called Magnuseen soot model that combines the kinetic theory formation of Tesner et al. (1971a, 1971b) and oxidation rates in the respective Eqs. (79) and (80) is widely applied in many field modeling investigations (for example, room fire simulations by Luo and Beck, 1996). It is also a standard feature in majority of commercial CFD codes such as ANSYS Inc., Fluent and ANSYS Inc., CFX. Note that the same model constants tabulated above have been successfully used to produce the flame data for other fuels, including methane.

Moss et al. method

Moss et al. (1988) has proposed a soot model that incorporates the essential physical processes of soot nucleation, coagulation and surface growth influencing the soot volume fraction and particulate number density. Whilst the kinetic theory of soot formation by Tesner et al. (1971a, 1971b) has been extensively employed, its focus solely on particle number density neglects the important role of surface growth in relation to soot mass addition. Particle size evolution could not be tracked and without the knowledge of particle size, the aerosol surface area cannot be determined. Heterogeneous chemical process like oxidation would therefore be unsatisfactorily ascertained. This alternative model which is represented by the soot volume fraction f_v and particulate number density n, important to the description of the post-flame burnup, permit nonetheless the soot aerosol surface area to be estimated from the average particle diameter. The Favre-averaged conservation equation for the soot particulate number density and soot volume fraction in terms of normalized variables can be written as:

$$\frac{\partial}{\partial t}\left(\bar{\rho}\tilde{\zeta}_n\right) + \frac{\partial}{\partial x_j}\left(\bar{\rho}\tilde{u}_j\tilde{\zeta}_n\right) = D^{th}_{num\_dens} + \frac{\partial}{\partial x_j}\left[\frac{\mu_T}{Sc_T}\frac{\partial \tilde{\zeta}_n}{\partial x_j}\right] + \bar{R}^+_{num\_dens} + \bar{R}^-_{num\_dens} \tag{81}$$

$$\frac{\partial}{\partial t}\left(\bar{\rho}\tilde{\zeta}_s\right) + \frac{\partial}{\partial x_j}\left(\bar{\rho}\tilde{u}_j\tilde{\zeta}_s\right) = D^{th}_{vol\_frac} + \frac{\partial}{\partial x_j}\left[\frac{\mu_T}{Sc_T}\frac{\partial \tilde{\zeta}_s}{\partial x_j}\right] + \bar{R}^+_{vol\_frac} + \bar{R}^-_{vol\_frac} \tag{82}$$

where $\tilde{\zeta}_n = \tilde{n}/\left(\bar{\rho}N_o\right)$ and $\tilde{\zeta}_s = \left(\rho_s\tilde{f}_v\right)/\bar{\rho}$ with $D^{th}_{num\_dens}$ and $D^{th}_{vol\_frac}$ expressed in the form analogy to Eq. (73). The influence rate processes of nucleation, coagulation and surface growth on the particulate number density and volume fraction are modeled as:

$$\bar{R}^+_{num\_dens} = \underbrace{\bar{\alpha}}_{nucleation} - \underbrace{\bar{\rho}^2\bar{\beta}\tilde{\zeta}^2_n}_{coagulation} \tag{83}$$

$$\overline{R}^{+}_{vol\_frac} = \underbrace{\overline{\delta}}_{nucleation} + \underbrace{N_o^{1/3}\overline{\rho\gamma}\tilde{\zeta}_s^{2/3}\tilde{\zeta}_n^{1/3}}_{surface\ growth} \tag{84}$$

The first term in Eq. (83) represents the increase in soot particle number density due to particle inception, which is given by:

$$\overline{\alpha} = C_\alpha \overline{\rho}^2 \tilde{T}^{1/2} \tilde{X}_{fu} \exp\left(-\frac{T_\alpha}{\tilde{T}}\right) \tag{85}$$

The second term in equation (83) describes the loss of particles as a result of coagulation. According to Moss et al. (1988), it can be described by the Smoluchowski (Fuchs, 1964) expression as:

$$\overline{\beta} = C_\beta \tilde{T}^{1/2} \tag{86}$$

Increase of the soot volume fraction in Eq. (82) is represented by nucleation of new particles in the first term and the result of surface growth in the second term. Based on the Eq. (85), the accompanying mass growth of the former may be represented (for 12 carbon atoms initially) by:

$$\overline{\delta} = 144\overline{\alpha} \tag{87}$$

The latter is modelled according to the surface growth of soot suggested by Syed *et al.* (1990), which contained a linear dependence on aerosol surface area, is controlled by the rate relationship:

$$\overline{\gamma} = C_\gamma \overline{\rho} \tilde{T}^{1/2} \tilde{X}_{fu} \exp\left(-\frac{T_\gamma}{\tilde{T}}\right) \tag{88}$$

Methane combustion was studied in Syed *et al.* (1990). The pre-exponential constants C_α, C_α and C_α and activation temperatures T_α and T_γ determined empirically for methane are:

$C_\alpha = 65400\ m^3\ kg^{-2}\ K^{-1/2}\ s^{-1}$

$C_\beta = 1.3 \times 10^7\ m^3\ K^{-1/2}\ s^{-1}$

$C_\gamma = 0.1\ m^3\ kg^{-2/3}\ K^{-1/2}\ s^{-1}$

$T_\alpha = 46100$ K; $T_\beta = 12600$ K

From a detailed chemical kinetic model of fuel pyrolysis, it is possible to identify the specific hydrocarbon species that are precursors to drive the nucleation and surface growth expressions through the mean fuel mole fraction $\tilde{X}_{fu}$. Depending on the parent fuel, the choice of acetylene and benzene as critical species for these processes is strongly supported by experimental evidence, for example, Harris and Weiner (1983a, 1983b) and Smyth (1985). Nonetheless, as will be demonstrated later, the above soot parameters are applicable to other flames that are lightly sooting in nature like methane.

Leung et al. method

The soot formation model of Leung et al. (1991) differs from those of Tesner et al. (1971a, 1971b) and Moss et al. (1988) in the aspect of which assumed specifically acetylene as the precursor for soot nucleation and growth. Hence, the rates of soot nucleation and growth are directly proportional to the acetylene concentration rather than to the parent fuel concentration. This concept in retrospect is physically more plausible although it complicates the modelling to some extent since there is a concerted need to determine the acetylene mass fraction. The model solves for the particulate number density n and soot mass fraction Y_s as follow:

$$\frac{\partial}{\partial t}(\overline{\rho}\tilde{n})+\frac{\partial}{\partial x_j}(\overline{\rho}\tilde{u}_j\tilde{n})=D^{th}_{num\_dens}+\frac{\partial}{\partial x_j}\left[\frac{\mu_T}{Sc_T}\frac{\partial\tilde{n}}{\partial x_j}\right]+\frac{2N_o}{C_{\min}}\overline{\rho}\overline{R}_1+\overline{\rho}\overline{R}_3 \quad (89)$$

$$\frac{\partial}{\partial t}(\overline{\rho}\tilde{Y}_s)+\frac{\partial}{\partial x_j}(\overline{\rho}\tilde{u}_j\tilde{Y}_s)=D^{th}_{soot}+\frac{\partial}{\partial x_j}\left[\frac{\mu_T}{Sc_T}\frac{\partial\tilde{Y}_s}{\partial x_j}\right]+2M_C\overline{\rho}(\overline{R}_1+\overline{R}_2)+M_C\overline{\rho}\overline{R}_4 \quad (90)$$

where C_{min} is the number of carbon atoms in the incipient carbon particle. Fairweather et al. (1992) who adopted the soot reaction mechanism of Leung et al. (1991) for a methane air jet flame have assumed a value of 9×10^4 carbon atoms. The rates appearing on the LHS of equation are modelled based on a simple kinetic mechanism, which entails the following four reaction steps:

nucleation	$2C_2H_2 \rightarrow 2C(s)+H_2$
surface growth	$nC(s)+C_2H_2 \rightarrow (n+2)C(s)+H_2$
coagulation	$nC(s) \rightarrow C_n(s)$
oxidation	$C(s)+\frac{1}{2}O_2 \rightarrow CO$

From the reaction steps described above, the notation C(s) represents soot, and a mole of soot is taken as a mole of carbon atoms. The reaction step defined by the particle nucleation is similar to that outlined by Tesner et al. (1971a) for premixed acetylene-air flames. For simplicity, the rate of nucleation is assumed to be first order in acetylene concentration, which can be expressed in terms of mass fraction so that:

$$\bar{R}_1 = 1\times 10^4 \exp\left(-\frac{21100}{\tilde{T}}\right)\left(\frac{\bar{\rho}\tilde{Y}_{C_2H_2}}{M_{C_2H_2}}\right) \tag{91}$$

where $\tilde{Y}_{C_2H_2}$ and $M_{C_2H_2}$ are the corresponding mean mass fraction and molecular weight of acetylene. For the surface growth, an *ad hoc* assumption is made that the number of active sites is taken to be proportional to the square root of the total surface area available locally in the flame. The soot growth rate is given as:

$$\bar{R}_2 = 6\times 10^3 \exp\left(-\frac{12100}{\tilde{T}}\right)\sqrt{\pi\left(\frac{6M_c}{\pi\rho_{soot}}\right)^{2/3}}\left(\frac{\bar{\rho}\tilde{Y}_{C_2H_2}}{M_{C_2H_2}}\right)\left(\frac{\bar{\rho}\tilde{Y}_s}{M_C}\right)^{1/3}\tilde{n}^{1/6} \tag{92}$$

Here, M_C refers to the molecular weight of carbon. The decease of the particle number density is accounted by the particle agglomeration of which this step is modeled using the normal square dependence, i.e.

$$\bar{R}_3 = -2C_a\left(\frac{6M_C}{\pi\rho_{soot}}\right)^{1/6}\left(\frac{6\kappa\tilde{T}}{\rho_{soot}}\right)^{1/2}\left(\frac{\bar{\rho}\tilde{Y}_s}{M_C}\right)^{1/6}\tilde{n}^{11/6} \tag{93}$$

where C_a is the agglomeration rate constant taken to have a value of 9.0 and κ is the Boltzmann constant given as 1.381×10^{-23} J K^{-1}. Instead of the Nagle and Strickland-Constable (1962) model, Leung et al. (1991) used the rate of soot oxidation proposed by Lee et al. (1962), which is due by the limiting mechanism of oxidation by O_2. The rate of soot oxidation where the dependence on local surface area has been retained as:

$$\bar{R}_4 = -1\times 10^4 T^{1/2} \exp\left(-\frac{19680}{\tilde{T}}\right)\pi\left(\frac{\bar{\rho}6\tilde{Y}_s}{\pi\rho_{soot}\tilde{n}}\right)^{2/3}\left(\frac{\bar{\rho}\tilde{Y}_{O_2}}{M_{O_2}}\right)\tilde{n} \tag{94}$$

It is noted that the above oxidation rate has been adjusted by a factor of 14 to provide adequate agreement with the measurements of Garo et al. (1990). Again, such *ad hoc* adjustment is required in order to necessitate the neglect of OH radical as an oxidant of soot. Alternatively, an oxidation step involving OH could readily have been formulated in the context of present model. Following investigations by Fenimore and Jones (1967), Puri et al. (1994) and Garo et al. (1990), the rate of soot by the OH radical, assuming a collision efficiency of 0.04 according to Brookes and Moss (1999), may be simply written as:

$$\bar{R}_4 = -4.2325T^{1/2}\pi\left(\frac{\bar{\rho}6\tilde{Y}_s}{\pi\rho_{soot}\tilde{n}}\right)^{2/3}\left(\frac{\bar{\rho}\tilde{Y}_{OH}}{M_{OH}}\right)\tilde{n} \tag{95}$$

2.5.3. Soot modelling with detailed chemistry

Attempting to attack the shortcomings of the semi-empirical models, Frenklach and co-workers attempted to develop a soot formation model with detail chemistry consideration which resulted in a number of contributory paper over the last decades (Frenklach *et al.*, 1984; Frenklach and Wang, 1990, 1994; Wang and Frenklach, 1997; Frenklach, 2002).

The development of their models started from modelling soot formation in the shock tube pyrolysis of acetylene (Frenklach *et al.*, 1984). A pyrolysis mechanism, as proposed by Tanzawa and Gardiner (1980), was employed in their study. Studies of the first aromatic ring were preformed. They found that the formation of the first aromatic ring in flames of non-aromatic fuel begins usually with vinyl addition to acetylene Afterwards, the aromatic ring grew through the so-called "H Abstraction- C_2H_2 Addition (HACA)" procedures to become polycyclic aromatic hydrocarbon (PAH) species such as acenaphtalene and coronene. They concluded that the formation of the first aromatic ring was a major "bottleneck" in the soot formation of acetylene pyrolysis. Therefore, a sophisticated model with a detailed modelling of the sequenced chemical reaction of the formation aromatic ring and HACA procedures is inevitable.

In 1990, Frenklach and Wang (1990) attempted to model the soot particle nucleation and growth by using the basis of earlier formation mechanism studies (Frenklach *et al.*, 1984). They introduced a detailed reaction mechanism for the initial PAH formation with 337-reaction and 70-species mechanism. The Sandia burner code (Kee *et al.*, 1985) was employed for solving the profiles of H, H_2, C_2H_2, O_2, OH, H_2O and of a prescribed-size PAH. All these profile were then used as inputs for the particle nucleation and growth simulation of an in-house kinetic code. In the kinetic code, the method of moments was incorporated for accounting several simultaneously occurring processes. The computational results presented agreed quantitatively with the experimental data. Although the computational results were not in perfect agreement with the measured value, such a model depicted a new direction in the modelling soot formation and demonstrated the feasibility of incorporating the detailed reaction mechanism in the model.

One of significant breakthrough of the proposed moments method is the capability in simulating the occurring coagulate processes and the surface growth processes simultaneously. The method of moments was first introduced by Frenklach and Harris (1987), which they further explained in detail in later articles (Frenklach and Wang, 1994) (Frenklach, 2002). In the moments method, two moments were defined: concentration moments and size moments. By using these two moments, the evolution of individual PAHs or soot particles can be grouped into an infinite set of differential equations. In theory, the knowledge of all moments is equivalent to knowing the distribution function itself. Therefore, by solving all these differential equations, all the distribution functions of the PAHs and soot particles can be obtained without any approximation. As the set of equation contained an infinite number of differential equations, solving all the differential equations is near impossible and impractical. The authors, therefore, calculated the first few moments of the differential equation as an approximation of the solution. Moreover, the author also addressed a problem that the moments method might encounter. In evaluating the time derivatives of the soot particles moments, the coagulation terms should be taken into account. According to the formulation of the coagulation terms (Frenklach and Harris, 1987), the coagulation terms involved the evaluation of fraction-order positive moments which are undeterminable.

Therefore, the authors proposed to use Lagrange quadratic interpolation method as a closure of the model. It can be revealed that the proposed method of moments provides a new trend and direction for incorporating the sequential chemical reaction details into the soot formation model. However, extension of such a model to a turbulent diffusion flame is rare. Thus, further development in this area should be done in future.

3. FIRE MODELLING APPLICATION FOR ENCLOSURE FIRES

In previous section, theoretical concepts and modelling approaches of some essential physical sub-models for fire modelling are briefly reviewed. To demonstrate the application and performance of these models, in the following sections, some practical fire modelling problems will be presented and validated against full-scale experimental data. In this section, to exemplify the coupling effects of combustion, radiation and soot formation, discussion is firstly centred at the fire modelling application for enclosure fires where combustion, radiation and soot formation processes are taken in confined room or space.

3.1. Combustion Modelling of a Single Compartment Fire

In this numerical study, field modelling that incorporates increasingly complex illustration of the chemical process to characterize the Steckler's single-compartment fire is described through the application of different combustion modelling approaches. The parametric study includes systematic representations of the fire source by different combustion models. Numerical simulations are performed through an in-house computer code FIRE3D. For the purpose of validating and verifying the different combustion models that could be applied in fire engineering, comparison of the computed results is made not only against measurements made by Steckler et al. (1984) but also the numerical results obtained from Lewis et al. (1997) for the same heat release rate of 62.9 kW. Special emphasis on the range of fire studies performed by Lewis et al. (1997) allows the feasibility of direct comparison of similarly applied combustion models in order to establish confidence in the usage of these models in fire safety investigations.

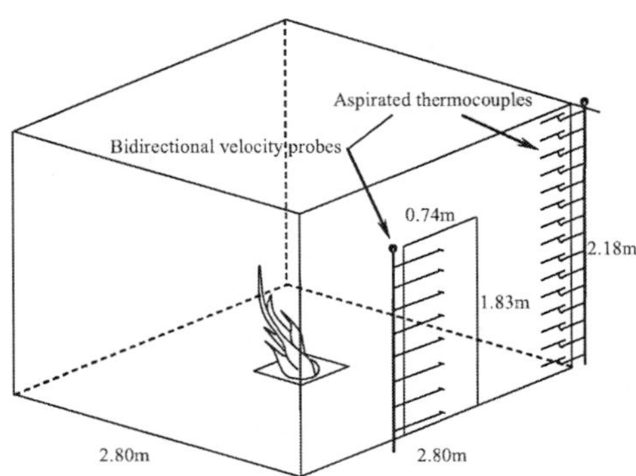

Figure 16. Schematic drawing Steckler's burn room

Figure 16 shows the schematic drawing of the particular geometry of the compartment. The non-spreading fire was fuelled by commercial grade methane, having a circular gas burner diameter D of 0.3 m centrally located in the room on a square enclosure of side 2.8 m and height 2.18 m. Air was drawn into the burn room through a doorway opening of 1.83 m high and 0.74 m wide located in one of the walls as depicted in Figure 16. Compartment walls and ceiling were covered with ceramic fiber board insulation to establish near steady state conditions within 30 minutes. Detailed measurements of temperature, using aspirated thermocouples, and velocity by bi-directional probes reported in Steckler et al. (1984) at the doorway are employed to validate the model predictions. A heat release rate $\dot{Q}$ of 62.9 kW was selected. Special emphasis on the range of fire studies performed by Lewis et al. (1997) allows the feasibility of direct comparison of similarly applied combustion models in order to establish confidence in the usage of these models in fire safety investigations.

Numerical features: Numerical solutions to a system of three-dimensional Favre-averaged equations for the transport of mass, momentum and enthalpy with the addition of mass fractions of gas species as well as mixture fraction accompanied by its fluctucation scalar are ascertained. Instead of the specific need of imposing a physical volume of the fire source within the computational domain, the fuel flow rate of 0.0013 kg/s corresponding to the heat release rate of 62.9 kW is now specified at the surface of the gas burner. Since the room configuration is symmetrical about the vertical plane bisecting the doorway and burner, mesh generation for the compartment fire was only carried out on half of the room, thus improving the resolution of the flow and thermal fields. A uniform rectangular mesh is generated spanning the length, width and height of the reduced configuration. In order to eliminate any errors that could arise due to mesh generation, exact mesh distribution comprising of a total of 83160 grid nodes is used for the numerical calculations performed in both the in-house and commercial computer codes. As the doorway temperature and velocity distributions are of significant interest for model assessments, a large extended region away from the doorway is constructed to reduce the end effects of the extended boundaries affecting the flow and thermal characteristics at the doorway. A fixed pressure boundary condition is imposed on all the external boundaries (open boundaries). All the compartment walls are taken to be adiabatic

For the comparative study against Lewis et al. (1997) numerical solutions, every attempt has been made to adopt similar or identical pressure-velocity linkage method, numerical discretisation scheme and turbulence model. Simulations for the single compartment fire are carried out using the SIMPLE pressure correction algorithm alongside with the *hybrid differencing scheme* and standard k-ε turbulence model. The eddy dissipation combustion (EDM) model of Magnussen and Hjertager (1976) based on single step chemistry of methane and the conserved scalar approach employing the Sivathanu and Faeth (1990) state relationships are assessed against similar combustion models employed in Lewis et al. (1997). To demonstrate the performance of combustion models, numerical simulation was also carried out by representing the fire as a volumetric heat source with a specified volume of 0.3 $\times$ 0.3 $\times$ 0.3 m^3. Furthermore, since this room configuration is symmetrical about the vertical plane bisecting the doorway and burner, mesh generation for the compartment fire is only carried out on half of the room, thus improving the resolution of the flow and thermal fields.

The mesh density of 83,160 grid nodes comparable to the finest mesh used by Lewis et al. (1997) of 70,432 grid nodes is employed for the comparison of the attained results.

The inclusion of an extended region away from the doorway usually demands an important modelling consideration in order to correctly predict the migration of combustion products and entrainment of ambient air through the doorway. In retrospect, the size of the extended region attached to the compartment fire is usually not known *a priori*. To determine the extent of the open boundaries of the extended boundaries affecting the flow and thermal characteristics at the doorway, two regions having a size of 3 m × 2.8 m in plan and 6.8 m in height and with the same plan area and a lower height of 3.8 m are investigated. Numerical experiments have revealed that no appreciable differences could be found for the predicted temperature profiles in the two simulation cases. It is nevertheless demonstrated in Figure 17that the predicted velocity profiles just above the floor level for the large extended region appear to be closer to the experimental data than those of the small extended region. The entrainment characteristic at the doorway can thus be inferred to be better accommodated through the requirement of a large extended region.

Numerical results: Predicted line graphs for the doorway temperature and velocity profiles utilizing different combustion models based on the eddy dissipation and conserved scalar by the in-house computer code are compared against those of Lewis et al. (1997) numerical results employing the eddy dissipation model and Steckler et al. (1984) experimental data. The numerical results obtained for the volumetric heat source approach from the previous worked example are also re-plotted in the same Figures in order to assess the relative merits in characterizing the fire combustion as a result of modeling through the simpler or more complicated approach to field modeling. All relevant numerical results of the doorway temperature and velocity profiles and experimental data are shown in Figure 18 and Figure 19 respectively.

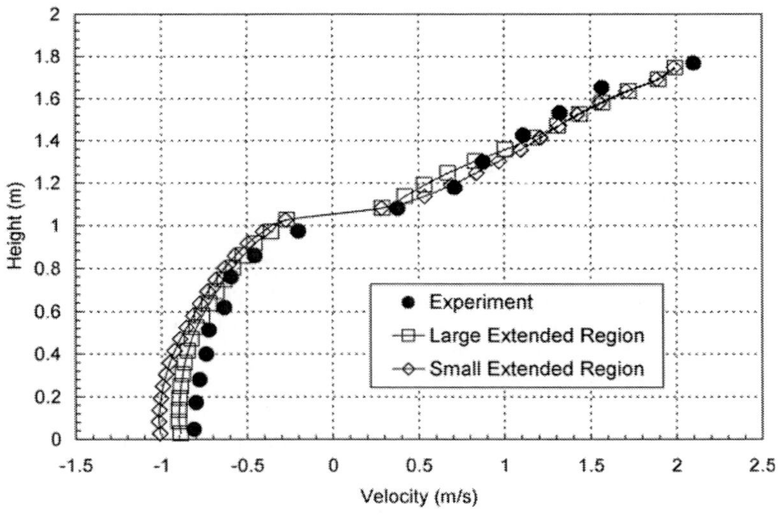

Figure 17. Comparison of velocity profiles at the doorway for different extended regions

Comparing the volumetric heat source and combustion modeling predictions in Figure 19, it is evident that the distinct separation of the hot and cold layers present within the compartment is better predicted by the latter than the former, which incidentally confirms the distinct two-layer structure observed during experiment. On the basis of the combustion simulations made by Lewis et al. (1997) and the in-house computer code for the temperature predictions near the top edge of the doorway, deviations observed in the numerical results could be attributed to a number of factors: (i) Neglect of soot formation and its radiation and combustion efficiency to account for the incomplete combustion of the methane fuel, and (ii) Consideration of adiabaticity at the compartment walls of which the only probable path of the heat escaping from the compartment is through the doorway. Based on these, it is to be expected that the in-house predicted temperatures are likely to be higher than those predicted by Lewis et al. (1997).

In contrast to the predicted temperature profiles, the velocity profiles in Figure 19 appear not to be strongly sensitive to the different approaches adopted. Here again, the combustion modeling predictions appear to fair marginally better than the volumetric heat source approach to illustrate the distinct demarcation between the hot and cold layers that is evidently present within the compartment; cold air is entrained into the burn room at the bottom half while combustion products are seen exhausting at the top half of the doorway.

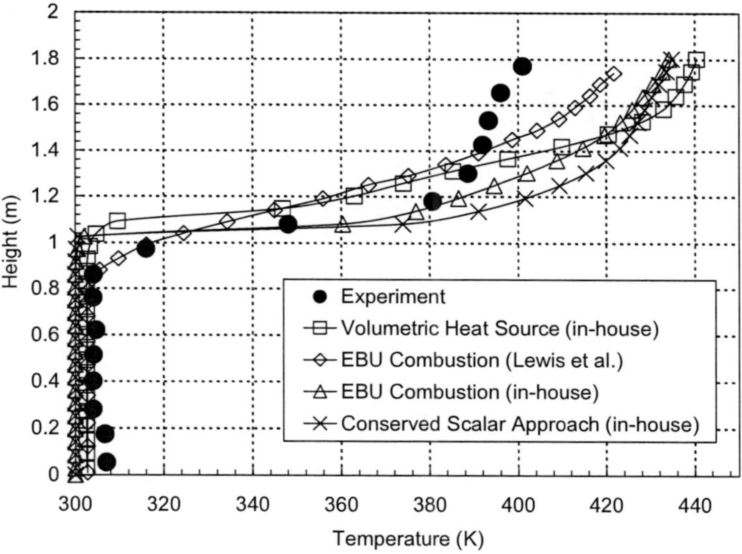

Figure 18. Comparison of different temperature profiles at the doorway

Major species predictions of the concentration contours of H_2O and CO at the vertical symmetry plane dissecting the gas burner are reported in Figure 20 and Figure 21 respectively. It is noted that actual measurements of the species in the Steckler et al. (1984) experiment are not available for comparison. Nevertheless, the capability of the conserved scalar approach using the Sivathanu and Faeth (1990) state relationships could still be checked against Lewis et al. (1997) results to verify whether similar values or trends could be obtained. Since the compartment is well-ventilated and the ambient levels of water vapour H_2O and carbon monoxide CO, remote from the fire source and plume regions, are as expectedly found to be very low. For H_2O, in-house predictions yield a peak level of 0.006 whereas a similar contour structure of 0.01 was obtained by Lewis et al. (1997). For CO, a peak level of 0.026 was predicted by Lewis et al. (1997) while a peak level of 0.01 is achieved through the in-house calculations. It is nevertheless noted that our computed CO levels are still within the same order of magnitude with those predicted by Lewis et al. (1997). Previously in Figure 19, doorway velocity profiles have indicated higher inflow and outflow mass fluxes than those predicted by Lewis et al. (1997). Since substantially more cold air is stipulated to be entering into the compartment, leaner fire combustion in turn is encouraged and promoted

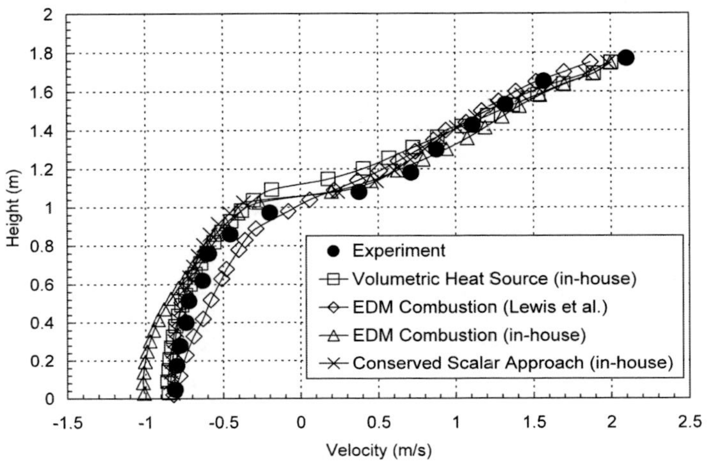

Figure 19. Comparison of different velocity profiles at the doorway

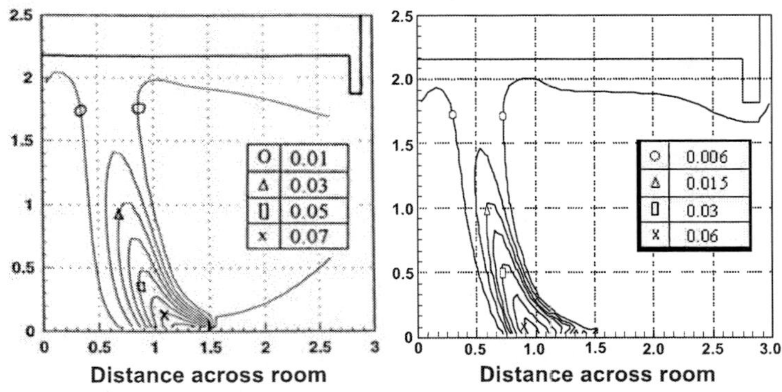

Figure 20. Concentration contours of H_2O at the vertical symmetry plane dissecting the gas burner

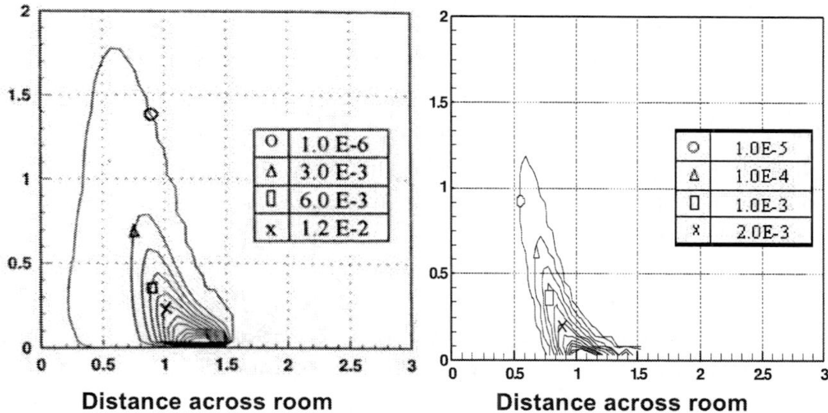

Figure 21. Concentration contours of CO at the vertical symmetry plane dissecting the gas burner

Conclusions: Coupled with a suitable two-equation turbulence model, application of the eddy dissipation combustion model of Magnussen and Hjertager (1976) and the conserved scalar approach employing the Sivathanu and Faeth (1990) state relationships to a full-scale compartment fire is demonstrated through this section. This particular approach does not depend on the *a priori* knowledge of the shape and size of the fire source such as required by the volumetric heat source approach but instead depends only on the specification of the fuel flow rate at the burner surface corresponding to the required heat release of the fire to be simulated. More importantly, this section aptly shows the prevalence of a two-layer structure which generally typifies a compartment fire through the additional consideration of turbulent combustion models. The feasibility of combustion modelling via the laminar flamelet representations of the turbulent fire in order to allow predictions of intermediate chemical species should suffice as a viable model for field modelling investigations if the detailed reaction mechanism is known for the fuel

3.2. Radiative Effects in Enclosure Fires

3.2.1. Single compartment fire

Thermal radiation represents an important mode of heat transfer since some proportion of the energy released by the combustion of fire is radiated to other parts of the flame and to external objects within the compartment. The influence of radiation heat transfer in the development of a fire in an enclosure is further demonstrated utilizing the benchmark case of the Steckler's single-room compartment fire in this section. Numerical results are obtained via the in-house computer code FIRE3D. For the purpose of assessing the model predictions, comparison of results is made for the same heat release rate of 62.9 kW against measurements made by Steckler et al. (1984) and the numerical results obtained from Lewis et al. (1997).

Numerical features: The system of governing equations and all other boundary conditions are identical to those described in the previous section 0. Numerical solutions have been attained by invoking the SIMPLE pressure correction algorithm to link the velocity and pressure, the *hybrid differencing scheme* to approximate the advective term in the governing equations and the standard two-equation k-ε model for turbulence. The eddy dissipation combustion (EDM) model of Magnussen and Hjertager (1976) based on single step chemistry of methane and the conserved scalar approach employing the Sivathanu and Faeth (1990) state relationships are employed to characterize the flaming fire. A mesh density consisting of 83160 grid nodes overlaying half of the burn room attached to an extended region having a size of 3 m × 2.8 m × 6.8 m is employed.

The radiant exchange between fluid elements and the compartment boundaries are handled through the S_4 discrete ordinates model (DOM). In Lewis et al. (1997), the flame radiation has nonetheless been characterized by the application of the discrete transfer radiative method (DTRM). Their room fire simulations considered a total of 16 discrete rays emanated from the solid surface on each boundary cell. The number of rays that were chosen represented a compromise between computational economy and uniform coverage. For direct comparison of the model predictions against those of Lewis et al. (1997), a constant absorption coefficient of 0.2 is prescribed. This is in accordance with the approach adopted in their room fire simulations

Numerical results: Predicted line graphs for the doorway temperature and velocity profiles utilizing the eddy dissipation combustion and conserved scalar model with and without the consideration of radiation exchange are depicted in Figure 22 and Figure 23 alongside with the measured data of Steckler et al. (1984). When compared against predictions without the effects of radiation, the inclusion of a radiation heat transfer model is however seen to significantly reduce and considerably sharpen the temperature profiles in the upper region at the doorway. In practical fires, radiation heat loss could account as high as 30% of the total heat release rate of the fire. Subsequently lower flame temperatures that are experienced above the fire source will result in lower advected temperatures away from the fire source. In spite of the significant discrepancies between the predicted and measured temperatures, the predicted velocity profiles are nevertheless not strongly sensitive to the effects of radiation as shown by the results without radiation in Figure 22a and with radiation in Figure 22b

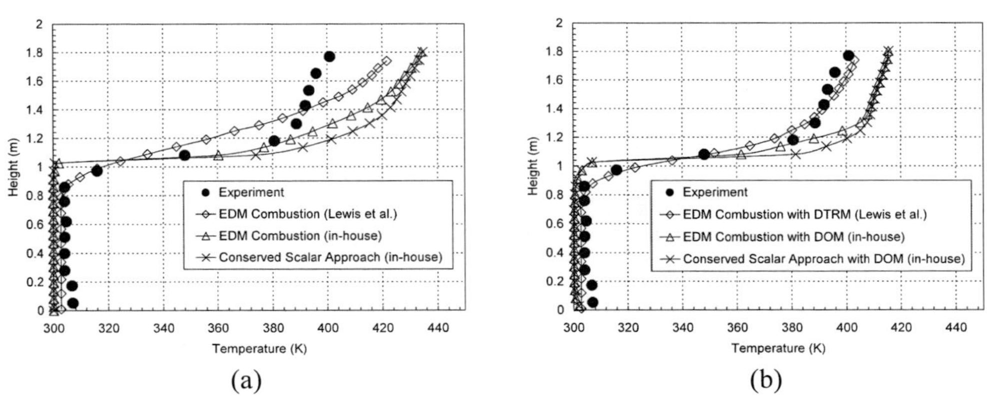

Figure 22. Comparison of doorway temperature profiles using the EBU combustion and conserved scalar models: (a) without and (b) with radiation exchange

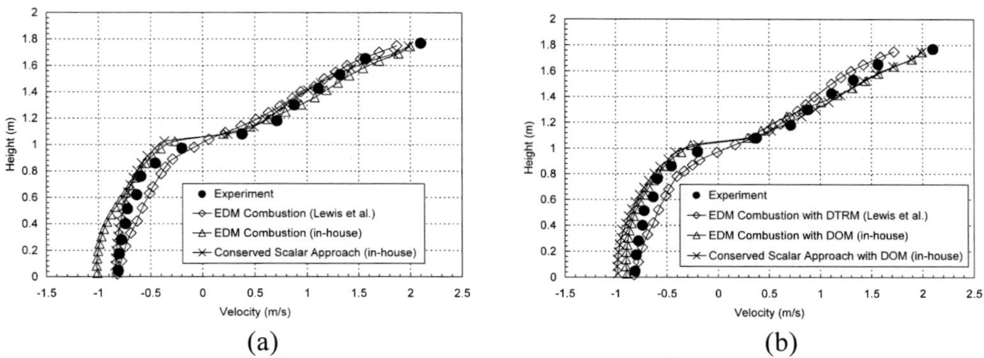

Figure 23. Comparison of doorway velocity profiles using the EBU combustion and conserved scalar models: (a) without and (b) with radiation exchange

Table 3 highlights the predictions of the height of the neutral layer, h_N, (relative to that of the doorway, h_O), the respective inflow and outflow mass fluxes, m_{in} and m_{out}, and the upper layer temperature, $T_{upper\ layer}$, against Lewis et al. (1997) computational results and Steckler et al. (1984) experimental measurement. The definition of the neutral height and calculation of mass fluxes can be found in Lewis et al. (1997) and will not be repeated here. Discrepancies against experimental data for h_N / h_O, m_{in}, m_{out}, $T_{upper\ layer}$ are respectively 3%, 6%, 9% and 1% based on the numerical results by Lewis et al. (1997). Present model predictions with radiative exchange yield discrepancies of 1%, 2%, 2% and 1% for the solutions employing the EDM combustion model and 1%, 2%, 2% and 4% for the conserved scalar model respectively. Comparing the case where the EDM combustion model is employed in the numerical simulations, predictions of the inflow and outflow mass fluxes made by the present model fair better. One possible explanation for the improvement in our predictions could be the large extended region to isolate the end effects of the extended boundaries on the doorway. Better predictions of the neutral height are obtained as can be seen in Table 3. The conserved scalar model is found to yield similar results to those of the EDM combustion model

Table 3. Comparison between model predictions against Lewis et al. (1997) and Steckler et al. (1984)

Grid/ No. of nodes	Combustion	Radiation	h_N / h_O	Inflow (kg/s)	Outflow (kg/s)	Temp. Upper Layer(°C)
Experiment			0.561	0.554	0.571	129
Lewis et al. (1997)						
70432	EDM	DTRM	0.546	0.521	0.523	128
Current						
83160	EDM	DOM	0.557	0.567	0.558	130
	Conserved Scalar	DOM	0.556	0.565	0.558	134

Illustrative temperature profiles of the predicted room corner temperatures are shown in Figure 24. It is clearly observed that predicted temperature profile by Lewis et al. (1997) using the conserved approach without radiation fails to predict the distinct separation of the hot and cold layers as observed during the experiment. Present model predictions without the consideration of radiation behave otherwise. With the inclusion of radiation, the predicted temperature profile correctly replicates the behaviour of the hot layer gas radiation heating the cold layer gas temperatures above the floor level.

A closer examination at the fire source is the apparent deflection of the fire plume towards the back wall due to the incoming flow through the doorway at the bottom as evidently observed during the experiments. The temperature contours for the predicted fire plume at the vertical symmetry plane of the compartment are illustrated in Figure 25. On the basis of these two contour plots, the inclination of the fire for both of the two combustion models with radiative exchange produce plumes at an angle of approximately 45°. These encouraging results are in good qualitative agreement with the observation of Quintiere et al (1981) where they have observed a flame angle which lies between 33° and 43° in their experiments.

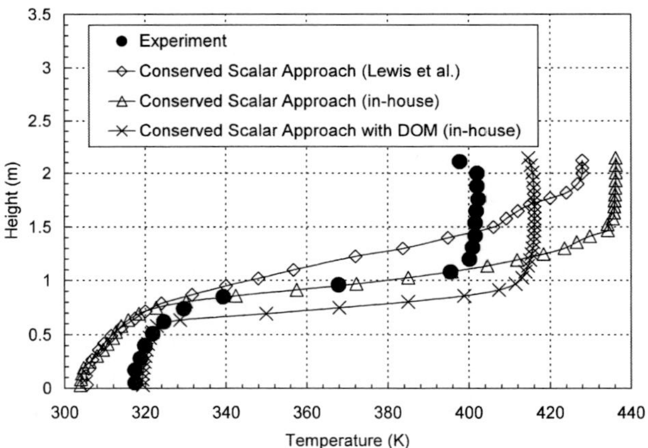

Figure 24. Comparison of predictions against measurements for corner rack temperature profiles using the conserved scalar appoarch with and without the consideration of radiation heat exchange

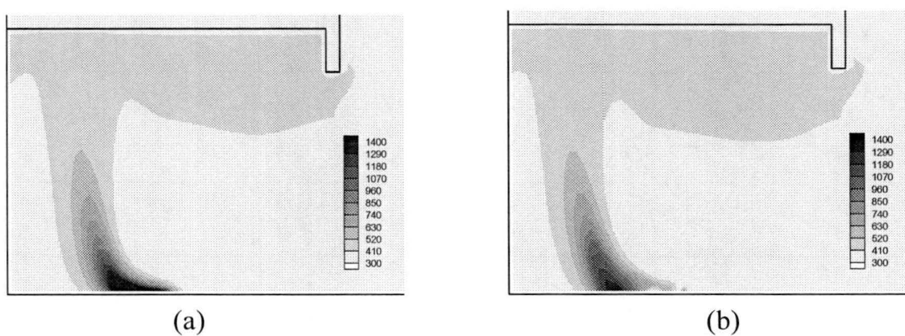

Figure 25. Temperature contours through the centre of fire source and doorway: (a) Eddy diispation model and (b) Conserved scalar model

Conclusions: It has been demonstrated through from the above numerical results that flame radiation can play a significant role and should be properly accounted in field modelling. Through the consideration of radiation heat exchange via the DOM and when coupled with a two-equation turbulence model, eddy dissipation combustion model of Magnussen and Hjertager (1976) and the conserved scalar approach employing the Sivathanu and Faeth (1990) state relationships to characterize the turbulence and combustion of the turbulent buoyant fire, the case with radiation has been shown to yield much better agreement with the experimentally measured profiles when compared to the case without radiation as exemplified by not only the temperatures profiles at the doorway but also at the corner location of the compartment.

3.2.2. Two-compartment fire

To further exemplify the influence of radiation heat transfer, numerical study has bee carried out in two-room compartment fire in this section. Comparison of field modelling predictions with radiation alongside with the computed temperature profiles without radiation are assessed against the experimental data of turbulent buoyant diffusion flames measured by Nielsen and Fleischmann (2000). Numerical simulations are performed through an in-house computer code FIRE3D with the same heat release rate of 110 kW.

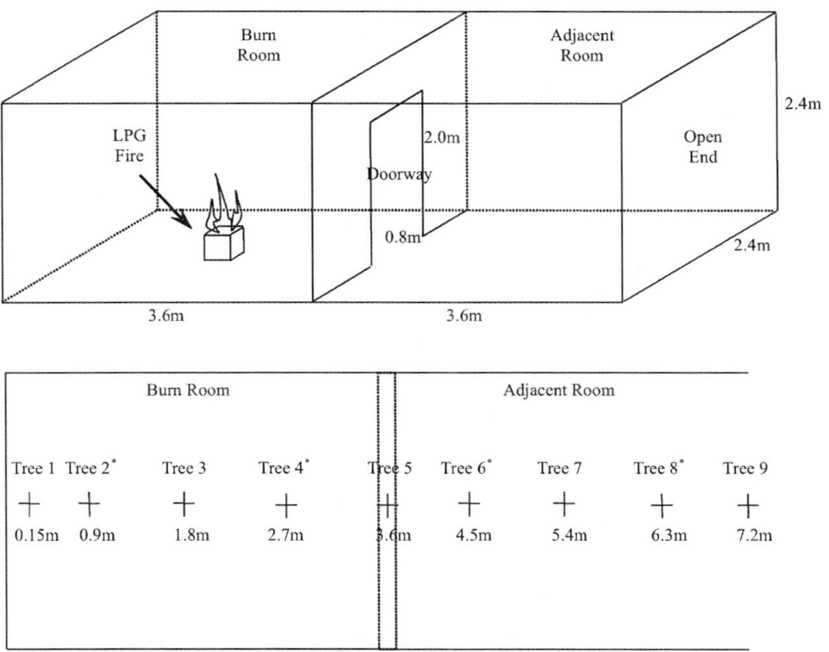

Figure 26. Schematic drawing of the two-compartment structure and distribution of thermocouple tree positions

Figure 26 illustrates the schematic drawing of the particular geometry being investigated. The non-spreading buoyant fire was fueled by a square sand-box LPG burner of 0.3 m wide and elevated 0.3 m above the floor centrally located in the burn room. Air was drawn into the two-room compartment through the open end of the adjacent room and entered the burn room through a doorway opening of 2.0 m high and 0.8 m wide as depicted in Figure 3.23. Compartment walls and ceiling were insulated with Gib® Fyreline and Intermediate Service Board to minimize heat transfer and damage to the structure and instrumentation. Aspirated thermocouples were placed evenly throughout the compartment to enable the validation of vertical temperature profiles. Figure 26 also shows the spatially distributed thermocouple tress within the burn and adjacent rooms. More detailed information regarding the setup can be referred in Nielsen and Fleischmann (2000). An in-house computer code FIRE3D is employed to generate the required numerical results. A heat release rate $\dot{Q}$ of 110 kW has been selected for the validation exercise

Numerical features: The three-dimensional Favre-averaged equations for the transport of mass, momentum and enthalpy are solved. A *hybrid differencing scheme* is employed for the convection terms. The velocity and pressure linkage is achieved through the SIMPLE algorithm. The eddy-viscosity concept is employed for the representation of the turbulent diffusivities in the governing equations due to turbulence. This is expressed by the solution of the standard k-ε turbulent model with additional source terms to account for buoyancy effects. An explicit equation for the mass fraction of fuel is solved for the eddy dissipation combustion model (EDM) of Magnussen and Hjertager (1976). LPG (Liquefied Petroleum Gas) as described in Nielsen and Fleischmann (2000) comprises of approximately 80% of propane (C_3H_8) and 20% of butane (C_4H_{10}). The global, one-step description of LPG combustion is assumed and can be written as:

$$0.8C_3H_8 + 0.2C_4H_{10} + 4.5O_2 \rightarrow 3.2CO_2 + 4.2H_2O \tag{96}$$

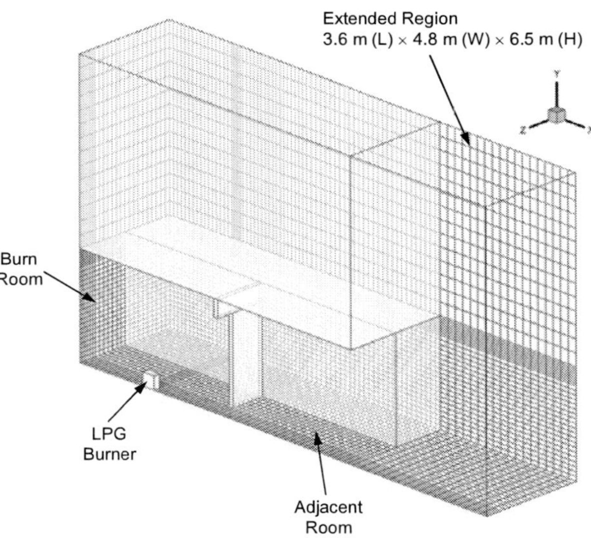

Figure 27. Comparison of predicted and measured temperature profiles in the burn room with and without gaseous radiation

Figure 3.25 shows the grid distribution of the two-compartment geometry. Experimental observations by Nielsen and Fleischmann (2000) for the central fire in the burn room revealed that this configuration was symmetrical about the vertical plane bisecting the burner, doorway and open end. Similar to the single-room compartment fire, mesh generation is thus carried out on half of the room, improving the resolution of the flow and thermal fields. Also, numerical experiments for the single-room compartment fire revealed that the inclusion of an extended region away from the open end was important to correctly model the flow through the open end. An extended region of 3.6 m × 4.8 m in plan and 6.5 m in height is attached to the two-compartment structure to isolate the end effects of the extended boundaries from the open end of the geometry. The computational grid is 85 × 44 × 25 (i.e. a total of 93,500 control volumes).

Radiant heat exchange within the two-room compartment is handled through DOM via the S_4 numerical quadrature approximation. The mean absorption concept of Hubbard and Tien (1978) is adopted to calculate the absorption coefficients of the gaseous combustion products. For soot, the absorption coefficient is based on an expression from Kent and Honnery (1990).

Numerical results: For comparison, results from the simpler strategy of employing the volumetric heat source approach to fire problems by specifying a fixed volume above the fire source to represent the flaming fire are also obtained. On the basis of a prescribed volume size of 0.9 m × 0.3 m × 0.3 m is taken, the volumetric heat capacity of (110 W/ 0.081 m^3) is used as an input parameter for the source term of the energy equation. A constant value of 5.97 as recommended by Hubbard and Tien (1978) is used to account for the radiation heat loss due to combustion products.

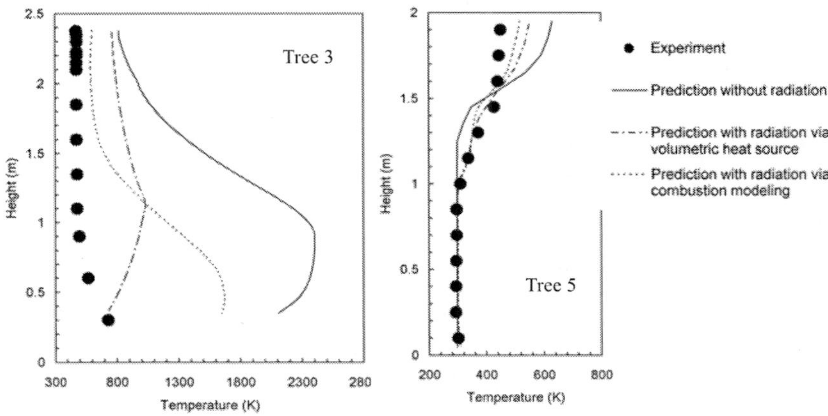

Figure 28. Comparison of predicted and measured temperature profiles above the fire source (Tree 3) and at the doorway (Tree 5) with and without gaseous radiation adopting the volumetric heat source and combustion modeling approaches

Figure 28 illustrates the vertical temperature distribution above the fire source (Tree 3) and at the doorway (Tree 5) with the additional results provided by the inclusion of the effects of radiation in the field model through the volumetric heat source and combustion modeling approaches. Substantially lower predicted temperatures above the fire source clearly identify the significance of radiation heat loss experienced by the fire. Nevertheless, it is observed that the use of the volumetric heat source approach grossly misrepresents the temperature behaviour above the fire source whilst the consideration of a combustion model is otherwise shown to aptly emulate the consistent burning behaviour of the fire as evidenced by the predicted vertical temperature distribution. Away from the fire source, the post-combustion temperatures recorded at the doorway via the volumetric heat source approach are however found to be of satisfactory agreement with the experimentally measured temperatures. This again confirms the unduly predicted high temperatures being confined to the region just above the fire source and exerts only marginal influence on the temperature distributions away from the fire source. Significant improvements to the temperature prediction are however achieved by the consideration of combustion as well as radiation in the field model not only above the fire source but also at the doorway joining the burn room and adjacent room.

Similar to the consideration of radiation in the single-room compartment fire as discussed previously, the surrounding temperatures as demonstrated in Figure 29 are also better predicted within the burn room when radiation heat transfer is accommodated in the field model. Owing to the lower temperatures predicted in the burn room, the over-spilling of high temperatures are not as significantly felt in the adjacent room thereby resulting in the predicted temperatures being much closer to the experimental profiles, as seen in Figure 30. The predicted temperature profiles in the burn room and adjacent room correspond to the solutions attained from the field model adopting the combustion modelling approach. Figure 31 presents the line contour plots of the temperature at the symmetry plane for the cases with and without the consideration of radiation. Radiation contribution by the combustion products is seen to significantly reduce the thermal plume, confirming the lower than expected temperatures found in the adjacent room in contrast to higher predicted temperatures in the case where radiation is neglected.

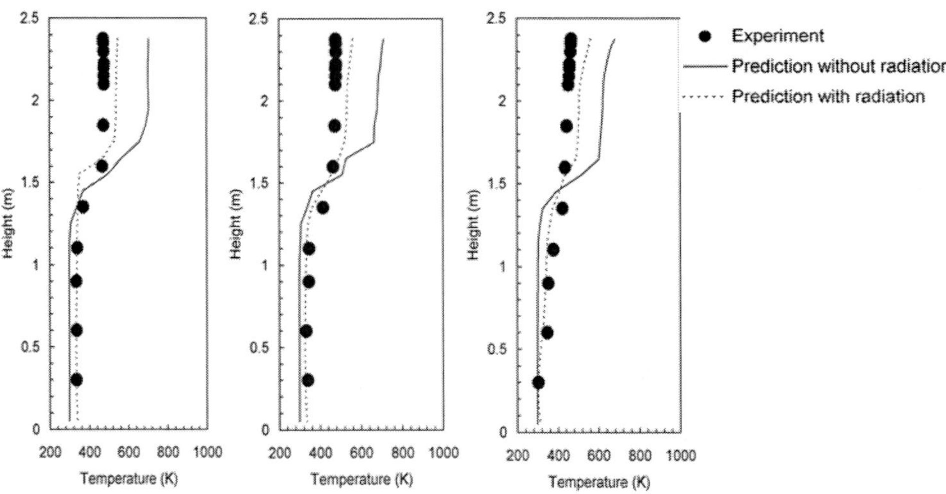

Figure 29. Comparison of predicted and measured temperature profiles in the burn room with and without gaseous radiation

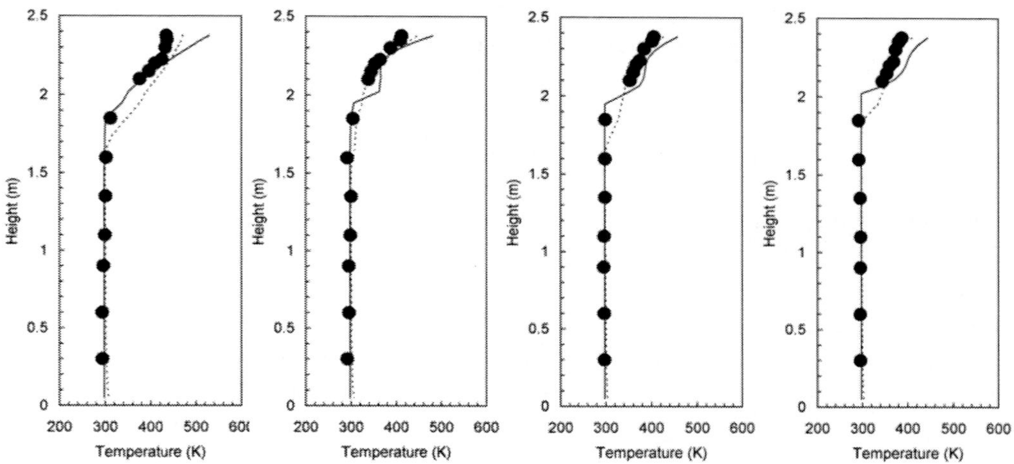

Figure 30. Comparison of predicted and measured temperature profiles in the adjoining room with and without gaseous radiation

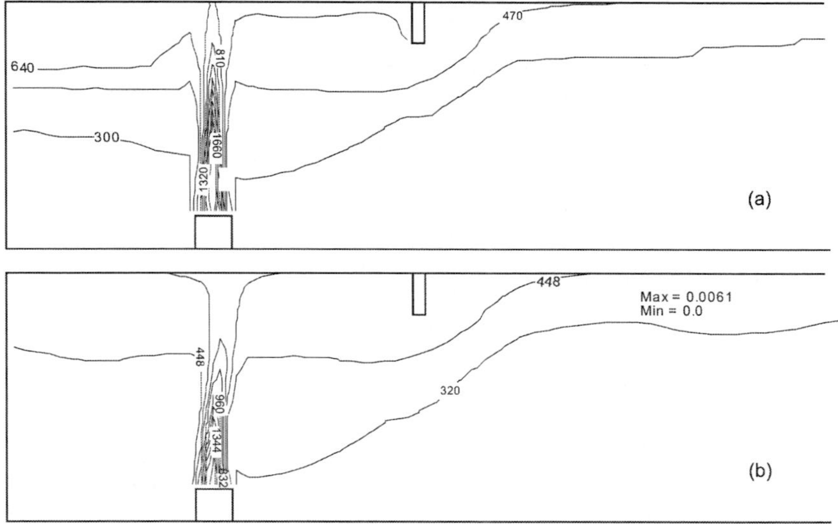

Figure 31. Predicted temperature distribution: (a) Without gaseous radiation and (b) With gaseous radiation

Conclusions: Two important aspects are demonstrated through this section. Firstly, the simple approach based on representing the fire as a volumetric heat source is not recommended if knowledge of temperatures or other local burning characteristics within the flaming fire are required. Combustion modelling remains the only effective way of treating the burning fire. Consideration of combustion yields temperature distribution that is more consistent with measurements as seen by the predicted temperature profiles in Figure 28. Secondly, the inclusion of radiation heat transfer appears to be an integral component in field modelling. Like in single-room compartment fire, flame radiation plays a significant role and should be incorporated in field modelling investigations.

3.3. Soot Modelling in Two-Compartment Fire

After discussing the importance of radiative heat transfer modelling, the two-compartment fire case is used to further explore the importance of luminous soot radiation. A heat release rate of 110 kW for the fire is also adopted as the basis of comparison. Numerical simulations are performed through an in-house computer code FIRE3D. Results incorporating soot and gaseous radiation are assessed against previous field modelling predictions with the consideration of only gaseous radiation presented in above section and experimental data of turbulent buoyant diffusion flames measured by Nielsen and Fleischmann (2000).

Numerical features: Following the study presented in above section, the identical system of governing equations and boundary conditions alongside with the numerical models of turbulence, combustion and radiation models were adopted in the calculaion. For soot formation, three models are considered. They are: (i) the single-step empirical model of Khan and Greeves (1974), (ii) semi-empirical model of Tesner et al. (1971a, 1971b) and (iii) semi-empirical model of Moss et al. (1988). For convenience, these models are hereby referred as: (i) Model 1, (ii) Model 2 and (iii) Model 3. These empirical soot models are attractive since soot production is described simply in terms of concentration of the parent fuel and local temperature. The constants used for the soot formation equations in Model 1, Model 2 and Model 3 can be found in section 4.3.2. A constant C_s = 0.01 kg N^{-1} m^{-1} s^{-1} is assumed in Model 1 while a mean particle diameter of 22.5×10^{-9} m is prescribed in Model 2. In all the calculations, the soot density is taken to be 2000 kg·m^{-3}. The computational grid is 85 × 44 × 25 (i.e. a total of 93,500 control volumes).

Numerical results: The predicted and measured vertical temperature distributions above the fire source (Tree 3) are shown in Figure 32. In spite of the additional consideration of soot radiation, the computed vertical temperature profiles are still appreciably higher than the measured data. As discussed before, Wen et al. (2001) have demonstrated that temperatures measured through bare-wire thermocouples do not actually reflect the real fluid temperatures. Heat transfer processes that generally occur by convection and radiation taking place within the sensor, surrounding surfaces and fluid balance each other to record temperatures between the surface and surrounding fluid temperatures. This error is particularly amplified in high temperature regions especially temperatures above the fire source. On the basis of the analysis carried out by Wen et al. (2001), the thermocouple readings on this tree could be corrected by assuming the following heat transfer equilibrium equation

$$\dot{q}_{convective\ to\ and\ from\ thermocouple} = \dot{q}_{radiative\ to\ and\ from\ thermocouple}$$

$$h\left(T_{gas} - T_{th}\right) = e_{th}\left(\sigma T_{th}^4 - \dot{q}_r\right) \tag{97}$$

where $\dot{q}_r$ is the local radiative flux which may be obtained from the predicted values evaluated via DOM. The convective heat transfer coefficient, h, is estimated using correlation taken from Holman (1992) for a sphere in cross-flow while the cross-flow velocity is set to according to the predicted mean flow velocity across the sensor. The emissivity of the thermocouple e_{th} can be set to 0.2 according to the data in Perry and Chilton (1997). Applying Eq. (97), considering the local radiative flux evaluated from the soot model of Model 3, the corrected temperatures as shown in Figure 32 are seen to behave more consistently with typically observed temperatures (Drysdale, 1999).

Figure 33 compares the predictions of the vertical temperature profiles in the burn room against the spatial measurements carried out for thermocouple tress 1, 2, 4 (away from the fire source) and at the doorway by the thermocouple tree 5. Comparing against the results where only gaseous radiation is considered, the presence of soot radiation is seen to significantly augment the global radiation exchange by significantly improving the temperature predictions.

Table 4 presents the measured floor temperatures against temperatures predicted by the various soot models. Floor temperatures in the burn room increase because of the backward radiation from the hot smoke layer below the ceiling of the compartment. Predicted floor temperatures by considering only gaseous radiation in the computational analysis does not yield satisfactory comparison to the experimental data measured at the locations of the thermocouple tress 1, 2 and 4 within the burn room. It is apparent that the radiation heat transfer is not only due to radiation contribution by the combustion products alone. By accounting further the luminous soot radiation, the model predicts temperatures that are inadvertently much closer to the measured temperatures. In Figure 33, temperatures at the upper part of the doorway are also found to be more comparable to the measured data due to the effect of soot radiation. From a modelling viewpoint, the presence of soot radiation improves the accuracy of predictions.

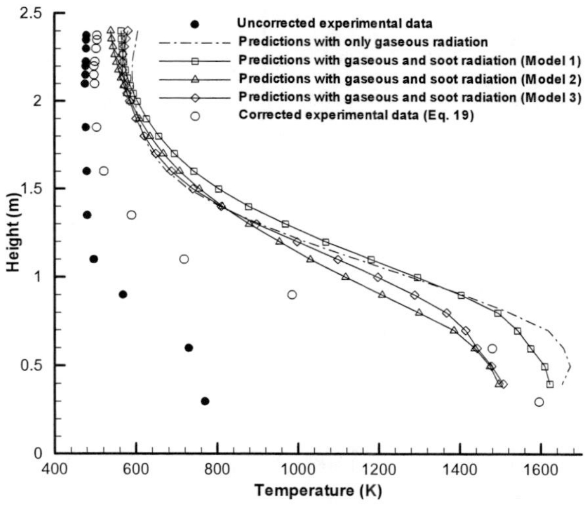

Figure 32. Temperature distribution above the fire source: comparison of models predictions against uncorrected and corrected experimental data

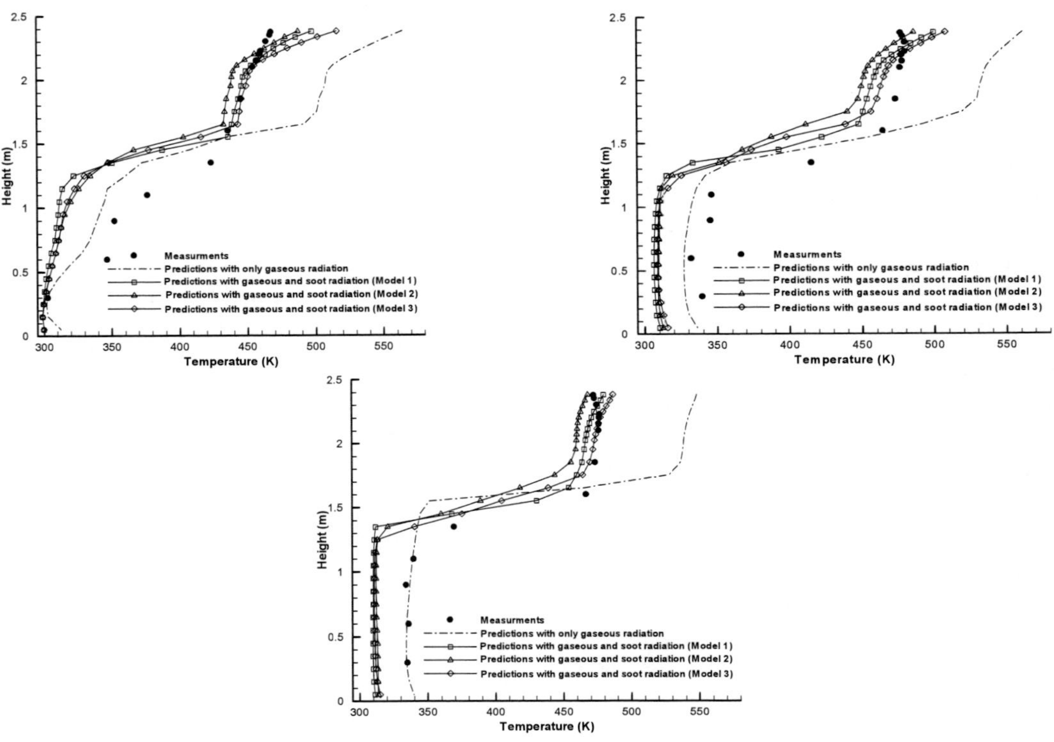

Figure 33. Comparison of temperature profiles measured and predicted at thermocouples Tree 1, Tree 2 and Tree 4 in the burn room

Table 4. Comparison of floor temperatures in the burn room

	Tree 1	Tree 2	Tree 4
Experiment (K)	398.95	413.65	423.75
Only gaseous radiation (K)	495.76	515.91	499.04
Normalized Prediction Error (%)	24.27%	24.72%	17.77%
Soot and gaseous radiation – Model 1 (K)	352.47	369.83	365.75
Normalized Prediction Error (%)	-11.65%	-10.59%	-13.69%
Soot and gaseous radiation – Model 2 (K)	353.74	368.95	364
Normalized Prediction Error (%)	-11.33%	-10.81%	-14.10%
Soot and gaseous radiation – Model 3 (K)	356.16	374.16	369.49
Normalized Prediction Error (%)	-10.73%	-9.55%	-12.80%

Numerical predictions of the vertical temperature profiles in the adjacent room and spatially measured temperatures carried out for thermocouple tress 6, 7, 8 and 9 are illustrated in Figure 35. All models considering soot radiation marginally under-predict the temperatures below the ceiling within the adjacent room. This could be attributed to the neglect of soot burnout or oxidation in the computational analysis. The measured and predicted floor temperatures in this adjacent room are tabulated in Table 5. As a consequence of a thinner ho smoke layer, the effect of backward radiation is not as pronounced as in the burn room. Predicted floor temperatures employing soot Model 3 are observed to be marginally closer to the measured data.

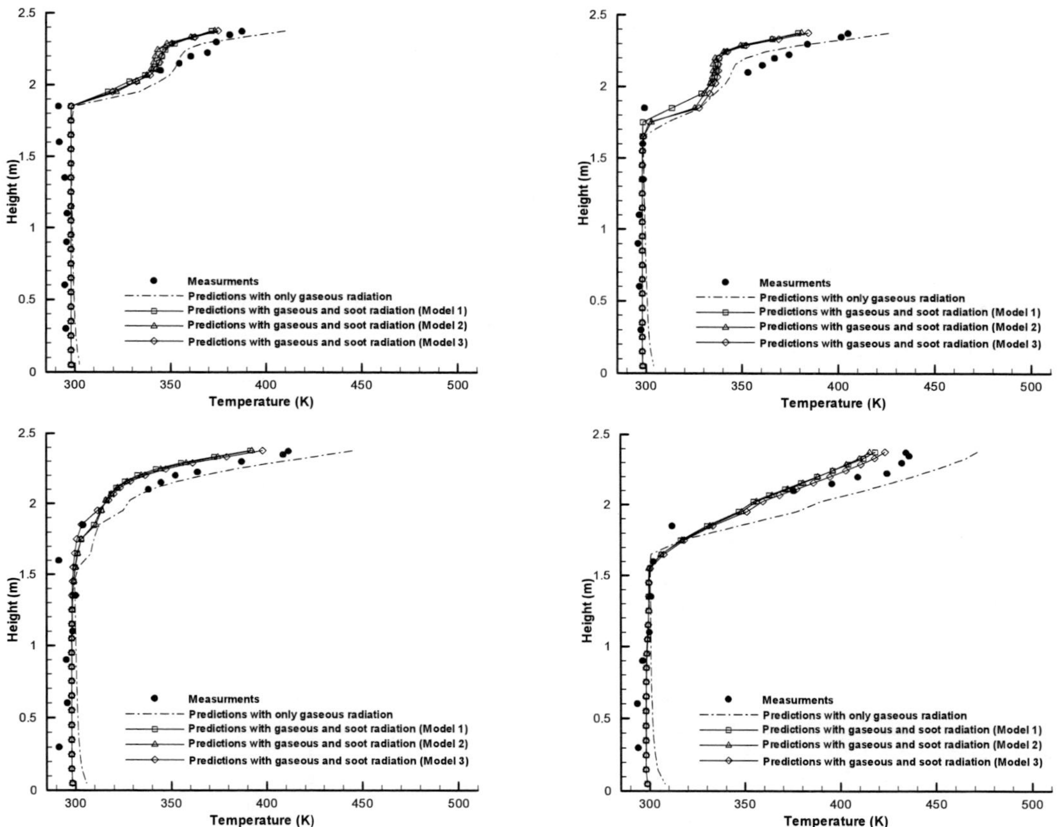

Figure 34. Comparison of predicted and measured temperature profiles at thermocouples Tree 6, Tree 7, Tree 8 and Tree 9 in the adjacent room

Line contour plots of the soot distribution at the symmetrical plane for the respective three soot models are shown in Figure 35. Among all the three models, Model 1 gives the lowest soot distribution while higher soot yield is evidenced in Model 2. This is not entirely surprising since Model 2 is based on pre-determined constants derived from experimental data fitted for acetylene flames. The consistently of low floor temperatures predicted in

Table 4 and Table 5 confirms such assertion. In hindsight, the pre-exponential constant in Model 1 could have been set higher to produce more soot. Increasing the soot loading may however significantly compromise the solutions of already lower floor temperatures thereby contradicting measurements. For such lightly sooty flame, Model 1 which only considers the soot inception rate as the dominant mechanism for the generation of soot is clearly inadequate in predicting reasonable soot concentrations. Model 3 employs constants that were derived from methane combustion, a weakly sooting flame. Since LPG comprises of fuels of predominantly weakly sooting in nature, it is not surprising that reasonable soot levels are predicted when the same constants are applied.

Table 5. Comparison of floor temperatures in the adjoining room

	Tree 6	Tree 7	Tree 8	Tree 9
Experiment (K)	326.05	318.45	317.55	295.75
Only gaseous radiation (K)	403.47	381.67	367.34	343.37
Normalized Prediction Error (%)	23.74%	19.85%	15.68%	16.10%
Soot and gaseous radiation – Model 1 (K)	319.64	315.87	312.09	306.17
Normalized Prediction Error (%)	-1.97%	-0.81%	-1.72%	3.52%
Soot and gaseous radiation – Model 2 (K)	316.26	315.28	311.84	306.15
Normalized Prediction Error (%)	-3.00%	-1.00%	-1.80%	3.52%
Soot and gaseous radiation – Model 3 (K)	319.97	316.17	312.47	306.48
Normalized Prediction Error (%)	-1.86%	-0.72%	-1.60%	3.63%

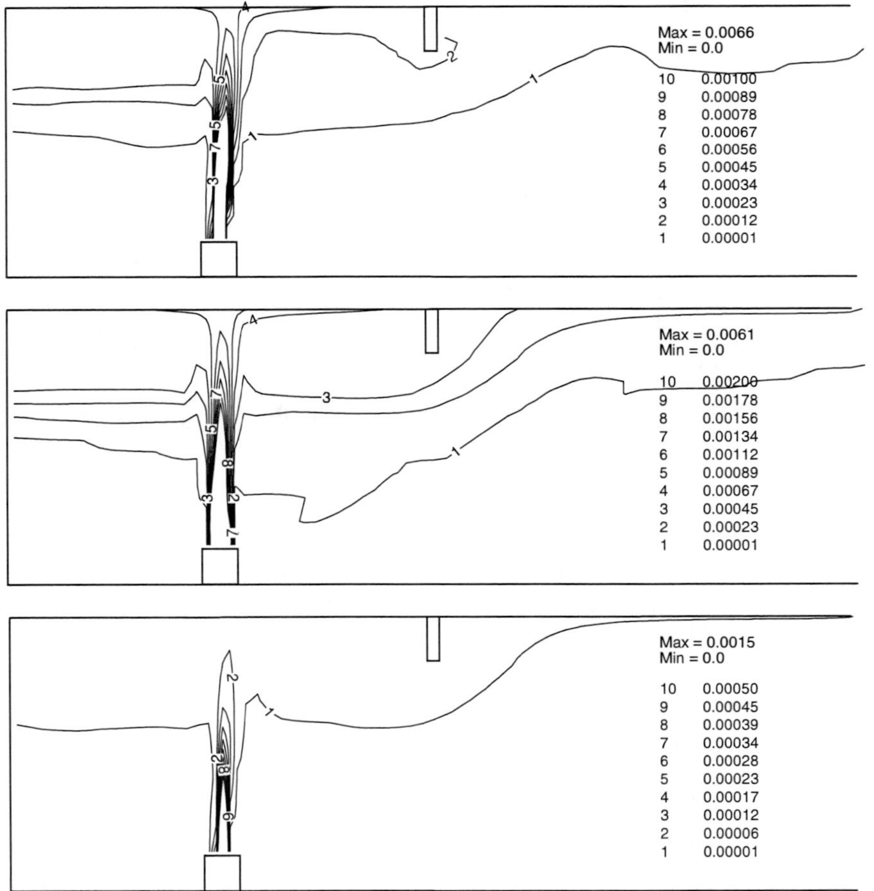

Figure 35. Predicted soot particulate volume fraction distribution: (a) Model 1, (b) Model 2 and (c) Model 3

Conclusions: The consideration of soot for field modelling investigation of a two-compartment fire is demonstrated in this section. On the basis of the above analysis, the inclusion of thermal radiation has shown to reduce the size of the fire plume where the maximum temperatures are located. More importantly, the presence of luminous soot radiation in conjunction with the radiation contribution by combustion products significantly improves the numerical predictions.

4. Capturing Pulsation Frequency of Buoyant Pool Fire

In previous section, the coupling effects of combustion, radiation and soot formation has been examined parametrically through numerical simulations. Nonetheless, one should notice that the time-averaged technique was employed for turbulence closure in all enclosure fire problems. In reality, fire/flame are always unstable or so-called "puffing". This section describes the physical phenomenon of the pulsation behaviour of a buoyant pool fire and the application of LES modelling to capture its temporal vortical structures.

4.1. "Puffing" Effect of Buoyant Pool Fire

From the period of 1970s to the 1980s, many investigators have reported the behaviour of buoyant pool fires exhibiting a periodic oscillatory motion close to their origin which has often been referred to as the "puffing" effect (Bryan and Nelson, 1970; Protscht, 1975; McCaffrey, 1983; Zukoski et al., 1984). By inferring to their observatory studies, the pulsation of fires can be taken to be caused by the formation of large vortical structures with the corresponding length scales on the order of the fire radius. These rotational flow motions create instability to the combustion of fuel with the air entrainment and induce subsequent development of alternative "necking" and "bulging" of the flame surface.

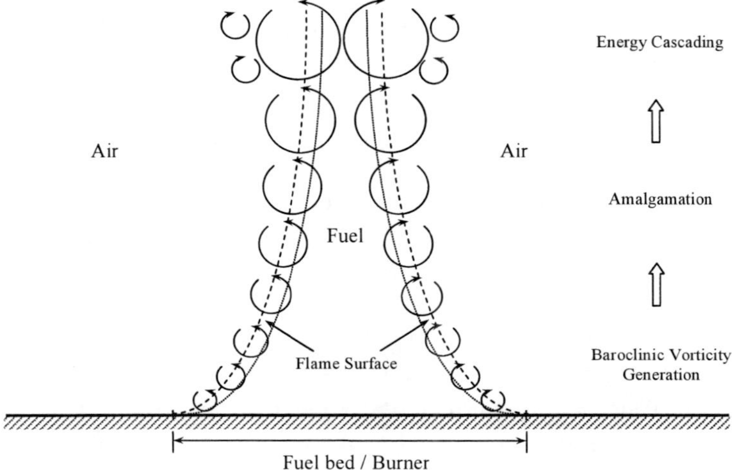

Figure 36. A schematic of the mechanisms contribute to the three stages of vortical structure development

On the basis of the phenomenological reasoning and observational data of various large-scale fire experiments, Tieszen et al. (1996) have further postulated three mechanisms primarily responsible for the vortical structures in fire. The three mechanisms that contribute to the three stages of vortical structure development are depicted in Figure 36. Within the fire, static pressure gradient exists in the vertical direction as a result of the gravitational force, and the density gradient is formed by the rapid temperature change between hot gaseous and surrounding ambient air in the horizontal direction. The misalignment of these two gradients causes the hot and cold fluid to be twisted into rotational motions which thereby initiates the formation of vortical structures. This mechanism also known as the "baroclinic vorticity generation" which can be expressed in terms of the vector cross of the two gradients as: $(\nabla\rho\times\nabla P)/\rho^2$. As the thickness of this density layer controls the strength of the vorticity generation, the strongest vortex formation usually occurs at the smallest scale of the flame surface (i.e. locations of the highest density gradient). These small-scale vorticities are then raised up by the presence of buoyancy forces within the fire. While they are travelling upwards, some may well combined with other eddies that are also rotating in the same sense. The occurrence of this vortices-pairing phenomenon often referred as "amalgamation" leads to the growth of larger flaming vortices and then results in the oscillatory characteristic of necking and bulging of the fire. These large-scale vortical structures undergo subsequent processes of energy cascade, which they collapse back to the smaller eddies, until they burn out at the top of the flame.

4.2. Large Eddy Simulation (LES) for the Vortical Structures of Free-Standing Fire

Based on the description of the aforementioned mechanisms, it must be confessed that vortical structures of fire are generated in a wide range of length scales where exothermic combustion, fluid motions and the spatial distribution of density and pressure occur in highly non-linear unsteady conditions. Modelling the periodic oscillation of pool-fire (also known as pulsation frequency) is thus an extremely challenging task for fire engineers. Macroscopic fire simulations based on time-averaged turbulence models (i.e. turbulence models which has been adopted in section 3) have been known to be unable to capture the scale dependent dynamic behaviours which are prevalent throughout the pulsation cycle of buoyant fires. Alternatively, the Large Eddy Simulation (LES) approach, which involves direct numerical simulation of the large-scale turbulence and modelling the small-scale turbulence, has recently become the central focus of fire modelling (Xin et al., 2002,2005; McGrattan and Forney, 2004; Kang and Wen, 2004). Based on the spatial filtering technique, LES can provide information to match macroscopic observables (scales that are resolved on computational mesh) while the microscopic (unresolved) information are indirectly reflected by the formulation of subgrid-scale (SGS) turbulence model. As the dynamic behaviours of large eddies are resolved directly, the temporal vortical structures are expected to be better captured by the LES as opposed to the RANS approach (Tieszen et al., 1996; Baum et al., 1994).

An in-house developed large eddy simulation fire model, which involves direct numerical simulation of the large-scale turbulence and subgrid scale modeling of the mixture fraction based combustion model, radiation heat transfer via the discrete ordinates method and finite-rate soot chemistry model of Moss et al. (1988) and Syed et al. (1990), is demonstrated in this section to the temporal vortical structure of a buoyant pool fire. The pulsating instability via the frequency of pulsation is also examined. Numerical results from the model as reported in Cheung et al. (2007a) and Cheung et al. (2007b) are validated against a large-scale (i.e. 1.0m diameter) methane pool fire measured by Tieszen et al. (2002).

The fire experiment by Tieszen et al. (2002) used for the comparison exercise centres on a buoyant fire in an open environment. In their experiment, a 1m diameter burner fuelled with methane was placed at the centre. Different heat release rates of the buoyant fire were investigated by adjusting the flow rate of methane; the case of 2.07M is adopted in the present investigation. Throughout the experiment, both instantaneous and time-averaged velocity field were measured using Particle Image Velocimetry (PIV) system.

Numerical features: The low-Mach-number Favre-filtered mass, momentum, energy and species (mixture fraction, scalar variance of the mixture fraction, soot particulate number density and soot volume fraction) conservation equations which have been described in previous sections are solved. In these equations, the molecular viscosity μ is assumed to be a function of the temperature such that $\mu = \mu_{ref} \left(\tilde{T}/T_{ref}\right)^{0.7}$. The molecular Prandtl number is set to a value of 0.7 while the molecular Schmidt numbers for the mixture fraction, the mixture fraction variance, soot particulate number density and soot volume fraction are prescribed at values of 0.7, 0.7, 700 and 700 respectively. Schmidt numbers for the soot have been attained from Sivathanu and Gore (1994).

On the basis of the application of the standard Smagorinsky-Lilly model, the Smagorinsky constant C_s is prescribed at 0.2 while the turbulent Prandtl and all the scalar turbulent Schmidt numbers of 0.3 are imposed. Zhou et al. (2001) have indicated that a little larger C_s is generally used for many thermal flows; C_s for cold jets is usually taken to lie within a range of 0.1 – 0.13. They have employed a value of 0.23 for their large eddy simulation study. The scalar turbulent Schmidt numbers correspond to the combustion studies on turbulent diffusion flame recently performed by Yaga et al. (2002) and confirmed through direct numerical simulation data in Jiménez et al. (2001).

In order to realise a true predicative capability of the fire model, it is imperative to understand the range of length scales that are required to be resolved for a large eddy simulation simulation. For a fire plume, the characteristic length scale can be related to the total heat release rate $\dot{Q}$ (W) by the following relationship as suggested by McGrattan et al. (1998):

$$L^* = \left(\frac{\dot{Q}}{\rho_{ref} T_{ref} C_p \sqrt{g}} \right)^{2/5} \tag{98}$$

In general, the large-scale structure that is controlled by the inviscid terms can be completely described when this characteristic length L^* is adequately resolved. For the heat release considered in this present investigation, the characteristic length L^* is approximately in the order of 1.3m. McGrattan et al. (1998) have ascertained that the large-scale structure can be completely described when L^* is spanned by roughly ten computational cells. This implies that adequate resolution of the fire plume particularly above the porous square burner can be achieved with a spatial resolution of about 0.13 m. With the spatial value of 0.13 m used as reference, two non-uniform mesh distributions of 96 × 96 × 96 and 116 × 116 × 116 cells have been tested within the computational domain with finer grid cells centred above the burner to better capture all the necessary macroscopic large-scale features of the flaming fire. No significant difference of the predicted results is observed when simulations are performed on the two grid resolutions. For the best trade-off between numerical accuracy and cost, the mesh of 96 × 96 × 96 cells is thus employed. Finer grid cells with the minimum spacing of 1.4 cm were generated above the burner to better capture all the necessary macroscopic features of the vortical flame structure. A transient analysis is preformed via the two-stage predictor-corrector approach for low Mach number compressible flows to account for the strong coupling between the density and fluid flow equations. The computational is set to 35 seconds to ensure that it reaches the stable and converged status. The time step is determined by employing a CFL number of 0.35 to achieve time-accurate solution via:

$$dt = \frac{0.35}{\max\left(\left|\frac{\tilde{u}}{\Delta x}\right| + \left|\frac{\tilde{v}}{\Delta y}\right| + \left|\frac{\tilde{w}}{\Delta z}\right|\right)} \tag{98}$$

Numerical results: A quasi-steady state solution was obtained when the physical time arrived at 35 seconds. Time-averaged field quantities were then extracted by performing time-weighted averaging calculation over 10 seconds of instantaneous solutions. Figure 37 illustrates the predicted and measured time-averaged vertical velocity profiles at different centre-line locations (i.e. Y=0.2, 0.4, 0.6 and 0.8m) above the methane burner. In general, the predicted velocity profiles agreed reasonably well with the measurement. Especially at the location Y=0.8m, the maximum error of prediction is 13% which is well within the ±20% uncertainly bounds of the measurements. Larger discrepancies were observed at the vicinity of burner surface (i.e. Y=0.2 and Y=0.4) in Figure 37a and Figure 37b. As predicted in the figures, the distance between two velocity peaks was considerably over-predicted.

A more comprehensive depiction could be found in Figure 38 here time-averaged horizontal and vertical velocity contour plots were compared with measured results. Overall, the predicted time-averaged velocity contours of the present fully-coupled LES model are in good agreement with the measurements. In Figure 38a, aligned with the observation in Figure 37, the predicted gap between two velocity regions can be observed to be wider than measurement. This suggested that the present model slightly over-estimated the spreading rate of velocities and the width of the fire plume as evident in vertical contours shown in Figure 38b.

Similar numerical errors were also reported by Rawat et al. (2002) and Desjardin (2005) where they both concluded that these numerical errors were attributed to the insufficient grid resolution for LES models to resolve the microscopic baroclinic vorticity generation near the burner surface. As this mechanism was observed to be responsible for the sequential up-scale turbulent energy transfer to large scale eddies, filtering of these small-scale turbulent motions by the LES models might cause under-prediction of the local turbulent mixing rate resulting in an inaccurate estimation of the local heat release, combustion and velocity spread rates. Furthermore, the single-step combustion reaction assumption adopted in the numerical simulation could be another source of error coupling with the velocity field.

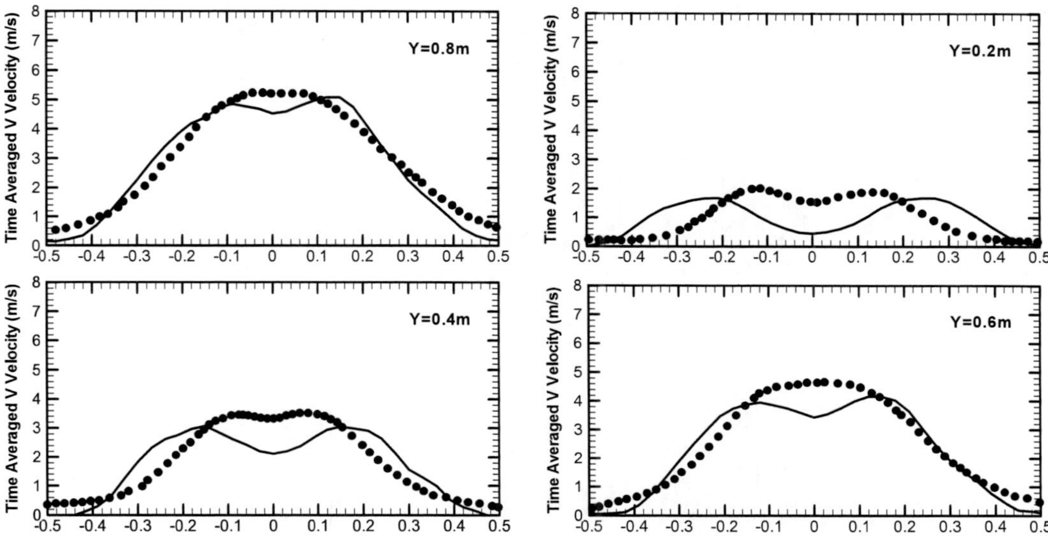

Figure 37. Predicted and measured time-averaged vertical velocity at different centre-line locations: (a) Y=0.2m; (b) Y=0.2m; (c) Y=0.2m and (d) Y=0.8m

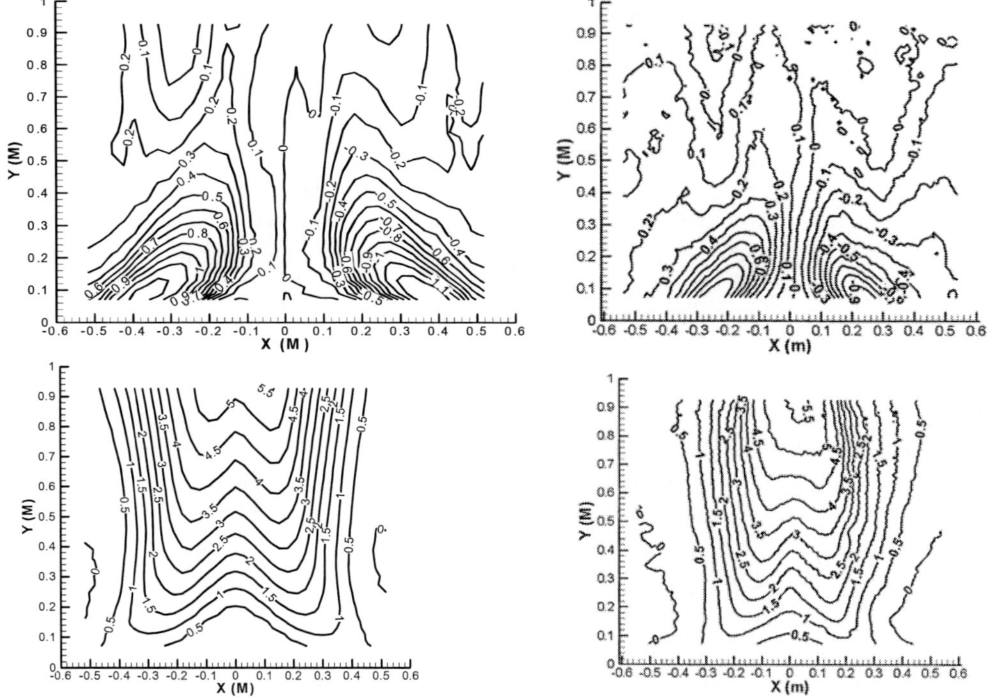

Figure 38. Comparison of time-averaged velocity component contours captured from PIV measurement of Tieszen et al. (2002) (left column) and the present LES model (right column) at the centre-plane of the fire: (a) U velocity (horizontal) and (b) V velocity (vertical)

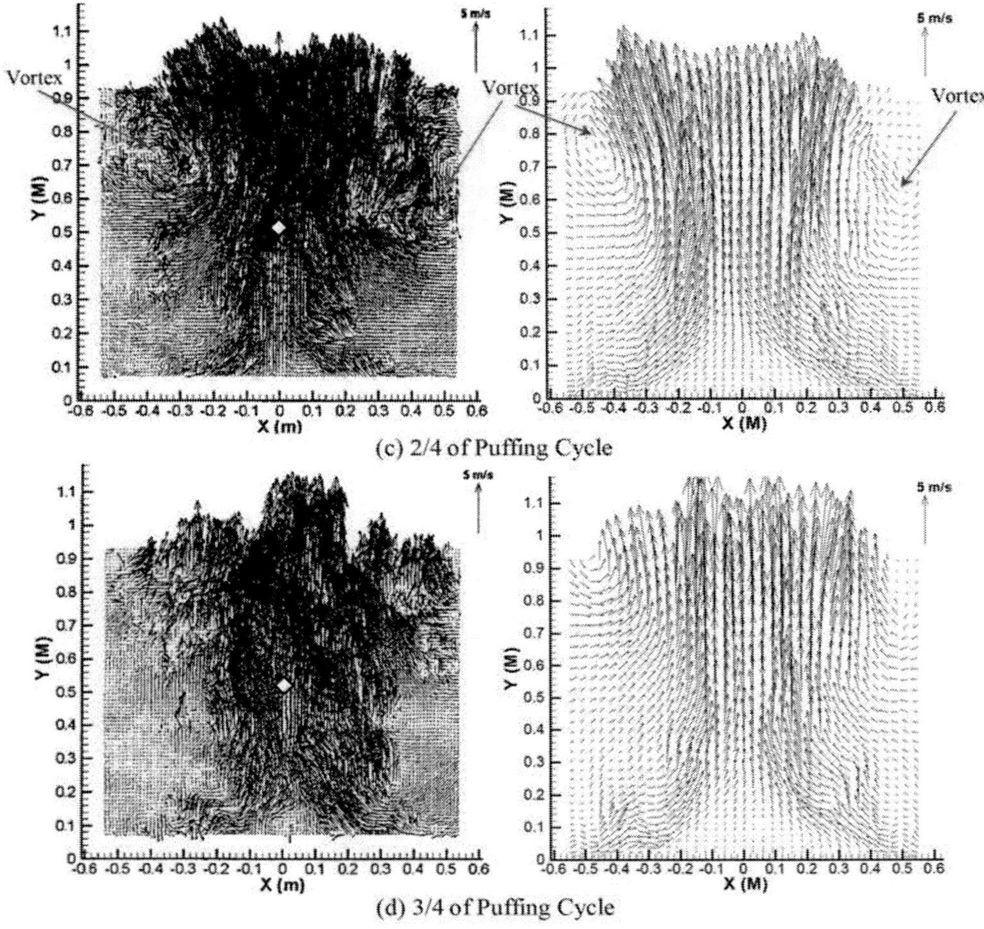

Figure 38c-d. Four instantaneous velocity field at the centre plane of the fire plume captured by PIV measurement of Tieszen et al. (2002) (left column) and the present LES model (right column): (c) 3/4 and (d) end puffing cycle

The temporal vortical structures and its coupled combustion behaviour of the large scale pool fire are discussed herein. Figure 39 shows the instantaneous velocity field at the centre plane of the fire plume captured by the PIV measurement of Tieszen et al. (2002) and the present LES model. The four sequential PIV results on the left column clearly illustrated the relationship between large vortical structures and the puffing cycle. At the start of the cycle, turbulent eddies were firstly stemmed from the base of the fire which has been caused by baroclinic vorticity generation. Owing to the amalgamation of eddies and buoyancy forces, the size of vortical structures continuously increased and accelerated in vertical direction until it has been advected out of the image (see Figure 39b-c). In comparison to the PIV measurements, four sequential predicted vector plots were also extracted within one arbitrary puffing cycle from the present model (as shown in the right column of Figure 39). A broad overview of the figure suggested that the predicted instantaneous velocity fields were in excellent agreement with the PIV measurements.

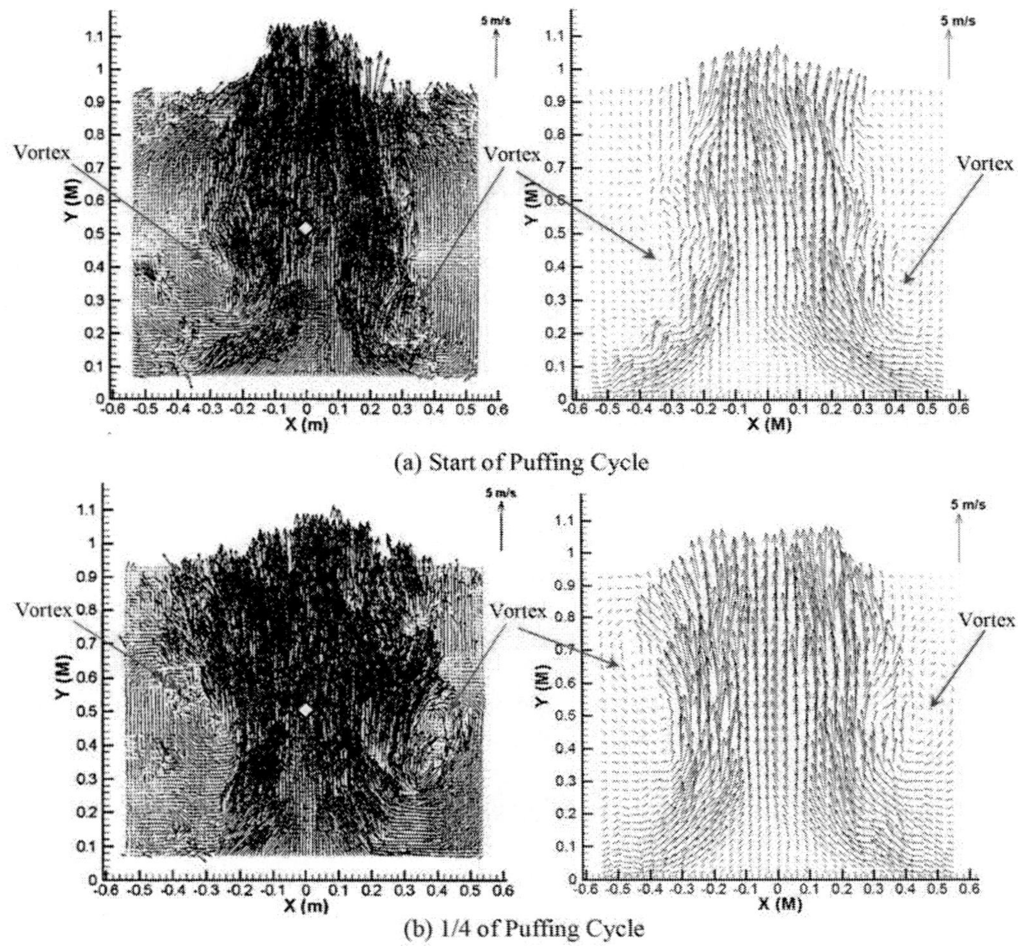

Figure 39a-b. Four instantaneous velocity field at the centre plane of the fire plume captured by PIV measurement of Tieszen et al. (2002) (left column) and the present LES model (right column): (a) start and (b) 2/4 puffing cycle

Velocity vectors of predicted results also exhibited assembly similar behaviour / distribution to the experimental data. Special attention was paid on the simulated vortical structures at both sides of the fire. As depicted in the figure, the vortical structures were successfully captured by the present model. From the start to the end of puffing cycle, the simulated vortical structures were created at the fire base and continuously developed in size and convected upward away through the top of plot. The above development of the vortices was clearly aligned with the observations of PIV measurements. Vortex locations of the numerical results were also comparable to the measurement.

In connection with the above qualitative comparison, a closer examination of the pulsating behaviour of the pool fire can be analysed by tracing the time history of the vertical velocity at the centre-point of fire (X = 0, Y = 0.505 m) in comparison to the record of PIV measurement in Figure 40. As summarized in Tieszen et al. (2002), the PIV record indicated that the periods of the puffing cycles preserved a somewhat regular pattern but varied slightly from time to time. In general, the predicted vertical velocity history exhibited similar pattern with the measurement. Nevertheless, it must be confessed that the predicted time history line appears much "smoother" than the scattering measurement points where a more chaotic behaviour and sharp peaking of the velocity fluctuations have been observed. Such smoothing effect on the numerical results was contributed by nature of the LES models. Based on the spatial filtering technique, LES models only resolve large eddy motions above the subgrid length. Subgrid scale eddy motions were filtered and modelled by the SGS model. Re-capturing these chaotic flow behaviours would require significant improvement of the current grid resolution or even Direct Numerical Simulation scheme which demands excessive computational resources. The frequency spectrum obtained from above time history line of instantaneous vertical velocity by Fast Fourier Transform (FFT) shows a dominant frequency at 1.526 Hz which is only 8% faster than the measured pulsation frequency (1.65Hz). Based on the above thorough quantitative comparison, it can be concluded that the present LES model is capable to resolve the temporal effects from the vortical structures and its resultant pulsation frequency of large scale pool fire.

Conclusion: The fire model based on large eddy simulation has shown to be capable of reproducing the temporal and spatial evolution of the self-excited large toroidal flaming vortex structures that correspond to the experimentally observed vortex shedding phenomenon of a typical buoyant diffusion fire. The time-averaged velocity and temperature profiles from the present model were in good agreement in comparison to experiment data and other numerical results. Instantaneous vector plots were compared alongside with PIV measurements. The predicted velocity magnitudes and vortex locations were found to be qualitatively comparable with experimental data suggesting that the simulation is capable reproducing the typical temporal and spatial evolution of the large-scale vortical structures and its puffing behaviour of buoyant pool fire. Quantitative validations of the time history of velocity fluctuations and the pulsation frequency were also presented with excellent agreement with experimental data.

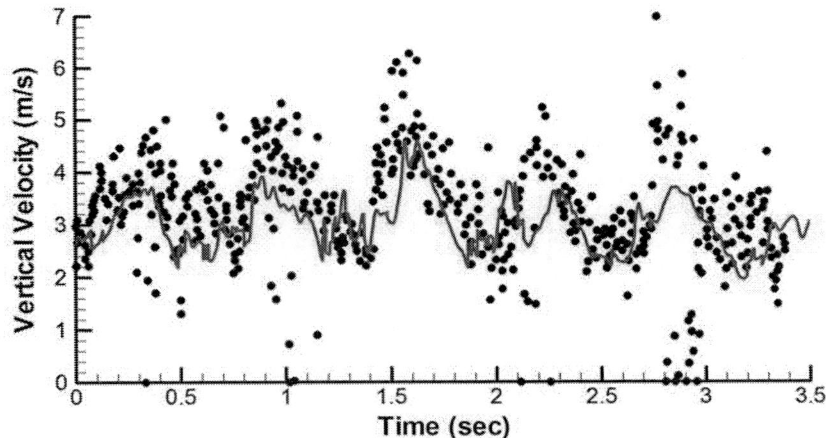

Figure 40. Time history of the vertical velocity at the centre-point of fire (X=0, Y=0.505) predicted by present model (line) and recorded by PIV measurement (points)

5. Conclusions and Further Developments

In the previous sections, theoretical concepts and mathematical formulation of combustion, radiation and soot sub-models have been discussed. Furthermore, to demonstrate the necessity and assess the capability of the sub-models, a number of numerical studies have been also presented. From the presented validation results, one should be able to appreciate the capability and maturity of the existing fire models in replicating the sophisticated coupling effects of both enclosure and free-standing fires. This is also the reason why fire field model is now becoming a viable and prevalent predictive tool for fire safety assessment or fire engineering approach.

Nevertheless, advancements to the models are still required especially to better resolve complicated flaming conditions in practical fire scenarios. Current combustion models that have been applied to solve a whole range of fire problems need to be further improved beyond the fast chemistry assumption. Development of combustion models able to accommodate a wide range of chemical and turbulent time scales will certainly assist in the predictive capability of the field model in simulating the growth and spread of fires of not only in a well-ventilated environment but also in an under-ventilated environment such as the depletion of oxygen supply in a room with the door shut. The latter aspect has serious implication towards the possible dire consequence of a back draft fire incident, usually persisting only few seconds before exhausting its fuel supply.

As discussed in previous sections, soot formation and oxidation models in field modelling can still be regarded as very empirical in nature. Soot modelling with detailed chemistry is yet to be fully explored. Recently, the viability of the population balance approach to handle the soot formation and oxidation particulate processes has certainly received unprecedented attention from both academic as well as industrial quarters. The structure of the population balance equation, generally expressed as an *integrodifferential* form of the particle size distribution, is generally very complex and their solution by analytical means remains elusive for all but the most idealized situations (Bove et al., 2005). Several numerical techniques have been proposed in the literature to best handle the population balance equations not only achieving some considerable levels of accuracy but also obtaining the solutions in real time with moderate computational load.

The standard method of moments (SMM) by Frenklach (1985), which has been introduced in the section 0, can be considered one of the earliest algorithms developed in align with the population balance approach. Numerically speaking, the mathematical difficulty of SMM lies in obtaining the appropriate closure. The simplest way of accomplishing this is to presume the functional form of the PSD function (Dobbins and Mulholland, 1984, Pratsinis, 1988). To achieve closure, Frenklach (2002) have attested the validity of interpolation closure in obtaining these moments by first expressing the natural logarithmic of moments in terms of polynomial and later determining the required moments by separating the interpolation for positive-order and negative-order moments via the Lagrange interpolation among logarithms of the whole-order moments. As an attractive alternative to the SMM, the quadrature method of moment (QMOM) as proposed by McGraw (1997) could be employed to purposefully approximate the moment integrals. Here, nodes or abscissas and weights of the quadrature approximation can be determined from the moments of the distribution by using a very efficient algorithm (Gordon, 1968, McGraw, 1997,

McGraw and Wright, 2003). QMOM is a rather sound mathematical approach and represents an elegant tool in solving the population balance equations with limited computation burden. In essence, this method may be regarded as a presumed PSD method where the underlying distribution is assumed to be made of delta functions. It thus possesses many similarities with the conventional SMM where the PSD is assumed to be a monodisperse or lognormal distribution. Marchisio and Fox (2005) formulated a direct formulation (direct quadrature method of moment or DQMOM) for multi-dimensional problems, which this particular approach is able to cater for poly-disperse systems with two or more internal coordinates. Encouraging results attained by Barret and Webb (1998) and Marchisio et al. (2003a, 2003b) have certainly elevated DQMOM as a serious competing method that presents the main advantage of being extremely accurate in solving monovariate (i.e. one internal coordinate) problems and amenable for coupling with CFD calculations.

On the other hand, to resolve the embedded chemophysical nonlinearity in a more accurate sense, the density of mesh grid is of primary importance at the flame region. Especially for the LES simulations, such mesh requirement could become extremely critical since grid spacing governs the filtering width of subgrid scale eddy motions. To circumvent this problem, the Local Grid Refinement (LGR) technique could be one of the directions aiming at refining the mesh spacing without incurring significant computation burden within the calculation (Magnus and Yoshihara, 1970; Rai, 1986; Coelho *et al.*, 1991; Kao and Liou, 1995; Drikakis *et al.,* 2001)

The Nested Overlapping Grid (NOG) method is one the LGR techniques which has been recently proposed for fire zone where combustion, radiation and soot formation processes are closely coupled in the finest length scale. Figure 41 shows the comparison of the predicted temperature profiles based on the conventional structure grid and NOG method. As depicted, using the NOG method, significant reduction of mesh grid nodes (resulting over 50% saving of computational time) could be attained with the comparable prediction accuracy. The success of incorporating the LGR techniques into the LES fire modelling is expected to be a crucial step for enhancing the practicality of the current LES fire models.

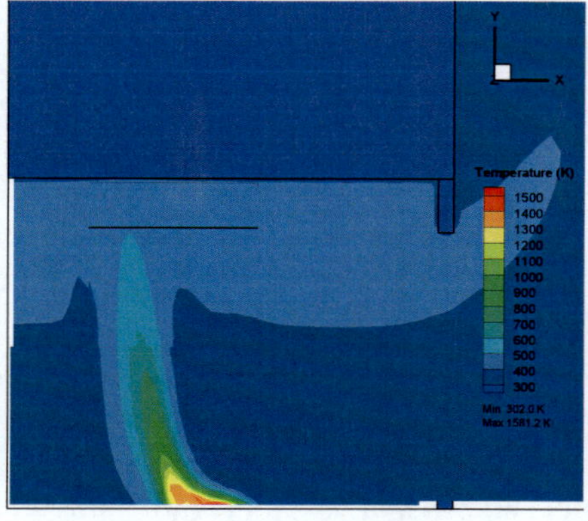

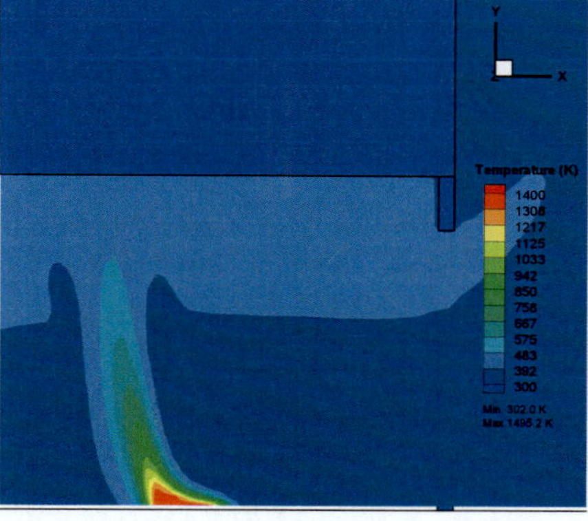

Figure 41. Comparison of the predicted temperature profiles based on: (a) conventional structure grid (with 151,200 nodes); (b) NOG method (with 63,218 nodes)

ACKNOWLEDGMENT

The financial support provided by the Australian Research Council (ARC project ID LP0882413) is gratefully acknowledged.

REFERENCES

Abramowitz, M. & Stegun, I. A (1964). *Handbook of Mathematical Functions*, Washington.

Barrett, J. C. & Webb, N. A. (1998). A Comparison of Some Approximate Methods for Solving the Aerosol General Dynamics Equation, *J Aerosol Sci*, *29*, 31-39.

Baum, H. R, Ezekoye, O. A, Mcgrattan, K. B. & Reum, R. G. (1994). Mathematical-Modelling and Computer-Simulation of Fire Phenomena. *Theoretical and Computational Fluid Dynamics*, *6*, 125-139.

Beer, J. M, Foster, P. J. & Siddall, R. G. (1971). Calculation methods of radiative heat transfer. *HFTS Design Report No. 22*.

Bilger, R. W. (1975). A Note on Favre Averaging in Variable Density Flows. *Combust. Sci. Tech. 11*, 215-217.

Bilger, R. W. (1980). *Turbulent Flows with Non-premixed Reactions*. New York: Academic Press.

Bilger, R. W. & Kent, J. H. (1972). *Measurements in Turbulent Diffusion Flames*, Report F41.

Bishop, S. R. Drysdale, D. D. (1998). Fire in Compartments: the phenomenon of flashover. *Philosophical Transaction of the Royal Society of London series A*, 1748, 2855-2872.

Bove, S, Solberg, T. & Hjertager, B. H. (2005). A Novel Algorithm for Solving Population Balance Equations: The Parallel Parent and Daughter Classes. Derivation, Analysis and Testing. *Chem. Eng. Sci*, *60*, 1449-1464.

Brookes, S. J, Moss, J. B. (1999). Predictions of Soot and Thermal Radiation Properties in Confined Turbulent Jet Diffusion Flames. *Combust. Flame*, *116*, 486-503.

Brown, F. B. & Martin, W. R. (1984). Monte-Carlo Methods for Radiation Transport Analysis on Vector Computers. *Progress in Nuclear Energy*, *14*, 269-299.

Bryan, G. M. & Nelson R. M. (1970). The modelling of pulsating fires, Fire Tech. *6* 102-110.

Bryan, G. M. & Nelson, R. M. (1970). The modelling of pulsating fires. *Fire Tech*, *6* 102-110.

Calcote, H. F. & Manos, D. M. (1983). Effect of molecular-Structure on Incipient Soot Formation. *Combustion and Flame*, *49*, 289-304.

Case, K. M. & Zweifel P. F. (1967). *Linear Transport Theory*, Addison-Wesley, Reading, MA.

Chai, J. C, Patankar, S. V, Lee, H. S. (1994). Evaluation of Spatial Differencing Practices for the Discrete Ordinates Method. *J. Thermophys. Heat Transfer*, *8*, 140-144.

Chandrasekhar, S. (1960). *Radiative Transfer*. New York: Dover.

Cheung, S. C. P. (2006). *Modelling of Building Fires coupled with Turbulent, Combustion, Soot Chemistry and Radiation Effects*. PhD Thesis. City University of Hong Kong, Hong Kong.

Cheung, S. C. P., Yeoh, G. H., Cheung, A. L. K. & Yuen, R. K. K. (2007a). Flickering Behavior of Turbulent Buoyant Fires Using Large-Eddy Simulation. *Numer. Heat Transfer, Part A*, *52*, 679-712.

Cheung, S. C. P., Yeoh, G. H. & Tu, J. Y (2007b). On the Numerical Study of Isothermal Vertical Bubbly Flow Using Two Population Balance Approaches. *Chem. Eng. Sci*, *62*, 4659-4674.

Collin, A., Boulet, P., Lacroix, D. & Jeandel, G. (2005). On Radiative Transfer in Water Spray Curtains using the Discrete Ordinate Method. *Journal of Quantitative Spectroscopy & Radiative Transfer*, *92*, 85-110.

Dembele, S., Delmas, A., Sacadura, J. F. (1997). A Method for Modelling the Mitigation of Hazardous Fire Thermal Radiation by Water Spray Curtains. *Journal of Heat Transfer*, *119*, 746-753.

Dembele, S. & Wen, J. X. (2000). Investigation of a Spectral Formulation for Radiative Heat Transfer in One-dimensional Fires and Combustion System. *International Journal of Heat and Mass Transfer*, *43*, 4019-4030.

Desjardin, P. E. (2005). Modeling of conditional dissipation rate for flamelet models with application to large eddy simulation of fire plumes. *Comb. Sci. & Tech*, *177*, 1883-1916.

Dobbins, R. A. & Mulholland, G. W. (1984). Interpretation of Optical measurements of Flame Generated Particles, *Combust. Sci. Tech*, *40*, 175–191.

Drysdale, D. (1999). *An Introduction to Fire Dynamics*, New York: Wiley.

Eddington, A. S. (1988). *The Internal Constitution of the Stars*, UK: Cambridge University Press.

Edwards, D. K. (1976). Molecular Gas Band Radiation, *Advances in Heat Transfer*, *12*, 100-193.

Edwards, D. K. (1976). Molecular Gas Band Radiation. *Advances in Heat Transfer*, *12*, 100-193.

El Ghobashi, E. E. (1974). *Characteristics of Gaseous Turbulent Diffusion Flames in Cylindrical Chambers, A Theoretical and Experimental Investigation*, London University.

Erlebacher, G., Hussaini, M. Y., Speziale, V. G. & Zang, T. A. (1992). Towards the large-eddy simulations of compressible turbulent flows. *J. Fluid Mech*, *238*, 155-185.

Fairweather, M., Jones, W. P. & Lindstedt, R. P. (1992). Predictions of Radiative Transfer from a Turbulent Reacting Jet in Cross-Wind. *Combust. Flame*, *89*, 45-63.

Favre, A. (1965). Equation des gaz Turbulents Compressibles. *J. Mechanique*, 4, 361-392.

Felske, J. D., Tien, C. L. (1974). A Theoretical Closed Form Expression for the Total Band Radiating Gases. *Int. J. Heat Mass Transfer*, *17*, 155-158.

Fenimore, C. P. & Jones, G. W (1967). *Journal of Physical Chemistry*, *71*, 593.

Fiveland, W. A. (1984). Discrete-Ordinates Solutions of the Radiative Transport Equations for Rectangular Enclosures. *Journal of Heat Transfer*, *106*, 699-706.

Fiveland, W. A. (1987). Discrete Ordinate Methods for Radiative Heat Transfer in Isotropically and Anisotropically Scattering Media. *Journal of Heat Transfer*, *109*, 809-812.

Fiveland, W. A. (1988). Three-Dimensional Radiative Heat-Transfer Solutions by the Discrete-Ordinates Method. *J. Thermophysics*, *2*, 309-316.

Fiveland, W. A. (1994). Finite Element Formulation of the Discrete-Ordinates Method for Multidimensional Geometries. *Journal of Thermophysics and Heat Transfer*, *8*, 426-433.

Fletcher, D. F., Kent, J. H., Apte, V. B. & Green, A. R. (1994). Numerical Simulations of Smoke Movement from A Pool Fire in a vertical Tunnel. *Fire Safety Journal*, *23*, 305-325.

Frenklach, M. (2002). Method of Moments with Interpolative closure. *Chemical Engineering Science*, *57*, 2229-2239.

Frenklach, M., Clary, D. W., Gardiner, J. & Stein, S. E. (1984). Detailed Kinetic Modelling of Soot Formation in Shock-Tube pyrolysis of Acetylene, In: *Twentieth Symposium (International) on Combustion*, The Combustion Institute, University of Michigan, USA.

Frenklach, M. & Harris, S. J. (1987). Aerosol Dynamics Modelling using The Method of Moments. *Journal of Colloid and Interface Science*, *118*, 252-261.

Frenklach, M. & Wang, H. (1990). Detailed Modelling of Soot Particle Nucleation and Growth, In: *Twenty-third Symposium (International) on Combustion*, University of Orlans, France.

Frenklach, M. & Wang, H. (1994). Detail Mechanism and Modelling of Soot Particle Formation IN: *Soot Formation in Combustion*, Bockhorn, H. ed., Springer-Verlag, PA, pp. 162-190

Fuchs, N. A. (1964). *The Mechanics of Aerosols*. Oxford: Pergamon Press.

Garo, A., Prado, G. & Lahaye, J. (1990). Chemical Aspects of Soot Particles Oxidation in a Laminar Methane-Air Diffusion Flame. *Combust. Flame*, *79*, 226-233.

Germano, M., Piomelli, U., Moin, P. & Cabot, W. H. (1991). A Dynamics Subgrid-scale Eddy Viscosity Model. *Physics of Fluids A*, *3*, 1760-1765.

Ghosal, S., Lund, T. S, Moin, P. & Akselvoil, K. (1995). A Dynamics Localization Model for Large-Eddy Simulation of Turbulent Flows. *J. Fluid Mech 286*, 229-255.

Gill, R. J., Olson, D. B. & Calcote, H. F. (1984). Corrections of Soot Formation in Turbojet Engines and Laboratory Flames. *Mechanical Engineering*, *106*, 87-87.

Goody, R. M. (1952). A Statistical Model for Water Vapor Absorption. *Quart. J. Roy. Meteor. Soc*, *78*, 165-169.

Gordon, R. G. (1968). Error Bounds in Equilibrium Statistical Mechanics, *J. Math. Phys*, *9*, 655-672.

Goussebaile, J. & Viollet, P. L. (1982). On the Modelling of Turbulent Flow under Strong Buoyant Effects in Cavities with Curved Boundaries. *Proc. of Symp. Refined Modelling of Flows.*

Hanjalic, K. & Jakirlic, S. (1993). A Model of Stress Dissipation in Second Moment Closures. *Applied Scientific Research*, *51*, 513-528.

Harris, S. J, Weiner, A. M. (1983). Determination of the Rate Constant for Soot Surface Growth. *Combust. Sci. Tech*, *32*, 267-275.

Harris, S. J, Weiner, A. M. (1983). Surface Growth of Soot Particles in Premixed Ethylene/Air Flames. *Combust. Sci. Tech*, *31*, 155-167.

Holman, J. P. (1992). *Heat Transfer*, *7th ed.* New York: McGraw-Hill.

Hossain, M. S. & Rodi, W. (1982). *A Turbulence Model for Buoyant Flows and Its Application to Vertical Buoyant Jets*, (p. 121-178). Oxford: Pergamon.

Hottel, H. C (1954). *Radiant Heat Transmission, Heat Transmission.* New York: McGraw-Hill.

Hottel, H. C. & Sarofim, A. F. (1967). *Radiative Transfer*, New York: McGraw-Hill.

Hubbard, G. L. & Tien, C. L. (1978). Infrared Mean Absorption Coefficients of Luminous Flames and Smoke. *ASME J. Heat Transfer*, *100*, 235-239.

Hutchinson, P., Khalil, E. E. & Whitelaw, J. H. (1977). Measurement and Calculation of Furnace Flow Properties. *J. Energy*, *1*, 212-219.

Ince, N. Z. & Launder, B. E. (1989). On the Computation of Buoyancy-driven Turbulent Flows in Rectangular Enclosures. *International Journal of Heat and Fluid Flow*, *10*, 110-117.

Jamaluddin, S. & Smith, P. J. (1988). Predicting Radiative Transfer in Rectangular Enclosures using the Discrete Ordinates Method. *Combust. Sci. Tech*, *59*, 321-340.

Jeans, J. H. (1917). The Equations of Radiative Transfer of Energy. *Mon. Not. Roy. Astr. Soc*, *78*, 28-36.

Jiménez, C., Ducros, F., Cuenot, B. & Bédat, B. (2001). Subgrid Scale Variance and Dissipation of a Scalar Field in Large Eddy Simulations. *Phys. Fluids*, *13*, 1748-1754.

Jones, W. P. (1980). Models for Turbulent Flows with Variable Density, VK-I Lecture Series 1979-2. In: W Kollmann, (Ed) *Prediction Methods for Turbulent Flows*, New York: Hemisphere Publishing.

Kang, Y. & Wen, J. X. (2004). Large Eddy Simulation of a Small Pool Fire. *Com. Sci. & Tech*, *176*, 2193-2223.

Kee, R. J., Grgar, J. F., Smooke, M. D. & Miller, J. A. (1985). A Fortran Program of Modelling Steady Laminar One-Dimensional Premixed Flames, *In*: *Sandia Report No. SAND85-8240*.

Kennedy, I. M. (1997). Models of soot formation and oxidation. *Prog. Energy Combust. Sci*, *23*, 95-132.

Kent, J. H. & Bilger, R. W. (1977). The Prediction of Turbulent Diffusion Flame Fields and Nitric Oxide Formation, *In*: *Sixteenth Symposium (International) on Combustion*, pp. 1643-1656.

Kent, J. H. & Honnery, D. R. (1990). A Soot Formation Rate Map for a Laminar Ethylene Diffusion Flame. *Combustion and Flame*, *79*, 287-298.

Keramida, E. P., Boudouvis, A. G., Lois, E., Markatos, N. C. & Karayannis, A. N. (2001). Evaluation of Two Radiation Models in CFD Fire Modelling. *Numerical Heat Transfer - Part A*, *39*, 711-722.

Khalil, E. E. (1977). *Flow and Combustion in Axisymmetric Furnaces*, London University.

Khan, M. & Greeves, G. A. Eds. (1974). *A Method for Calculating the Formation and Combustion of Soot in Diesel Engines*, Washington: Scripta Book.

Kim, T. K. & Lee, H. S. (1988). Effect of Anisotropic Scattering on Radiative Heat Transfer in Two-Dimensional Rectangular Enclosures. *Int. J. Heat Mass Transfer*, *31*, 1711-1721.

Kuo, K. K. (1986). *Principles of Combustion*, New York: A Wiley-Interscience Publication.

Lathrop, K. D. (1976). THREETRAN A Program to Solve the Multigroup Discrete Ordinates Transport Equation in (x,y,z) Geometry. *Report LA-4848-MS*.

Launder, B. E. & Spalding, D. B. (1972). *Lectures in Mathematical Models of Turbulence*, London: Academic Press.

Launder, B. E. & Spalding, D. B. (1974). The Numerical Computation of Turbulent Flows. *Comp. Meth. Appl. Mech. Eng*, *3*, 269-289.

Lautenberger, C. W., de Ris, J. L., Dembsey, N. A., Barnett, J. R. & Baum, H. R. (2005). A Simplified Model for Soot Formation and Oxidation in CFD Simulation of Non-premixed Hydrocarbon Flames. *Fire Safety Journal*, *40*, 141-176.

Lee, K. B., Thring, M. W. & Beer, J. M. (1962) *Combustion and Flame*, (p. 137).

Leung, K. M., Lindstedt, R. P. & Jones, W. P. (1991). A Simplified Reaction Mechanism for Soot Formation in Nonpremixed Flame. *Combustion and Flame 87*, 289-305.

Lewis, E. E. & Miller J. W. F. (1984). *Computational Methods of Neutron Transport*, New York: Wiley.

Lewis, M. J., Moss, M. B. & Rubini, P. A. (1997). CFD Modeling of Combustion and Heat Transfer in Compartment Fire, In: *Fire Safety Science - Proceedings of Fifth International Symposium*, Melbourne.

Lilly, D. K. (1966). On the Application of the Eddy Viscosity Concept in the Inertial Sub-Range of Turbulence. *NCAR Report No, 123*.

Lilly, D. K. (1967). The Representation of Small-Scale Turbulence in Numerical Simulation Experiments. *Proceedings of the IBM Scientific Computing Symposium and Enviromental Science*.

Liu, F. & Wen, J. (2002). The Effect of Turbulence Modelling on the CFD Simulation of Buoyant Diffusion Flames, *Fire Safety Journal*, *37*, 125-150.

Lockwood, F. C. & Naguib, A. S. (1975). The Prediction of the Fluctuations in the Properties of Free, Round-Jet, Turbulent, Diffusion Flame. *Combustion and Flame*, *24*, 109-124.

Lockwood, F. C. & Shah, N. C. (1980). A New Radiation Solution Method for Incorporation in General Combustion Prediction Procedures, In: *Eighteenth Symposium (International) on Combustion*, pp. 1405-1414). The Combustion Institute, University of Waterloo.

Luo, M. & Beck, V. (1996). A Study of Non-flashover and Flashover Fires in a Full-Scale Multi-room Building. *Fire Safety Journal*, *26*, 191-219.

Lutz, A., Kee, R. J., Grcar, J. F. & Rupley, F. M. (1997). *OPPDIF: A FORTRAN Program for Computing Opposed-Flow Diffusion Flames*, Livermore.

Magnussen, B. F. & Hjertager, B. H. (1976). On Mathematical Modelling of Turbulent Combustion with Special Emphasis on Soot Formation and Combustion, In: *Sixteenth Symposium (International) on Combustion, Massachusetts Institute of Technology*, pp. 719-729). USA.

Marchisio, D. L., Vigil, D. R. & Fox, R. O. (2003a). Quadrature Method of Moments for Aggregation-Breakage Processes. *J. Colloid Interface Sci*, *258*, 322-324.

Marchisio, D. L., Pikturna, J. T., Fox, R. O. & Vigil, R. D. (2003b). Quadrature Method of Moments for Population-Balance Equations. *AIChE J*, *49*, 1266-1276.

Marchisio, D. L. & Fox, R. O. (2005). Solution of Population Balance Equations Using the Direct Quadrature Method of Moments, *J. Aerosol Sci 36*, 43-73.

McCaffery, B. J. (1979). *Purely Buoyant Diffusion Flames: Some Experimental Results*. NBSIR 79-1910, Centre of Fire Research, National Bureau of Standards, Washington D.C. 20234.

McCaffrey, B. J. (1983). Momentum Implication for Buoyant Diffusion Flames. *Combustion and Flame*, *52*, 149-167.,

McGrattan, K. B. & Forney, G. (2004). *Fire Dynamics Simulator (Version 4) - User's Guide, NIST Special Publication 1019, National Institute of Standard and Technology.*

McGrattan, K. B., Rehm, R. G. & Baum, H. R. (1994). Fire-Driven in Enclosures. *Journal of Computational Physics*, *110*, 285-291.

McGrattan, K. B., Rehm, R. G. & Baum, H. R. (1996). Numerical Simulation of Smoke Plumes from Large Oil Fires. *Atmospheric Environment*, *30*, 4125-4136.

McGrattan, K. B., Rehm, R. G. & Baum, H. R. (1998). Large Eddy Simulation of Smoke Movement. *Fire Safety Journal*, *30*, 161-178.

McGraw, R. (1997). Description of Aerosol Dynamics by the Quadrature Method of Moments, *Aerosol Sci, Tech*, *27*, 255-265.

McGraw, E. & Wright, D. L. (2003). Chemically Resolved Aerosol Dynamics for Internal Mixtures by the Quadrature Method of Moments, *J. Aerosol Sci*, *34*, 189-209.

Meneveau, C, Lund, T. S. & Cabot, W. H. (1996). A Lagrangian Dynamic Subgrid-Scale Model of Turbulence. *J. Fluid Mech*, *319*, 353-385.

Métais, O. & Lesieur, M. (1992a). Spectral Large_eddy Simulation of Isotropic and Stably Stratified Turbulence. *J. Fluid Mech*, *256*, 475-503.

Métais, O. & Lesieur, M. (1992b). Spectral Large eddy Simulation of Isotropic and Stably Stratified Turbulence. *J. Fluid Mech* , *256*, 475-503.

Modak, A. T. (1979). Radiation from Products of Combustion. *Fire Res*, 1, 339-361.

Modest, M. F. (1991). The Weighted-Sum-of-Gray Gases Model for Arbitrary Solution Methods in Radiative Transfer. *ASME J. Heat Transfer*, *113*, 650-656.

Modest, M. F. (2003). *Radiative Heat Transfer*, London: Academic Press.

Moss, J. B., Stewart, C. D. & Syed, K. J. (1988). Flowfield Modelling of Soot Formation at Elevated Pressure. *Twenty-Second Symposium (International) on Combustion*, 413-423.

Motevalli, V. (1994). Numerical Prediction of Ceiling Jet Temperature Profiles during Ceiling Heating using Empirical Velocity Profiles and Turbulent Continuity and Energy Equations. *Fire Safety Journal*, *22*, 125-144.

Nagle, J. & Strickland-Constable, R. F. (1962). Oxidation of Carbon between 1000-2000°C, In: *Fifth Carbon Conference*, pp. 154-164). Pergamon, Oxford.

Nam, S. & Bill, B. G. (1993). Numerical-Simulation of Thermal Plume. *Fire Safety Journal*, *21*, 231-256.

Nielsen, C. & Fleischmann, C. (2000). An Analysis of Pre-Flashover Fire Experiments with Field Modeling Comparison, In: *Fire Engineering Research Report*, *ISSN*, *1173-5996*, University of Canterbury, NZ.

Novozhilov, V. (2001). Computational Fluid Dynamics Modelling of Compartment Fire. *Progress in Energy and Combustion Science*, *27*, 611-666.

Novozhilov, V., Harvie, D. J. E., Green, A. R. & Kent, J. H. (1997). A Computational Fluid Dynamics Model of Fire Burning Rate and Extinction by Water Sprinkler. *Combustion Science and Technology*, *123*, 227-245.

Olson, D. B., Pickens, J. C. & Gill, R. J. (1985). The Effects of Molecular-Structure on Soot Formation .2. Diffusion Flame. *Combustion and Flame*, *62*, 43-60.

Ozisik, M. N. (1973). *Radiative Transfer and Interactions with Conduction and Convection*, New York.

Perry, R. H. & Chilton, C. H. (1997). *Chemical Engineers Handbook, Seventh Edition* , New York: McGraw-Hill.

Peters, N. (1984). Laminar Diffsion Flamelet Models in Non-Premixed Turbulent Combustion. *Prog. Energy Combust. Sci*, *10*, 319-339.

Peters, N. (1984). Laminar Diffusion Flamelet Models in Non-Premixed Turbulent Combustion. *Prog. Energy Combust. Sci. 10*, 319-339.

Peters, N. (1986). Laminar Flamelet Concepts in Turbulent Combustion. *Prog. Combust. Inst*, *21*, 1231-1250.

Piomelli, U. & Moin, P. A. & Ferziger, J. H. (1988). Model Consistency in Large Eddy Simulation of Turbulent Channel Flow. *Phys. Fluids*, *31*, 1884-1891.

Portscht, R. (1975). Studies on Charactistic Fluctuation of Flame Radiation emitted by Fires. *Combustion Science and Technology, 10*, 73-84.

Pratsinis, S. E. (1988). Simultaneous Nucleation, Condensation, and Coagulation in Aerosol Reactors, *J. Colloid Interface Sci, 124*, 416–427.

Puri, R. Santoro, R. J. & Smyth, K. C. (1994). The Oxidation of Soot and Carbon Monoxide in Hydrocarbon Diffusioin Flames. *Combust. Flame, 97*, 125-144.

Quintiere, J., Rinkinen, W. J. & Jones, W. W. (1981). The Effect of Room Openings on Fire Plume Entrainment, *Combust. Sci. Tech, 26*, 193-201.

Rasbash, D. J., Drysdale, D. D. (1982). Fundamentals of Smoke Production. *Fire Safety J, 5*, 77-86.

Rawat, R., Pitsch, H. & Ripoll, J. F. (2002). Large-Eddy Simulation of Pool Fires with Detailed Chemistry using an Unsteady Flamelet Model-Studying Turbulence Using Numerical Simulation Database IX, In: *Proceedings of the Summer Program, 2002, Centre of Turbulence Research*, Stanford University.

Rodi, W. (1985). *Calculation of Stably Stratified Shear Layer Flows with a Buoyancy-extended Turbulence Model*, (p. 111-140). Oxford: Clarendon Press.

Rogallo, R. S., Moin, P. (1984). Numerical Simulation of Turbulent Flows. *Ann Rev. Fluid Mech, 16*, 99-137.

Rogg, B., Ed. (1993). *RUN-1DL: The Cambridge Universal Laminar Flamelet Code, Reduced Kinetic Mechanisms for Applications in Combustion Systems*, Berlin: Springer-Verlag.

Rogg, B. & Wang, W. (1997). *The Laminar Flame and Flamelet Computer Code, User Manual*, Bochum: Lehrsthul Strömungsmechanik, Institut für Thermo-und Fluiddynamik, Ruhr-Universität Bochum.

Sagaut, P. (1996). Numerical Simulations of Separated Flows with Subgrid Models. *Rech. Aéro, 1*, 51-63.

Scotti, A., Meneveau, C. & Lilly, D. K. (1993). Generalized Smagorinsky Model for Anisotropic Grids. *Phys. Fluids A, 5*, 1229-1248.

Shah, N. G. (1979). *New Method of Computation of Radiation Heat Transfer in Combustion Chambers*, London University.

Siegel, R00. & Howell, J. (2002). *Thermal Radiation Heat Transfer .4th edition.* New York: Taylor & Francis.

Sivathanu, Y. R., Faeth, G. M. (1990). Generalized State Relationships for Scalar Properties in Nonpremixed Hydrocarbon/Air Flames. *Combustion & Flame, 82*, 211-230.

Smagorinsky, J. (1963). General circulation experiment with the primitive equations: Part I. The basic experiment. *Monthly Weather Rev.*, 91-99.

Smith, T. F., Shen, X. F. & Friedman, J. N. (1982). Evaluation of Coefficients for the Weighted Sum of Gray Gases. *ASME J. Heat Transfer, 104*, 602-608.

Spalding, D. B. (1976). Development of the Eddy Break-Up Model of Turbulent Combustion, In: *Sixteenth Symposium (International) on Combustion*, Pittsburgh, PA.

Steckler, K. D., Quintiere, J. G., Rinkinen, W. J. (1984). *Flow Induced by Fire in a Compartment, National Bureau of Standards (US)*, Washington.

Steward, F. R. & Cannon, P. (1971). Calculation of Radiative Heat Flux in a Cylindrical Furnace using Monte-Carlo Method. *International Journal of Heat and Mass Transfer, 14*, 245.

Syed, K. J., Stewart, C. D. & Moss, J. B. (1990). Modelling Soot Formation and Thermal Radiation in Buoyant Turbulent Diffusion Flames, *In*: *Twenty-Third Symposium (International) on Combustion*, France.

Takahashi, F. & Glassman, I. (1984). Sooting Correlations for Premixed Flames. *Combustion Science and Technology*, *37*, 1-19.

Taniguchi, H., Kudo, K., Hayasaka, H., Yang, W. J. & Tashiro, H. (1984). Fundamentals of Thermal Radiation Heat Transfer, *In*: *ASME HTD - Vol 40*, pp. 29-36).

Tanzawa, T. & Gardiner, W. C. J. (1980). Reaction-Mechanism of the Homogeneous Thermal-Decomposition of Acetylene. *Journal of physical Chemistry*, *83*, 236-239.

Taylor, P. B. & Foster, P. J. (1974). Some Gray Weighting Coefficients for CO2-H2O-Soot Mixtures. *Int. J. Heat Mass Transfer*, *18*, 1331-1332.

Tesner, P. A., Snegiriova, T. D. & Knorre, V. G. (1971a). Kinetics of Dispersed Carbon Formation. *Combust. Flame*, *17*, 253-260.

Tesner, P. A., Tsygankova, E. I., Guilazetdinov, E. I., Zuyev, V. P. & Loshakova, G. V. (1971b). The Formation of Soot from Aromatic Hydrocarbons in Diffusion Flames of Hydrocarbon-Hydrogen Mixtures. *Combust. Flame*, *17*, 279-285.

Than, C. F. & Savilonis, B. J. (1993). Modelling Fire Behaviour in an Enclosure with a Ceiling Vent. *Fire Safety Journal*, *20*, 151-174.

Tieszen, S. R., Nicolette, V. F., Gritzo, L. A. & Holen, J. K. D. M., J.L. M (1996). Vortical structure in pool fires: observation, speculation and simulation. *Sandia Report*, *96-2607*.

Tieszen, S. R., O'hern, T. J., Schefer, R. W., Weckman, E. J. & Blanchat, T. K. (2002). Experimental study of the flow field in and around a one meter diameter methane fire. *Comb. & Flame*, *129*, 378-291.

Tu, J. Y., Yeoh, G. H. & Lui, C. (2008) *Computational Fluid Dynamics: A Practical Approach*, & Sydney, Elsevier.

Versteeg, H. K. & Malalasekera, W. (1995) *An Introduction to Computational Fluid Dynamics: The Finite Volume Method*, New York, Addison-Wesley.

Viskanta, R., Menguc, M. P. (1987). Radiation Heat Transfer in Combustion System. *in Energy and Combustion Science*, *13*, 97-160.

Wang, H. & Frenklach, M. (1997). A Detailed Kinetic Modelling Study of Aromatic Formation in Laminar Premixed Acetylene and Ethylene Flames. *Combustion and Flame*, *110*, 173-221.

Wang, H. Y. & Joulain, P. (1996). Three-dimensional Modelling for Prediction of Wall Fires with Buoyancy-induced Flow along a Vertical Rectangular Channel. *Combustion and Flame*, *105*, 391-406.

Wang, H. Y., Joulain, P. (2000). Numerical Study of the Turbulent Burning between Vertical Parallel Walls with a Fire-induced Flow. *Combustion Science and Technology*, *154*, 119-161.

Wang, H. Y., Joulain, P. & Most, J. M. (1995). Three-dimensional Modelling and Parametric Study of Turbulent Burning Along the Walls of a Vertical Rectangular Channel. *Combustion Science and Technology*, *109*, 287-308.

Wang, H. Y., Joulain, P. & Most, J. M. (1999). Modelling on Burning of Large-scale Vertical Parallel Surfaces with Fire-induced Flow. *Fire Safety Journal*, *32*, 241-271.

Wen, J. X., Huang, L. Y. (2000). CFD modelling of confined jet fires under ventilation-controlled conditions. *Fire Safety Journal*, *34*, 1-24.

Wen, J. X., Huang, K. Y. & Roberts, J. (2001). The effect of microscopic and global radiative heat exchange on the field predictions of compartment fires. *Fire Safety Journal, 36*, 205-223.

Williams, F. A. (1975). *Turbulent Mixing in Non-reactive and Reactive Flows*, New York: Plenum Press.

Woodburn, P. J. & Britter, R. E. (1996). CFD Simulation of a Tunnel Fire - Part II. *Fire Safety Journal, 26*, 63-90.

Woodburn, P. J. & Drysdale, D. D. (1998). Fires in Inclined Trenches: the Dependence of the Critical Angle on the Trench and Burner Geometry, *Fire Safety Journal, 31*, 143-164.

Woodburn, P. J. & Britter, R. E. (1996). CFD Simulation of a Tunnel Fire - Part I. *Fire Safety Journal, 26*, 35-62.

Xin, Y., Gore, J. P., Mcgrattan, K. B., Rehm, R. G. & Baum, H. R. (2002). Large eddy simulation of buoyant turbulent pool fires. *Proc. 29th Symp. (International) Comb.*

Xin, Y, Gore, J. P., Mcgrattan, K. B., Rehm, R. G. & Baum, H. R. (2005). Fire dynamic simulation of a turbulent buoyant flame using a mixture-fraction-based combustion model. *Comb. & Flame, 141*, 329-335.,

Xue, H., Ho, J. C. & Cheng, Y. M. (2001). Comparison of Different Combustion models in Enclosure Fire Simulation. *Fire Safety Journal, 36*, 37- 54.

Yaga, M., Endo, H., Yamamoto, T., Aoki, H., Miura, T. (2002). Modeling of Eddy Characteristic Time in LES for Calculating Turbulent Diffusion Flame. *Int. J. Heat Mass Transfer, 45*, 2343-2349.

Yan, Z. H. & Holmstedt, G. (1996). CFD and Experimental Studies of Room Fire Growth on Wall lining Materials. *Fire Safety Journal, 27*, 201-238.

Yeoh, G. H., Yuen, R. K. K., Lee, E. W. M. & Cheung, S. C. P (2002). Fire and smoke distribution in a two-room compartment structure. *Int. J. Numer. Meth. Heat Fluid Flow, 12*, 178-194.

Yeoh, G. H., Yuen, R. K. K, Lo, S. M. & Chen, D. H. (2003). On numerical comparison of enclosure fire in a multi-compartment building. *Fire Safety Journal, 38*, 85-94.

Yeoh, G. H. & Yuen, K. K. (2009) *Computational Fluid Dynamics in Fire Engineering: Theory, Modelling and Practice*, Sydney, Elsevier. (In press)

Yuen, R., Davis, G,. de. V., Leonardi, E. & Yeoh, G. H. (2000). The Influence of Moisture on the Combustion of Wood. *Numerical Heat Transfer, Part A 8*, 257-280.

Zhou, X., Luo, K. H. & Williams, J. J. R. (2001). Numerical studies on vortex structures in the near-field of oscillating diffusion flames. *Heat & Mass Transfer, 137*, 101-110.

Zukoski, E. E., Cetegen, B. M., Kubota, K. (1984). Visible Structures of Buoyant Diffusion Flames, *In*: *Twenty Symposium (International) on Combustion*, Pittsburgh, PA.

In: Buildings and the Environment
Editors: Jonas Nemecek and Patrik Schulz
ISBN: 978-1-60876-128-9

Chapter 2

ENERGY EFFICIENCY, CLIMATE CHANGE, BUILDINGS AND THE NEED FOR DEVELOPMENT IN RENEWABLE ENERGY USE

Abdeen Mustafa Omer
Nottingham NG7 4EU, United Kingdom

ABSTRACT

The use of renewable energy sources is a fundamental factor for a possible energy policy in the future. Taking into account the sustainable character of the majority of renewable energy technologies, they are able to preserve resources and to provide security, diversity of energy supply and services, virtually without environmental impact. Sustainability has acquired great importance due to the negative impact of various developments on environment. The rapid growth during the last decade has been accompanied by active construction, which in some instances neglected the impact on the environment and human activities. Policies to promote the rational use of electric energy and to preserve natural non-renewable resources are of paramount importance. Low energy design of urban environment and buildings in densely populated areas requires consideration of wide range of factors, including urban setting, transport planning, energy system design and architectural and engineering details. The focus of the world's attention on environmental issues in recent years has stimulated response in many countries, which have led to a closer examination of energy conservation strategies for conventional fossil fuels. One way of reducing building energy consumption is to design buildings, which are more economical in their use of energy for heating, lighting, cooling, ventilation and hot water supply. Passive measures, particularly natural or hybrid ventilation rather than air-conditioning, can dramatically reduce primary energy consumption. However, exploitation of renewable energy in buildings and agricultural greenhouses can, also, significantly contribute towards reducing dependency on fossil fuels. Therefore, promoting innovative renewable applications and reinforcing the renewable energy market will contribute to preservation of the ecosystem by reducing emissions at local and global levels. This will also contribute to the amelioration of environmental conditions by replacing conventional fuels with renewable energies that produce no air pollution or greenhouse gases. This article presents review of energy sources, environment and sustainable development. This includes all the renewable energy technologies, energy savings, energy efficiency systems and measures necessary to reduce climate change.

Keyword: Renewable energies, wind power, solar energy, biomass energy, geothermal power, environment, sustainable development.

1. INTRODUCTION

Spaces without northerly orientations have an impact on the energy behaviour of a building. For sustainable development, the adverse impacts of energy production and consumption can be mitigated either by reducing consumption or by increasing the use of renewable or clean energy sources [1]. Bioclimatic design of buildings is one strategy for sustainable development as it contributes to reducing energy consumption and therefore, ultimately, air pollution and greenhouse gas emissions (GHG) from conventional energy generation. Bioclimatic design involves the application of energy conservation techniques in building construction and the use of renewable energy such as solar energy and the utilisation of clean fossil fuel technologies.

Most businesses could make savings of up to 20% by introducing basic improvements in energy efficiency. Meeting the target of a 60% reduction in carbon dioxide (CO_2) emissions on environmental pollution is both technologically feasible and financially viable. A genuine investment of energy and resources to meet the environmental challenges the world at equity for a small planet. One compelling reason why businesses should reduce emissions:

- It is right to reduce emissions and to use energy efficiency. There are inevitably concerned about costs. They want to provide goods and services at prices their customers can afford and without a competitive detriment.
- To reduce emissions in businesses is that customers care about the environment and would give a choice and support environmentally conscious business.

Renewable energy markets, industry and investment have never grown faster than they did in 2007. Countries with the largest amounts of new capacity investment were Germany, China, the United States, Spain, Japan and India. Sources of finance and investment now come from a diverse array of private and public institutions. From private sources, mainstream and venture capital investment is accelerating, for both proven and developing technologies.

Between 1980 and 2000 governmental awareness of wind energy mainly concentrated in Denmark and Germany, where a large number of wind turbines were manufactured and installed. Nowadays, most European governments are well aware of the potential of wind energy. Generally, the development and operation of a wind farm can be subdivided into four phases:

- Initiation and feasibility.
- Pre-building (conducted by go/no-go).
- Building.
- Operation and maintenance.

2. Renewable Energy

In the majority of cities that have installed significant amounts of renewable energy over the last 10 years. The local municipal government has played a key role in stimulating projects. When it comes to the installation of large amounts of renewables, these cities have several important factors in common. These factors include:

- Information provision about the possibilities of renewables.
- The presence of municipal departments or offices dedicated to the environment, sustainability or renewable energy.
- A strong local political commitment to the environment and sustainability.
- Obligations that some or all buildings include renewable energy.

Energy efficiency is the most cost-effective way of cutting carbon dioxide emissions and improvements to households and businesses. It can also have many other additional social, economic and health benefits, such as warmer and healthier homes, lower fuel bills and company running costs and, indirectly, jobs. Britain wastes 20 per cent of its fossil fuel and electricity use. This implies that it would be cost-effective to cut £10 billion a year off the collective fuel bill and reduce CO_2 emissions by some 120 million tones. Yet, due to lack of good information and advice on energy saving, along with the capital to finance energy efficiency improvements, this huge potential for reducing energy demand is not being realised. Traditionally, energy utilities have been essentially fuel providers and the industry has pursued profits from increased volume of sales. Institutional and market arrangements have favoured energy consumption rather than conservation. However, energy is at the centre of the sustainable development paradigm as few activities affect the environment as much as the continually increasing use of energy. Most of the used energy depends on finite resources, such as coal, oil, gas and uranium. In addition, more than three quarters of the world's consumption of these fuels is used, often inefficiently, by only one quarter of the world's population. Without even addressing these inequities or the precious, finite nature of these resources, the scale of environmental damage will force the reduction of the usage of these fuels long before they run out.

Throughout the energy generation process there are impacts on the environment on local, national and international levels, from opencast mining and oil exploration to emissions of the potent greenhouse gas carbon dioxide in ever increasing concentration. Recently, the world's leading climate scientists reached an agreement that human activities, such as burning fossil fuels for energy and transport, are causing the world's temperature to rise. The Intergovernmental Panel on Climate Change has concluded that "the balance of evidence suggests a discernible human influence on global climate". It predicts a rate of warming greater than any one seen in the last 10,000 years, in other words, throughout human history. The exact impact of climate change is difficult to predict and will vary regionally. It could, however, include sea level rise, disrupted agriculture and food supplies and the possibility of more freak weather events such as hurricanes and droughts. Indeed, people already are waking up to the financial and social, as well as the environmental, risks of unsustainable energy generation methods that represent the costs of the impacts of climate change, acid rain and oil spills. The insurance industry, for example, concerned about the billion dollar costs of hurricanes and floods, has joined sides with environmentalists to lobby for greenhouse gas emissions reduction. Friends of the earth are campaigning for a more sustainable energy policy, guided by the principal of environmental protection and with the objectives of sound natural resource management and long-term energy security. The key priorities of such an energy policy must be to reduce fossil fuel use, move away from nuclear power, improve the efficiency with which energy is used and increase the amount of energy obtainable from sustainable, renewable sources. Efficient energy use has never been more crucial than it is today, particularly with the prospect of the imminent introduction of the climate change levy (CCL). Establishing an energy use action plan is the essential foundation to the elimination of energy waste. Alternatively energy sources can potentially help fulfil the acute energy demand and sustain economic growth in many regions of the world.

3. Water Resources

The world was blank, white and unformed. In the system of water, clouds are a bucket brigade, not storage. The atmosphere around the planet carries only about a ten day supply of fresh water – about one inch of rain. Each day on earth almost 250 cubic miles of water evaporates from the sea and the land. Its stay in the air is short; it is always seeking particles to stick to and fall with as rain or snow.

Climatic and environmental changes and a rising water demand have increased the competition over water resources and have made cooperation between countries that share a transboundary river an important issue in water resources management and hydropolitics. Yet, in river basins around the world, international conflict and cooperation are influenced by different factors and general conclusions about forces driving conflict and cooperation have been difficult to draw. Rivers are an essential natural resource closely linked to a country's wellbeing and economic success. But rivers ignore political boundaries and competition over the water resources has lead to political tension between countries with transboundary rivers. Integrating international cooperation and conflict resolution into the water management of transboundary rivers has therefore become an important issue in water resources management and hydropolitics. The problem requires a good understanding of the history and patterns of conflict and cooperation among nations sharing international basins worldwide and of the different factors that have influenced their international relations.

Increasing water scarcity in the downstream areas of several river basins demands improved water management and conservation in upper reaches. Improved management is impossible without proper monitoring at various levels. It is well known that all existing sewage treatment plants are overloaded. Hence, the treated effluents do not comply with international effluent quality guidelines. The main reasons behind this are:

- Weak management and absence of environmental awareness.
- Public-sector institutional problems.
- Failure in process design, construction and operation.
- Lack of skilled operating staff and insufficient monitoring programmes.
- Poor maintenance and weak financial resources.
- Low level of public involvement and lack of financial commitment.

4. WIND ENERGY

Wind energy is one of the fastest growing industries nowadays. The development in wind turbine (WT) technology is not limited to the significant increase in the size of the modern units, but also includes the high reliability and availability of the current machine. Therefore, a great competition among the manufactures established on the market and newcomers in the field is witnessed nowadays. A rapid development in the wind energy technology has made it alternative to conventional energy systems in recent years. Parallel to this development, wind energy systems (WES) have made a significant contribution to daily life in developing countries, where one third of the world's people live without electricity [2].

Many developing nations need to expand their power systems to meet the demand in rural areas. However, extending central power systems to remote locations is too costly an option in most cases. Then, autonomous small-scale energy systems can meet the electricity demand in remote locations, even though they generate relatively little power. However, even little electricity would contribute greatly to the quality of life in some places of developing countries. Being one of the most promising autonomous power technologies, wind energy applications, in the power range from tens of Watts to kilowatts, are increasingly growing in rural areas of developing countries.

Technical and economical aspects of WESs should further be improved to sustain this growth. Techno-economically optimal designs are crucial for wind systems in competing with the conventional and more reliable power systems. High performance at the lowest possible cost will encourage the use of such systems and lead to more cost effective systems gradually (Figure 1). Design tools, allowing system performance assessment over a certain period of time, are therefore of great importance for sizing and optimisation purposes.

Wind power now accounts for the dominant share of global investment in renewable energy. Total wind power capacity grew by 28% worldwide in 2007 to reach an estimated 95 GW. Annual capacity additions by market size increased even more: 40% higher in 2007 compared to 2006. Wind markets have also become geographically broad, with capacity in over 70 countries. Even as turbine prices remained high, due in part to materials costs and supply-chain troubles, the industry saw an increase in manufacturing facilities in the United States, India and China, broadening the manufacturing base away from Europe with the growth of more localised supply chains. India has been exporting components and turbines for many years and it appeared that 2006 and 2007 marked a turning point for China as well, with deals announced for the export of Chinese turbines and components. The annual energy yield is calculated by multiplying the wind turbine power curve with the wind distribution function at the site:

$$E_y = \sum_{i=1}^{i=n} f_{wi} P_{wi} \tag{1}$$

Where:

E_y is annual energy yield in kWh.

w is the wind speed in m/s.

n is the number of data bins converting the wind speed range of the turbine (0.5 or 1 m/s intervals).

f_{wi} is the number of hours per year for which wind speed is w m/s.

P_{wi} is the power resulting from a wind speed of w m/s.

Based on power curve from Figure 2 and the Weibull wind speed distribution, with a shape factor of 2, and the gross energy yield corresponding to 7-8.5 m/s is 10 MW.

Unchanging for all wind turbines- big or small- is a number of crucial factors that together determine the annual energy-generating potential in kWh/m^2 of rotor swept area. Key factors that impact potential energy yield and their physical relationships are expressed in the formula:

$$P = \frac{1}{2} \rho C_p \eta_{me} \eta_{el} V^3 A \quad (2)$$

Where:

P is the wind turbine power performance fed into the grid (Watts).

C_p is an aerodynamic efficiency of conversion of wind power into mechanical power, often called the power coefficient.

η_{me} is the conversion efficiency of mechanical power in the rotor axis into mechanical power in the generator axis. Encompasses all combined losses in the bearings, gearbox and so on.

η_{el} is the conversion efficiency of mechanical power into electric power fed into the grid, encompassing all combined losses in the generator, frequency converter, transformer, switches, etc.

ρ is the air density in kg/m^3 depends on environmental conditions.

V is the wind speed some three rotor diameters upwind from the rotor plane in m/s.

A is the rotor swept area in m^2.

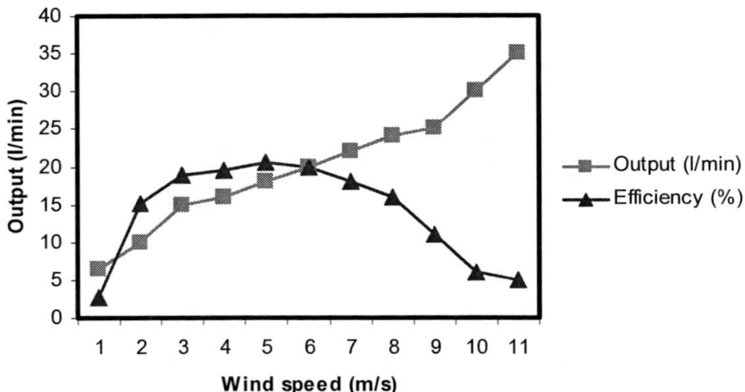

Figure 1. Performance of the wind pump

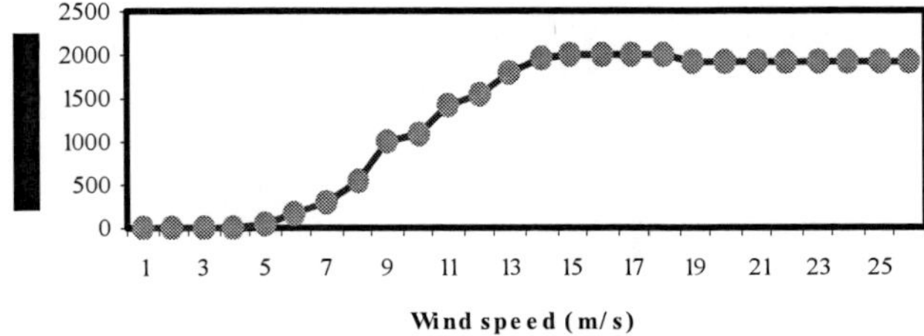

Figure 2. A power/wind speed curve

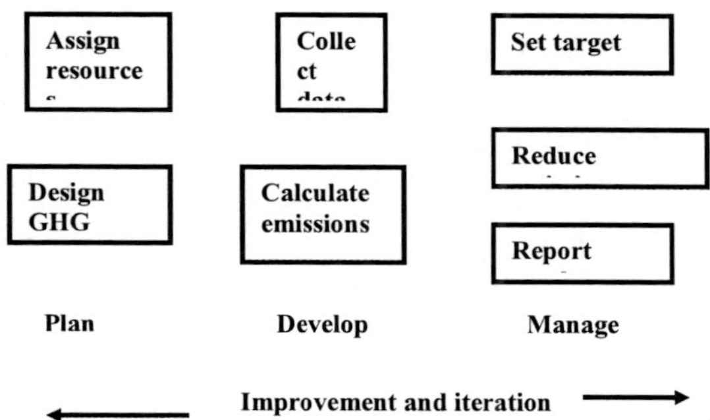

Figure 3. Shows office buildings can fight global warming

Each of the elements of the performance formula has its own distinct contribution to total wind turbine power output and resulting yearly energy yield. Traditionally wind turbines applied in an open field are horizontal-axis designs fitted with an upwind rotor. In the operational output range, wind power generated increases with wind speed cubed. Rotor swept area is a function of the rotor diameter squared and is the second key wind turbine output variable. The Boyle-Gay-Laussac Law shows the impact of temperature and pressure on density, whereby density is proportional to pressure divided by temperature. The influence of air density on wind turbine performance is therefore limited.

5. Energy Efficiency in Buildings

The world population is rising rapidly, notably in the developing countries. Historical trends suggest that increased annual energy use per capita is a good surrogate for the standard of living factors which promote a decrease in population growth rate. The term 'low energy' means achieving 'zero energy' requirements for a house or reduced energy consumption in an office (Figure 3). The main elements of energy concept are typical passive house components:

- Thermal bridges reduced to a minimum.
- Triple glazed windows with adequate frame and an optimised installation.
- A ventilation system with highly efficient heat recovery installed.
- Thermal solar collectors installed covering up 60% of the annual energy demand for domestic hot water.
- Excellent insulation level of opaque building elements: u-values range from 0.10 W/m^2K for walls and roof to 0.18 W/m^2K for basement ceilings.
- Highly efficient condensing gas boilers were installed; where possible, ducts have been insulated to a very good level; in other projects biomass boilers have been successfully tested.
- The air-tightness was improved by a factor of 6-10; the limiting value for new passive houses was achieved.

Compact development patterns can reduce infrastructure demands and the need to travel by car. As population density increases, transportation options multiply and dependence areas, per capita fuel consumption is much lower in densely populated areas because people drive so much less. Few roads and commercially viable public transport are the major merits. On the other hand, urban density is a major factor that determines the urban ventilation conditions, as well as the urban temperature. Under given circumstances, an urban area with a high density of buildings can experience poor ventilation and strong heat island effect. In warm-humid regions these features would lead to a high level of thermal stress of the inhabitants and increased use of energy in air-conditioned buildings.

However, it is also possible that a high-density urban area, obtained by a mixture of high and low buildings, could have better ventilation conditions than an area with lower density but with buildings of the same height. Closely spaced or high-rise buildings are also affected by the use of natural lighting, natural ventilation and solar energy. If not properly planned, energy for electric lighting and mechanical cooling/ventilation may be increased and application of solar energy systems will be greatly limited. Table 1 gives a summary of the positive and negative effects of urban density. All in all, denser city models require more careful design in order to maximise energy efficiency and satisfy other social and development requirements. Low energy design should not be considered in isolation, and in fact, it is a measure, which should work in harmony with other environmental objectives. Hence, building energy study provides opportunities not only for identifying energy and cost savings, but also for examining the indoor and outdoor environment.

Energy efficiency and renewable energy programmes could be more sustainable and pilot studies more effective and pulse releasing if the entire policy and implementation process was considered and redesigned from the outset. New financing and implementation processes are needed which allow reallocating financial resources and thus enabling countries themselves to achieve a sustainable energy infrastructure. The links between the energy policy framework, financing and implementation of renewable energy and energy efficiency projects have to be strengthened and capacity building efforts are required.

6. Solar Energy

Policies for solar hot water have grown substantially in recent years. In particular, mandates for solar hot water in new construction represent a recent trend at both national and local levels. Large-scale, conventional, power plant such as hydropower, has an important part to play in development. It does not, however, provide a complete solution. There is an important complementary role for the greater use of small-scale, rural based, power plant. Such plant can be used to assist development since it can be made locally using local resources, enabling a rapid built-up in total equipment to be made without a corresponding and unacceptably large demand on central funds. Renewable resources are particularly suitable for providing the energy for such equipment and its use is also compatible with the long-term aims. It is possible with relatively simple flat plate solar collectors (Figure 4) to provide warmed water and enable some space heating for homes and offices which is particularly useful when the buildings are well insulated and thermal capacity sufficient for the carry over of energy from day to night is arranged. Energy efficiency is related to the provision of the desired environmental conditions while consuming the minimal quantity of energy.

Table 1. Effects of urban density on city's energy demand

Positive effects	Negative effects
Transport: • Promote public transport and reduce the need for, and length of, trips by private cars. Infrastructure: • Reduce street length needed to accommodate a given number of inhabitants. • Shorten the length of infrastructure facilities such as water supply and sewage lines, reducing the energy needed for pumping. Thermal performance: • Multi-story, multiunit buildings could reduce the overall area of the building's envelope and heat loss from the buildings. • Shading among buildings could reduce solar exposure of buildings during the summer period. Energy systems: • District cooling and heating system, which is usually more energy efficiency, is more feasible as density is higher. Ventilation: • A desirable flow pattern around buildings may be obtained by proper arrangement of high-rise building blocks.	Transport: • Congestion in urban areas reduces fuel efficiency of vehicles. Vertical transportation: • High-rise buildings involve lifts, thus increasing the need for electricity for the vertical transportation. Ventilation: • A concentration of high-rise and large buildings may impede the urban ventilation conditions. Urban heat island: • Heat released and trapped in the urban areas may increase the need for air conditioning. • The potential for natural lighting is generally reduced in high-density areas, increasing the need for electric lighting and the load on air conditioning to remove the heat resulting from the electric lighting. Use of solar energy: • Roof and exposed areas for collection of solar energy are limited.

The encouragement of greater energy use is an essential component of development. In the short term it requires mechanisms to enable the rapid increase in energy/capita, and in the long term we should be working towards a way of life, which makes use of energy efficiency and without the impairment of the environment or of causing safety problems. Such a programme should as far as possible be based on renewable energy resources. As with any market, the benefit of experience is invaluable in the successful implementation of a growth strategy (Figure 5). The large-scale uptake of photovoltaics in urban areas potentially represents a vast market area that could be developed under the right conditions. A wide range of countries, project stages and stakeholders have been involved and this has led to the collection of a comprehensive set of lessons learnt and successful methods of promoting the implementation of photovoltaic (PV) within the urban planning process:

- Setting the stage- the impact of planning policy on renewables within urban areas.
- Implementation- from financing to design to construction.
- Occupation- when the real success of otherwise of a project can be seen.
-

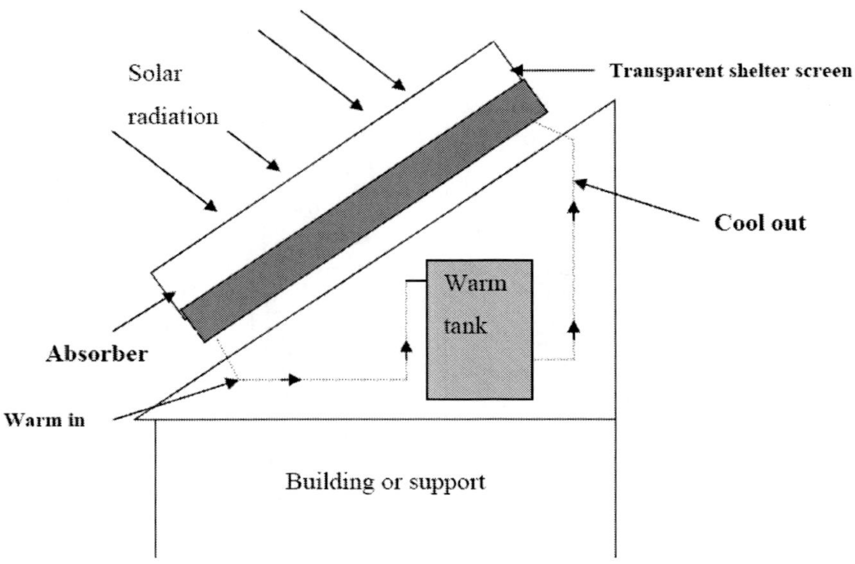

Figure 4. Solar water warmer

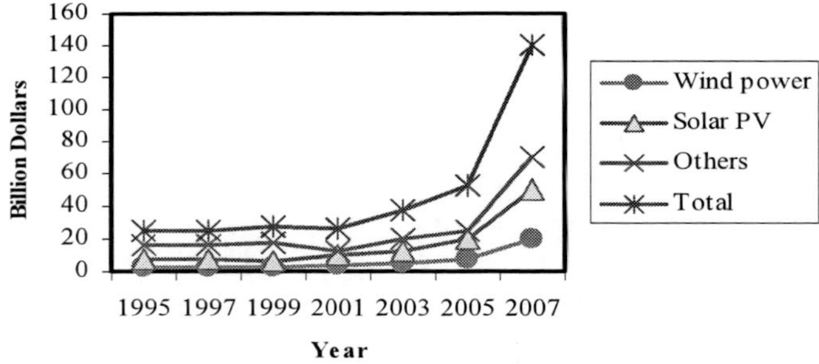

Figure 5. Annual investments in new renewable energy capacity 1995-2007 [3]

In many central European countries energy consumption for heating and domestic hot water causes around one third of national CO_2 emissions. For this reason the reduction of energy demand from buildings play an important role in efforts to control anthropogenic greenhouse gas emissions.

7. Biomass Energy

Biofuels are emerging in a world increasingly concerned by the converging global problems of rising energy demands, accelerating climate change, high priced fossil fuels, soil degradation, water scarcity and loss of biodiversity. For instance, the Intergovernmental Panel on Climate Change (IPCC) identified that in order to avoid more than an acceptable maximum 2-2.4°C rise in mean global temperature, greenhouse gas emissions will need to peak around 2015 and be reduced well below 50% of 2000 level by 2050. A lower figure is needed, which cannot be achieved by emissions reduction alone. Hence, there is a need for carbon removals, giving rise to enhanced supplies of biomass raw material and the potential of biofuels-related investments to show a profit from biofuels sales revenues.

Many nations have the ability to produce their own biofuels derived both from agricultural and forest biomass and from urban wastes, subject to adequate capacity building, technology transfer and access to finance. Trade in biofuels surplus to local requirements can thus open up new markets and stimulate the investment needed to promote the full potential of many impoverished countries. Such a development also responds to the growing threat of passing a tipping point in climate system dynamics. The urgency and the scale of the problem are such that the capital investment requirements are massive and more typical of the energy sector than the land use sectors. The time line for action is decades, not centuries, to partially shift from fossil carbon to sustainable biomass. Figure 6 shows the contribution of biomass to global primary and consumer energy supplies. Food and fodder availability is very closely related to energy availability.

8. Applications

There are various successful applications of renewable technologies:

8.1. Water Management

The current global usage of water identified to be 75% of an overall water consumption for agricultural purposes [4]. Excess heat and scarcity of water are the two major problems, which are usually encountered in irrigating land especially in the arid and semi-arid regions. It is predicted that a higher future demand on crops would take place as a result of a rapid increase in the world's population leading to a serious dilemma on the global water resources in the future. Furthermore, with the current environmental problems (i.e., global warming), many plants and trees started to be exposed to additional amount of heat, which resulted in hindering their growth. The current global requirements for optimising water usage for the agricultural purposes especially in the arid and semi-arid regions, considered being the most demanding task. Irrigation in these lands, encounters two major obstacles, which are lack of water and access heat that could damage plants and prevent them from a healthy growth and attainment of a maximum production of crops. There are numerous advantages that could be associated with proper water and heat management of agricultural lands especially in the arid and semi-arid regions. Beside water is being conserved; optimum water usage for irrigation at different meteorological conditions could result in a healthy growth and optimum production of crops. The amount of heat the plant is exposed to, could play a major role in the plant's growth and sometimes proper shading could be necessary to maintain a productive land. In general, a greater economical saving could be achieved if appropriate steps are taken to manage heat and water in these lands in order to attain a better production of crops.

8.2. Greenhouses Environment

Greenhouse cultivation is one of the most absorbing and rewarding forms of gardening for anyone who enjoys growing plants. The enthusiastic gardener can adapt the greenhouse climate to suit a particular group of plants, or raise flowers, fruit and vegetables out of their natural season. The greenhouse can also be used as an essential garden tool, enabling the keen amateur to expand the scope of plants grown in the garden, as well as save money by raising their own plants and vegetables. There was a decline in large private greenhouses during the two world wars due to a shortage of materials for their construction and fuel to heat them. However, in the 1950s mass-produced, small greenhouses became widely available at affordable prices and were used mainly for raising plants [5]. Also, in recent years, the popularity of conservatories attached to the house has soared. Modern double-glazing panels can provide as much insulation as a brick wall to create a comfortable living space, as well as provide an ideal environment in which to grow and display tender plants.

The comfort in a greenhouse depends on many environmental parameters. These include temperature, relative humidity, air quality and lighting. Although greenhouse and conservatory originally both meant a place to house or conserve greens (variegated hollies, cirrus, myrtles and oleanders), a greenhouse today implies a place in which plants are raised while conservatory usually describes a glazed room where plants may or may not play a significant role. Indeed, a greenhouse can be used for so many different purposes. It is, therefore, difficult to decide how to group the information about the plants that can be grown inside it.

Throughout the world urban areas have increased in size during recent decades. About 50% of the world's population and approximately 76% in the more developed countries are urban dwellers [6]. Even though there is an evidence to suggest that in many 'advanced' industrialised countries there has been a reversal in the rural-to-urban shift of populations. Virtually all population growth expected between 2000 and 2030 will be concentrated in urban areas of the world. With an expected annual growth of 1.8%, the world's urban population will double in 38 years [6]. This represents a serious contributing to the potential problem of maintaining the required food supply. Inappropriate land use and management, often driven by intensification resulting from high population pressure and market forces, is also a threat to food availability for domestic, livestock and wildlife use. Conversion to cropland and urban-industrial establishments is threatening their integrity. Improved productivity of peri-urban agriculture can, therefore, make a very large contribution to meeting food security needs of cities as well as providing income to the peri-urban farmers. Hence, greenhouses agriculture can become an engine of pro-poor 'trickle-up' growth because of the synergistic effects of agricultural growth such as [6]:

- Increased productivity increases wealth.
- Intensification by small farmers raises the demand for wage labour more than by larger farmers.
- Intensification drives rural non-farm enterprise and employment.
- Alleviation of rural and peri-urban poverty is likely to have a knock-on decrease of urban poverty.

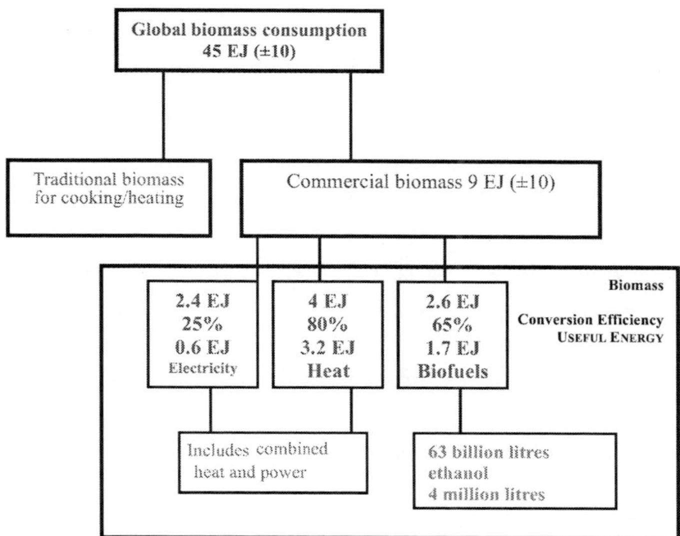

Figure 6. Contribution of biomass to global primary and consumer energy supplies

Despite arguments for continued large-scale collective schemes there is now an increasingly compelling argument in favour of individual technologies for the development of controlled greenhouses. The main points constituting this argument are summarised by [6] as follows:

- Individual technologies enable the poorest of the poor to engage in intensified agricultural production and to reduce their vulnerability.
- Development is encouraged where it is needed most and reaches many more poor households more quickly and at a lower cost.
- Farmer-controlled greenhouses enable farmers to avoid the difficulties of joint management.

Such development brings the following challenges:

- The need to provide farmers with ready access to these individual technologies, repair services and technical assistance.
- Access to markets with worthwhile commodity prices, so that sufficient profitability is realised.
- This type of technology could be a solution to food security problems. For example, in greenhouses, advances in biotechnology like the genetic engineering, tissue culture and market-aided selection have the potential to be applied for raising yields, reducing pesticide excesses and increasing the nutrient value of basic foods.

However, the overall goal is to improve the cities in accordance with the Brundtland Report [7] and the investigation into how urban green could be protected. Indeed, greenhouses can improve the urban environment in multitude of ways. They shape the character of the town and its neighborhoods, provide places for outdoor recreation, and have important environmental functions such as mitigating the heat island effect, reduce surface water runoff, and creating habitats for wildlife. Following analysis of social, cultural and ecological values of urban green, six criteria in order to evaluate the role of green urban in towns and cities were prescribed [7]. These are as follows:

- Recreation, everyday life and public health.
- Maintenance of biodiversity - preserving diversity within species, between species, ecosystems, and of landscape types in the surrounding countryside.
- City structure - as an important element of urban structure and urban life.
- Cultural identity - enhancing awareness of the history of the city and its cultural traditions.
- Environmental quality of the urban sites - improvement of the local climate, air quality and noise reduction.
- Biological solutions to technical problems in urban areas - establishing close links between technical infrastructure and green-spaces of a city.

The main reasons why it is vital for greenhouses planners and designers to develop a better understanding of greenhouses in high-density housing can be summarised as follows [7]:

- Pressures to return to a higher density form of housing.
- The requirement to provide more sustainable food.
- The urgent need to regenerate the existing, and often decaying, houses built in the higher density, high-rise form, much of which is now suffering from technical problems.

The connection between technical change, economic policies and the environment is of primary importance as observed by most governments in developing countries, whose attempts to attain food self-sufficiency have led them to take the measures that provide incentives for adoption of the Green Revolution Technology [8]. Since, the Green Revolution Technologies were introduced in many countries actively supported by irrigation development, subsidised credit, fertiliser programmes, self-sufficiency was found to be not economically efficient and often adopted for political reasons creating excessive damage to natural resources. Also, many developing countries governments provided direct assistance to farmers to adopt soil conservation measures. They found that high costs of establishment and maintenance and the loss of land to hedgerows are the major constraints to adoption [8]. The soil erosion problem in developing countries reveals that a dynamic view of the problem is necessary to ensure that the important elements of the problem are understood for any remedial measures to be undertaken. The policy environment has, in the past, encouraged unsustainable use of land [8]. In many regions, government policies such as provision of credit facilities, subsidies, price support for certain crops, subsidies for erosion control and tariff protection, have exacerbated the erosion problem. This is because technological approaches to control soil erosion have often been promoted to the exclusion of other effective approaches. However, adoption of conservation measures and the return to conservation depend on the specific agro-ecological conditions, the technologies used and the prices of inputs and outputs of production.

8.2.1. Heat balance

The greenhouse effect is one result of the differing properties of heat radiation generated at different temperatures. The high temperature sun emits radiation of short wavelength, which can pass through the atmosphere and through glass. Objects inside the greenhouse (or any other building), such as plants, absorb and then re-radiate this heat. Because the objects inside the greenhouse are at a lower temperature than the sun the radiated heat is of longer wavelengths and, hence, cannot re-penetrate the glass. This re-radiated heat is therefore trapped and causes the temperature inside the greenhouse to rise [9]. It has been observed that there is a significant rise of between 2.5 and 15°C in the enclosed room air temperature of a controlled environment greenhouse as reported by various authors [10-21]. This leads to a thermal energy saving for heating a greenhouse and maintaining a favourable environment for crop production during the off-season of up to 75%. At a certain prescribed depth an imposed temperature equal to a seasonal average ambient temperature, it is possible to identify the main parameters influencing the thermal behaviour of a greenhouse, which lead to a better understanding of the different processes inside the greenhouse.

Using the above concept with synthetic meteorological data (e.g., solar radiation, ambient temperature, wind speed, relative humidity, ground temperature, thermal and physical properties of the ground) will allow the prediction of energy consumption during a whole season in order to maintain the required inside air conditions for plant growth. Further modifications and introducing the more complex problem of plant growth, will allow programming the greenhouse for specific plants, thus improving the economic rentability of very specific types of solar collector.

8.2.2. Ventilation

Whereas heat loss in winter is a problem, it can be a positive advantage when greenhouse temperatures soar considerably above outside temperatures in summer. Table 2 illustrates typical greenhouse temperatures and gives an idea of temperature variation in a greenhouse. Therefore, ventilation is required to maintain the temperature at a level that plants can thrive, and to remove the still, humid air, which encourages the development of diseases. This can be achieved by natural ventilation, where warm air escapes and cool air enters through vents in the greenhouse roof (rigid vents), or by forced air ventilation, where a motorised fan designed specifically for greenhouses sucks warm air out of the greenhouse and pulling cool air in through openings on the opposite side.

Table 2. Typical greenhouse temperatures

Minimum winter temperature	Conditions
4°C	Frost-free
10°C	Temperate
15°C	Tropical

8.2.3. Relative humidity

The addition of side vents, which are optional extras on many small greenhouses, will, also, provide a more rapid movement of air. Cool air is drawn in from the side vents. As it heats up, it rises until it drawn out of the ridge vents. If only top ventilation is fitted, care needs to be taken not to open the vents too rapidly as rising warm air, particularly on cold days, will be quickly replaced by a block of cold air, which can prove a shock to plants. Where additional vents are available, these are best positioned on sidewalls. Table 3 gives typical ventilation space and vent numbers required for different greenhouses. Louvers will give adequate side ventilation and are less likely to cause draughts than standard ventilations [22]. Further, the door opening can provide additional ventilation, which is particularly valuable in small greenhouses. However, this is not always desirable where security is a problem. Generally, ventilators should be provided on all sides of the greenhouse, so that on windy days the windward side can be closed to prevent draughts, while ventilation is opened on the leeward end side. The temperature in small greenhouses can rise rapidly, and, hence, requires continuous monitoring and control. Note that manual control may lead to wide fluctuations in temperature. However, sufficient ventilation in greenhouse is essential for healthy plant growth. Hence, warm air gathering inside the roof may need to be released through a mop fan and replaced by cooler air.

Air humidity is measured as a percentage of water vapour in the air on a scale from 0% to 100%, where 0% being dry and 100% being full saturation level. The main environmental control factor for dust mites is relative humidity. The followings are the practical methods of controlling measures available for reducing dust mite populations:

- Chemical control.
- Cleaning and vacuuming.
- Use of electric blankets, and
- Indoor humidity.

Indoor relative humidity control is one of the most effective long-term mite control measures. There are many ways in which the internal relative humidity can be controlled including the use of appropriate ventilation, the reduction of internal moisture production and maintenance of adequate internal temperatures through the use of efficient heating and insulation [22]. Plants, usually, require a humidity level of 50%-60%, and succulents require 35%-40% [23]. For small greenhouses, simple, automatic controls would create ideal conditions for plants and ensure that ventilation is provided only when required. This prevents the rapid fluctuations in temperature and higher fuel bills that can occur with manual control. For larger ones, however, more sophisticated controls may need to be implemented.

As the sun comes through the glass and gives the plants the energy they need to grow, it will, also, quickly raise the temperature. With no shade and inadequate ventilation, temperatures up to 35°C can be reached very quickly. This is uncomfortable for both plants and people, although plants can tolerate high temperatures, if provided with good ventilation. Also, to allow the carbon dioxide needed for photosynthesis into the plant, the small pores in the leaf, called stomata, must be open and so water will be lost into the air. Keeping the atmosphere humid will help, but eventually the plant will have to shut its stomata, and then growth will come to a stop. All plants have a minimum, optimum and maximum temperature for growth. During the time of the year when they are actively growing, a temperature as close as possible to the optimum is preferred.

8.2.4. The ideal living area

Most people would maintain a steady temperature between 15.5°C and 21°C. For comfort, the level of atmospheric humidity should be low and indeed living areas should, generally, have a dry atmosphere [23]. Hence, artificial heating in autumn, winter and spring will, usually, be required. Plants cannot live comfortably in an area of high humidity like human beings. Also, most plants do not like the combination of high temperatures and dry air. Tropical and sub-tropical plants, for instance, would make very poor growth in such environment and the leaves may shrive and dry up, or turn brown at the edges. Table 4, which is reproduced from [23], classifies plants according to their climate preferences. The table, also, indicates that there are many plants that can be grown in a humid 'microclimate'. During the summer the greenhouse is generally warm and plants are sizzling. Good thermometers are good troubleshooters and enable the greenhouse operator to monitor the shifts in temperature and take appropriate action when necessary. Also, greenhouses can be very arid places in summer. Most greenhouse plants prefer a slightly humid atmosphere. A group of plants growing together will create heir own, more humid microclimate. Plants from the Mediterranean are a good choice as they are naturally adapted to hot, and dry summer.

Table 3. Ventilation needs

Greenhouse size	Ground area	Ridge ventilators space required	Number of 0.6 m x 0.6 m ridge-vents required
1.8 m x 2.5 m	4.5 m^2	0.9 m^2	3
2.5 m x 3 m	7.5 m^2	1.5 m^2	4
3 m x 3.7 m	11.1 m^2	2.2 m^2	6
3.7 m x 4.3 m	15.9 m^2	3.2 m^2	9

8.2.5. Transpiration

Transpiration is a process used by plants to maintain their temperature when the environment is too warm. It serves the same functions as sweating in some animals. In order to cool the area around it, plants release water through the leaves. High rates of transpiration will occur if temperature of the growing environment is too high, and this can seriously affect the nutrient solution. As a plant transpires, it will draw up more water from the solution to replace what has been lost. This will have an impact on the water ratio of the nutrient solution, which will affect both pH (potential of Hydrogen) and the level of conductivity of these nutrients from soil to the plant. Maintaining a basic nutrient solution of pH, which, in turn, affects how easy the plant, can take in the nutrient solution. In a good growing environment, the quantity of growth and yield primarily depends on the quality of the nutrient solution. A nutrient solution contains [24]:

- Water.

- A nutrient concentrate, which provides the basic food for plants.
- One or more nutrient additives, which provide supplementary nutrients, required to promoting specific processes within the plants development.
- Possibly pH solution.

Plants, like human beings, need tender loving care in the form of optimum settings of light, sunshine, nourishment, and water. Hence, the control of sunlight, air humidity and temperatures in greenhouses are the key to successful greenhouse gardening. The mop fan is a simple and novel air humidifier; which is capable of removing particulate and gaseous pollutants while providing ventilation. It is a device ideally suited to greenhouse applications, which require robustness, low cost, minimum maintenance and high efficiency. A device meeting these requirements is not yet available to the farming community. Hence, implementing mop fans aids sustainable development through using a clean, environmentally friendly device that decreases load in the greenhouse and reduces energy consumption. The effect of indoor (greenhouse) conditions and outdoor (ambient) conditions (temperature and relative humidity) on system performance is illustrated in Figure 7.

Table 4. Atmospheric conditions for plants [23]

Types/conditions	Warm conditions and high humidity	Warm conditions and thrive in a dry atmosphere	Cool
Shrubs	Aphelandra, Gardenia	Lantana, Cestrum	Acacia, Lantana
Climbers	Allamanda, Hoya	Plumbago	Jasminum, Plumbago
Flowering pot plants short-term	Capsicum, Celosia	Ipomoea	Solanum, Primula
Flowering pot plants long-term	Begonia, Euphorbia	Fuchsis, Cacti	Clivia, Erica
Foliage pot plants	Ferns, Ficus	Sansevieria	Coleus, Hedera
Bulbous and tuberous plants	Canna	Vallota, Gloriosa	Cyclamen, Lilium
Fruits	Citrus	-	Vitis Vinifera, Prunus Persica

8.2.6. Heat transfer

The total rate of heat transfer on a surface of a soil per unit area Q^{n}_{total}, shown in Figure 8, is a combination of radiation heat transfer Q^{n}_{rad}, convection heat transfer Q^{n}_{conv} and heat transfer due to evaporation Q^{n}_{evap}, illustrated in the following equation:

$$Q^{n}_{total} = Q^{n}_{rad} + Q^{n}_{conv} + Q^{n}_{evap} \tag{3}$$

Where:

Q''_{rad} Can be expressed in terms of emissivity ε, surface temperature T_s and the surrounding temperature T_{surr}:

$$Q''_{rad} = \varepsilon\, \delta\, (T_s - T_{surr}) \tag{4}$$

For the rate of convection heat transfer Q''_{conv}, combinations of both natural and forced convection are considered especially for low air velocities. The vapour pressure of air far away from the watered-surface $P_{v,\infty}$, is a function of relative humidity φ and saturated water vapour pressure $P_{T\infty, sat}$ [25]:

$$P_{v,\infty} = \varphi\, P_{T\infty, sat} \tag{5}$$

Treating the water vapour and air an ideal gas, the densities of water vapour, dry air and their mixture at air-water interface and far from the surface are determined in the following equations:

At the surface:

$$\rho_{v,s} = P_{v,s}/R_v T_s \tag{6}$$

$$\rho_{a,s} = P_{a,s}/R_a T_s \tag{7}$$

$$\rho_s = \rho_{v,s} + \rho_{a,s} \tag{8}$$

And away from the surface:

$$\rho_{v,\infty} = P_{v,\infty}/R_v T_\infty \tag{9}$$

$$\rho_{a,\infty} = P_{a,\infty}/R_a T_\infty \tag{10}$$

$$\rho_\infty = \rho_{v,\infty} + \rho_{a,\infty} \tag{11}$$

$$\begin{Bmatrix} \text{Gr}/R_e^2 < 0.1 \text{ forced convection} \\ < \text{Gr}/R_e^2 < 10 \text{ mixed (forced + natural) convection} \\ 10 < \text{Gr}/R_e^2 \text{ natural convection} \end{Bmatrix} \tag{12}$$

Where: G_r is Grashof number and R_e is Reynolds.

9. Ground Source Heat Pumps

Heat pumps function by moving (or pumping) heat from one place to another. Like a standard air-conditioner, a heat pump takes heat from inside a building and dumps it outside. The difference is that a heat pump can be reversed to take heat from a heat source outside and pump it inside. Heat pumps use electricity to operate pumps that alternately evaporate and condense a refrigerant fluid to move that heat. In the heating mode, heat pumps are far more "efficient" at converting electricity into usable heat because the electricity is used to move heat, not to generate it.

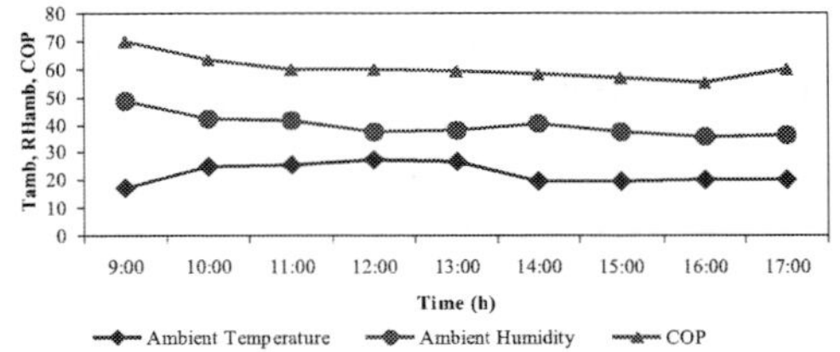

Figure 7. Ambient temperature, relative humidity and COP

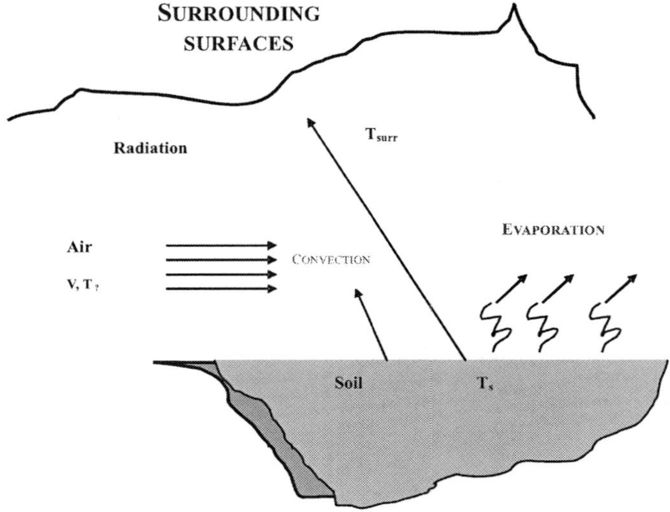

Figure 8. A combination of heat and mass transfer on the surface

The most common type of heat pump- air-source heat pump- uses outside air as the heat source during the heating season and the heat sink during the air-conditioning season. Ground-source and water-source heat pumps work the same way, except that the heat source/sink is the ground, groundwater, or a body of surface water, such as a lake. For simplicity, water-source heat pumps are often lumped with ground-source heat pumps, as in this case (Figure 9).

The efficiency or coefficient of performance (COP) of ground-source heat pumps (GSHPs) is significantly higher than that of air-source heat pumps because the heat source is warmer during the heating season and the heat sink is cooler during the cooling season. GSHPs are also known as geothermal heat pumps.

GSHPs are environmentally attractive because they deliver so much heat or cooling energy per unit of electricity consumed. The COP is usually 3 or higher. The best GSHPs are more efficient than high-efficiency gas combustion, even when the source efficiency of the electricity is taken into account.

GSHPs are generally most appropriate for residential and small commercial buildings, such as small-town post offices. In residential and small (skin-dominated) commercial buildings, GSHPs make the most sense in mixed climates with significant heating and cooling loads because the high-cost heat pump replaces both the heating and air-conditioning system.

Because GSHPs are expensive to install in residential and small commercial buildings, it sometimes makes better economic sense to invest in energy efficiency measures that significantly reduce heating and cooling loads, then install less expensive heating and cooling equipment. The savings in equipment may be able to pay for most of the envelope improvements. If a GSHP is to be used, planning the site work and project scheduling needed so carefully that the ground loop can be installed with minimum site disturbance or in an area that will be covered by a parking lot or driveway.

GSHPs are generally classified according to the type of loop used to exchange heat with the heat source/sink. Most common are closed-loop horizontal and closed-loop vertical systems (Figure 9). Using a body of water as the heat source/sink is very effective, but seldom available as an option. Open-loop systems are less common than closed-loop systems due to performance problems (if detritus gets into the heat pump) and risk of contaminating the water source or, in the case of well water, inadequately recharging the aquifer. GSHPs are complex. Basically, water or a nontoxic antifreeze-water mix is circulated through buried polyethylene or polybutylene piping. This water is then pumped through one of two heat exchangers in the heat pump. When used in the heating mode, this circulating water is pumped through the cold heat exchanger, where its heat is absorbed by evapouration of the refrigerant. The refrigerant is then pumped to the warm heat exchanger, where the refrigerant is condensed, releasing heat in the process. This sequence is reversed for operation in the cooling mode.

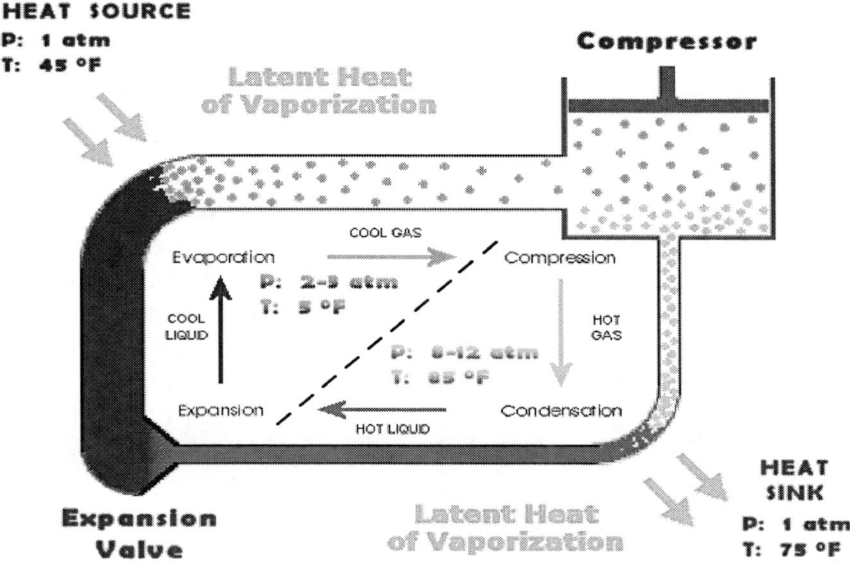

Figure 9. Heat pump works by promoting the evaporation and condensation of a refrigerant

Direct-exchange GSHPs use copper ground-loop coils that are charged with refrigerant. This ground loop thus serves as one of the two heat exchangers in the heat pump. The overall efficiency is higher because one of the two separate heat exchangers is eliminated, but the risk of releasing the ozone-depleting refrigerant into the environment is greater. Direct-exchange systems have a small market share.

An attractive alternative to conventional heating, cooling, and water heating equipment is the GSHP. The higher initial cost of this equipment must be justified by operating cost savings. Therefore, it is necessary to predict energy use and demand. However, there are no seasonal ratings for this type of equipment. The ratings for GSHPs calculate performance at a single fluid temperature (32°F) for heating COP and a second for cooling energy efficiency rating (EER) (77°F). These ratings reflect temperatures for an assumed location and ground heat exchanger type, and are not ideal indicators of energy use.

This problem is compounded by the nature of ratings for conventional equipment. The complexity and many assumptions used in the procedures to calculate the seasonal efficiency for air-conditioners, furnaces, and heat pumps (SEER, AFUE, and HSPF) make it difficult to compare energy use with equipment rated under different standards. The accuracy of the results is highly uncertain, even when corrected for regional weather patterns. These values are not indicators for demand since they are seasonal averages and performance at severe conditions is not heavily weighted.

The American Society of Heating, Refrigerating, and Air-Conditioning Engineers (ASHRAE) recommends a weather driven energy calculation, like the bin method, in preference to single measure methods like SEER, HSPF, EER, COP, and AFUE. The bin method permits the energy use to be calculated based on local weather data and equipment performance over a wide range of temperatures [29]. The bin method also calculates demand at the most severe conditions. This method was used to compare the energy use and demand of high efficiency equipment in Sacramento, California and Salt Lake City, Utah. The equipment considered was a high efficiency single speed air source, a variable speed air source heat pump and electric air-conditioner with a natural gas furnace, and a GSHP [26].

9.1. Heat Pump Principles

Heat flows naturally from a higher to a lower temperature. Heat pumps, however, are able to force the heat flow in the other direction, using a relatively small amount of high quality drive energy (electricity, fuel, or high-temperature waste heat). Heat pumps can transfer heat from natural heat sources such as the air, ground or water, to a building. By reversing the heat pump it can also be used for cooling. Heat is transferred in the opposite direction, from the application that is cooled, to surroundings at a higher temperature.

In order to transport heat external energy is needed to drive the heat pump. Theoretically, the total heat delivered by the heat pump is equal to the heat extracted from the heat source, plus the amount of drive energy supplied. Electrical powered heat pumps, for heating buildings, typically supply 100 kWh of heat with just 20-40 kWh of electricity. Because heat pumps consume less energy than conventional heating systems, there use will help to reduce the harmful emissions of carbon dioxide, sulphur dioxide and nitrogen oxides. However, the overall environmental impact of electric heat pumps depends very much on how the electricity is produced. Heat pumps driven by electricity generated by hydropower, wind power, photovoltaics or other renewable sources will reduce emissions more significantly than if the electricity is generated by coal, oil or gas-fired power plants.

The great majority of heat pumps work on the principle of the vapour compression cycle. The main components in such a heat pump are the compressor, the expansion valve and the two heat exchangers referred to as the evaporator and the condenser. A heat pump can take heat out of an interior space, or it can put heat into an interior space (Figures 10-11). A volatile liquid, known as the working fluid or refrigerant, circulates through the four components. In the evaporator the temperature of the refrigerant is kept lower than the temperature of the heat source. This allows heat to flow from the heat source (ground loops, air or loops in water, e.g., rivers etc.) to the refrigerant. As the refrigerant warms up it evaporates. This vapour is then compressed by the compressor to a higher pressure and temperature. The hot vapour then enters the condenser, where it condenses and gives off useful heat. Finally, the high-pressure working fluid is expanded to the evaporator pressure and temperature in the expansion valve. The refrigerant is returned to its original state and once again enters the evaporator.

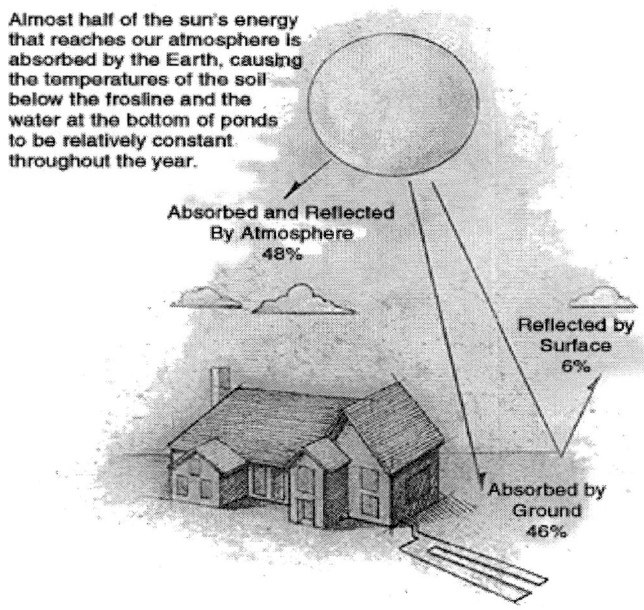

Figure 10. Earth heat

The compressor is driven by an electric motor and pumps circulate the water through (ground loops, or loops in water, e.g., rivers etc.). The domestic fridge uses the same technology. When putting food and drink into fridge the low-grade heat it carries (after all it is usually warmer than the inside of the fridge) is transferred from the icebox to the refrigerant in the unit. The refrigerant is then compressed and expanded to raise the heat; this high grade heat is then expelled from the back of the fridge. This is why the inside of the fridge remains cold whilst the back of the fridge gets hot.

In the cooling mode, cool vapour arrives at the compressor after absorbing heat from the building. The compressor compresses the cool vapour into a smaller volume, increasing its heat density. The refrigerant exits the compressor as a hot vapour, which then goes into the earth loop field. The loops act as a condenser condensing the vapour until it is virtually all-liquid. The refrigerant leaves the earth loops as a warm liquid. The flow control regulates the flow from the condenser such that only liquid refrigerant passes through the control. The refrigerant expands as it exits the flow control unit and becomes a cold liquid.

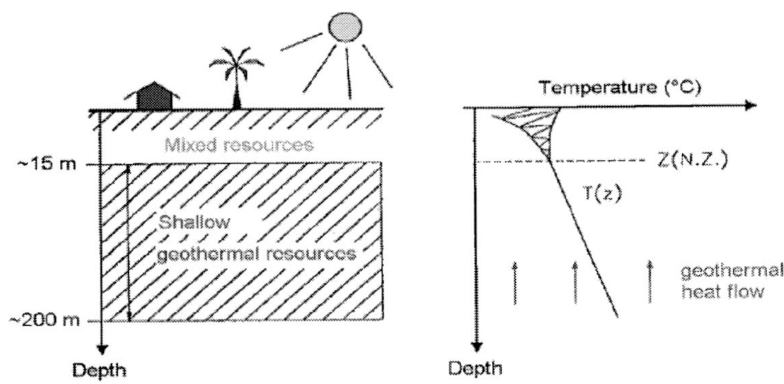

Figure 11. Geothermal energy, comprising geothermal and mixed resources in the shallow subsurface

The term "ground source heat pump" has become an all-inclusive term to describe a heat pump system that uses the earth, ground water, or surface water as a heat source and/or sink. The GSHP systems consist of three loops or cycles as shown in Figure 12. The first loop is on the load side and is either an air/water loop or a water/water loop, depending on the application. The second loop is the refrigerant loop inside a water source heat pump. Thermodynamically, there is no difference between the well-known vapour-compression refrigeration cycle and the heat pump cycle; both systems absorb heat at a low temperature level and reject it to a higher temperature level. The difference between the two systems is that a refrigeration application is only concerned with the low temperature effect produced at the evaporator, while a heat pump may be concerned with both the cooling effect produced at the evaporator as well as the heating effect produced at the condenser. In these dual-mode GSHP systems, a reversing valve is used to switch between heating and cooling modes by reversing the refrigerant flow direction. The third loop in the system is the ground loop in which water or an antifreeze solution exchanges heat with the refrigerant and the earth.

The GSHPs utilise the thermal energy stored in the earth through either vertical or horizontal closed loop heat exchange systems buried in the ground. Many geological factors impact directly on site characterisation and subsequently the design and cost of the system. The solid geology of the United Kingdom varies significantly. Furthermore there is an extensive and variable rock head cover. The geological prognosis for a site and its anticipated rock properties influence the drilling methods and therefore system costs. Other factors important to system design include predicted subsurface temperatures and the thermal and hydrological properties of strata. GSHP technology is well established in Sweden, Germany and North America, but has had minimal impact in the United Kingdom space heating and cooling market. Perceived barriers to uptake include geological uncertainty, concerns regarding performance and reliability, high capital costs and lack of infrastructure. System performance concerns relate mostly to uncertainty in design input parameters, especially the temperature and thermal properties of the source. These in turn can impact on the capital cost, much of which is associated with the installation of the external loop in horizontal trenches or vertical boreholes. The temperate United Kingdom climate means that the potential for heating in winter and cooling in summer from a ground source is less certain owing to the temperature ranges being narrower than those encountered in continental climates. This project will develop an impartial GSHP function on the site to make available information and data on site-specific temperatures and key geotechnical characteristics.

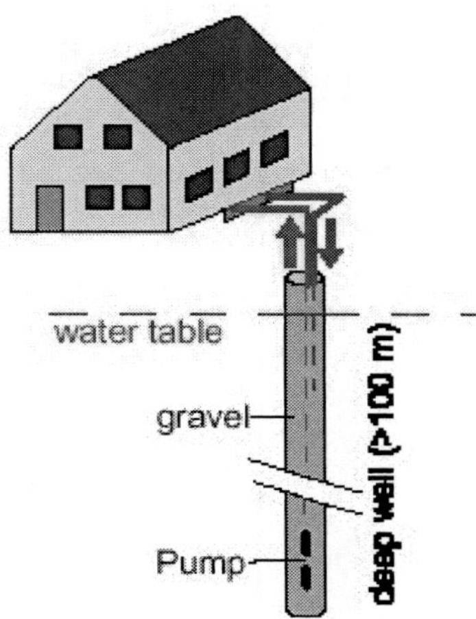

Figure 12. Standing column well

The GSHPs are receiving increasing interest because of their potential to reduce primary energy consumption and thus reduce emissions of greenhouse gases. The technology is well established in North Americas and parts of Europe, but is at the demonstration stage in the United Kingdom. The information will be delivered from digital geoscience's themes that have been developed from observed data held in corporate records. These data will be available to GSHP installers and designers to assist the design process, therefore reducing uncertainties. The research will also be used to help inform the public as to the potential benefits of this technology.

Geothermal energy use has a net positive environmental impact. Geothermal power plants have fewer and more easily controlled atmospheric emissions than either fossil fuel or nuclear plants. Direct heat uses are even cleaner and are practically non-polluting when compared to conventional heating. Another advantage, which differentiates geothermal energy from other renewables, is its continuous availability, 24 hours a day all year round. While production costs are at times competitive and in other cases marginally higher than conventional energy, front-end investment is quite heavy and not easily funded.

9.2. Heat Pumps

A heat pump can take low temperature heat and upgrade it to a higher and more useful temperature (Figure 13). If this heat comes from an ambient source, for example outside air or the ground, the use of a heat pump can result in savings in fossil fuel consumption and thus a reduction in the emission of the GHGs and other pollutants. The GSHPs in particular are receiving increasing interest and the technology is now well established with over 550,000 units (80% of which are domestic) installed worldwide and over 66,000 installed annually [27]. Despite increasing use elsewhere, the GSHPs are a relatively unfamiliar technology in the UK although the performance of systems is now such that, properly designed and installed, they represent a very carbon-efficient form of space heating.

The GSHPs can be used to provide space and domestic water heating and, if required, space cooling to a wide range of building types and sizes. The provision of cooling, however, will result in increased energy consumption and the efficiently it is supplied. The GSHPs are particularly suitable for new build as the technology is most efficient when used to supply low temperature distribution systems such as underfloor heating. They can also be used for retrofit especially in conjunction with measures to reduce heat demand. They can be particularly cost effective in areas where mains gas is not available or for developments where there is an advantage in simplifying the infrastructure provided. This application will concentrate on the provision of space and water heating to individual dwellings but the technology can also be applied to blocks of flats or groups of houses.

Figure 13. A photograph showing the connection of heat pump to the ground source

10. Bioenergy Utilisation

The increased demand for gas and petroleum, food crops, fish and large sources of vegetative matter mean that the global harvesting of carbon has in turn intensified. It could be said that mankind is mining nearly everything except its waste piles. It is simply a matter of time until the significant carbon stream present in municipal solid waste is fully captured. In the meantime, the waste industry needs to continue on the pathway to increased awareness and better optimised biowaste resources. Optimisation of waste carbon may require widespread regulatory drivers (including strict limits on the landfilling of organic materials), public acceptance of the benefits of waste carbon products for soil improvements/crop enhancements and more investment in capital facilities. In short, a significant effort will be required in order to capture a greater portion of the carbon stream and put it to beneficial use. From the standpoint of waste practitioners, further research and pilot programmes are necessary before the available carbon in the waste stream can be extracted in sufficient quality and quantities to create the desired end products. Other details need to be ironed out too, including measurement methods, diversion calculations, sequestration values and determination of acceptance contamination thresholds.

The internal combustion engine is a major contributor to rising CO_2 emissions worldwide and some pretty dramatic new thinking is needed if our planet is to counter the effects. With its use increasing in developing world economies, there is something to be said for the argument that the vehicles we use to help keep our inner-city environments free from waste, litter and grime should be at the forefront of developments in low-emissions technology. Materials handled by waste management companies are becoming increasingly valuable. Those responsible for the security of facilities that treat waste or manage scrap will testify to the precautions needed to fight an ongoing battle against unauthorised access by criminals and crucially, to prevent the damage they can cause through theft, vandalism or even arson. Of particular concern is the escalating level of metal theft, driven by various factors including the demand for metal in rapidly developing economies such as India and China.

10.1. Biogas Technology

Anaerobic digestion (AD) has, for some time, been considered an important technology in the treatment of waste and in the development of energy recovery solutions. Historically, many anaerobic digestion plants have tended to specialise in the treatment of manure or sludge. In today's market, the latest AD plants have to handle more complex substances and varying volume streams. As a result, the demands placed on this technology in terms of reliability, stability and robustness are significant. Also, significant is the potential contribution AD could make to solving our most pressing environmental concern- namely a reduction in the anthropogenic emission of GHGs. AD technology can reduce unwanted and uncontrolled emissions of methane by tapping the energy potential of this gas while reducing the volume of waste going to landfill. Anaerobic digestion is a biochemical process where, in the absence of oxygen, bacteria break down organic matter to produce biogas plus liquor and a fibre.

The biogas consists of 55-70% methane (CH_4) and 30-45% carbon dioxide (CO_2) and can be used to generate energy through a generator. The energy content of biogas is 20-25 MJm^3. Alternatively; the gas can be cleaned and then either compressed for use in vehicle transport (compressed natural gas) or injected into the gas distribution network. An average CH_4 yield per metric ton of treated waste (sludge, manure) ranges from 50-90 Nm^3 per ton and for municipal solid waste (MSW) the yield increases to 75-120 Nm^3 per ton. The liquid fraction, with a high nutrient content and the fibre fraction can be used as a soil improver. More modern plants have been developed to process MSW, industrial solid wastes and industrial wastewaters, but impurities and the varying content of lipids, proteins and carbohydrates can cause problems. These wastes can be characterised according to their COD concentration. COD refers to the total quantity of oxygen required for oxidation to carbon dioxide and water and is a measure of the organic content of the waste. COD loading rate is the daily quantity of organic matter, expressed COD, feed per m^3 digester volume per day, i.e., kg $COD/m^3/d$. Some systems have been invented to process substrates with a minor COD concentration (<25 gO_2/litre raw material), for example:

- Upflow anaerobic sludge blanket (UASB).
- Expanded granular sludge blanket (EGSB).
- Internal circulation (IC).

With a loading rate of $\geq$15 kg COD/m^3 fermenter/d, it possessed a sharp differentiation to traditional biogas plants. The advantages of the system are the following:

- Prevention of foam and floating layers- therefore high loading rates.
- No chemical requirement, no pH regulation- therefore cost savings.
- Low hydraulic retention time- therefore low demand for fermenter volume.
- Intense contact between substrate and microorganisms- therefore high degradation rates and rapid gas production.
- No accumulation of settling sediments (e.g., sand) in the system- thus supporting continuous operation.

The organic matter was biodegradable to produce biogas and the variation show a normal methanogene bacteria activity and good working biological process as shown in Figure 14. There are a number of factors that will give rise to greater interest in technologies such as AD. These include:

- Growing energy costs and import dependency within many countries.
- Decreasing capacity for landfill.
- Increasing world energy demand, in particular in China and India.
- Climate change needing urgent reactions and activities.
- 45% of European soils suffering from low organic matter content and reduced fertility.
- The most practical environmental solution will be deriving energy from waste, not only municipal solid waste but also the residues industry.

Anaerobic digestion has significant potential for industries with organic waste streams, such as food processing, the paper and textile industry, pharmaceutical industry and biofuels production. Anaerobic digestion combines several advantages. As a technology it can be regarded as being 'CO_2 neutral' because there is no net addition of CO_2 to the atmosphere. It degrades waste while producing biogas and a fertiliser product that contains a high nutrient content (nitrogen, phosphorous and potassium), but in order for the full potential of the waste/organic substrate/input to be realised, it is vital that the waste management industry is able to develop markets for all the by-products.

10.2. Sewage Sludge

Sewage sludge is rich in nutrients such as nitrogen and phosphorous. It also contains valuable organic matter, useful for remediation of depleted or eroded soils. This is why untreated sludge has been used for many years as a soil fertiliser and for enhancing the organic matter of soil. A key concern is that treatment of sludge tends to concentrate heavy metals, poorly biodegradable trace organic compounds and potentially pathogenic organisms (viruses, bacteria and the like) present in wastewaters. These materials can pose a serious threat to the environment. When deposited in soils, heavy metals are passed through the food chain, first entering crops, and then animals that feed on the crops and eventually human beings, to whom they appear to be highly toxic. In addition they also leach from soils, getting into groundwater and further spreading contamination in an uncontrolled manner. European and American markets aiming to transform various organic wastes (animal farm wastes, industrial and municipal wastes) into two main by-products:

- A solution of humic substances (a liquid oxidate).
- A solid residue.

The key to successful future appears to lie with successful marketing of the treatment by products. There is also potential for using solid residue in the construction industry as a filling agent for concrete. Research suggests that the composition of the residue locks metals within the material, thus preventing their escape and any subsequent negative effect on the environment.

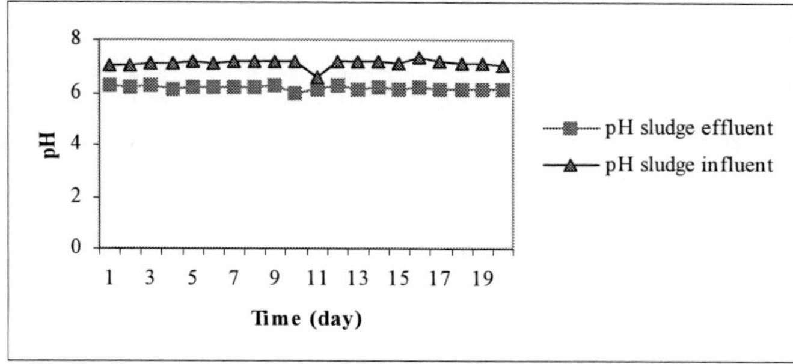

Figure 14. pH sludge before and after treatment in the digester

11. Energy Saving in Buildings

The admission of daylight into buildings alone does not guarantee that the design will be energy efficient in terms of lighting. In fact, the design for increased daylight can often raise concerns relating to visual comfort (glare) and thermal comfort (increased solar gain in the summer and heat losses in the winter from larger apertures). Such issues will clearly need to be addressed in the design of the window openings, blinds, shading devices, heating system, etc. In order for a building to benefit from daylight energy terms, it is a prerequisite that lights are switched off when sufficient daylight is available. The nature of the switching regime; manual or automated, centralised or local, switched, stepped or dimmed, will determine the energy performance. Simple techniques can be implemented to increase the probability that lights are switched off [28]. These include:

- Making switches conspicuous.
- Loading switches appropriately in relation to the lights.
- Switching banks of lights independently.

- Switching banks of lights parallel to the main window wall.

There are also a number of methods, which help reduce the lighting energy use, which, in turn, relate to the type of occupancy pattern of the building [28]. The light switching options include:

- Centralised timed off (or stepped)/manual on.
- Photoelectric off (or stepped)/manual on.
- Photoelectric and on (or stepped), photoelectric dimming.
- Occupant sensor (stepped) on/off (movement or noise sensor).

Likewise, energy savings from the avoidance of air conditioning can be very substantial. Whilst day-lighting strategies need to be integrated with artificial lighting systems in order to become beneficial in terms of energy use, reductions in overall energy consumption levels by employment of a sustained programme of energy consumption strategies and measures would have considerable benefits within the buildings sector. The perception often given however is that rigorous energy conservation as an end in itself imposes a style on building design resulting in a restricted aesthetic solution. It would perhaps be better to support a climate sensitive design approach, which encompassed some elements of the pure conservation strategy together with strategies, which work with the local ambient conditions making use of energy technology systems, such as solar energy, where feasible. In practice, low energy environments are achieved through a combination of measures that include:

- The application of environmental regulations and policy.
- The application of environmental science and best practice.
- Mathematical modelling and simulation.
- Environmental design and engineering.
- Construction and commissioning.
- Management and modifications of environments in use.

While the overriding intention of passive solar energy design is to achieve a reduction in purchased energy consumption, the attainment of significant savings is in doubt. The non-realisation of potential energy benefits is mainly due to the neglect of the consideration of post-occupancy user and management behaviour by energy scientists and designers alike. Buildings consume energy mainly for cooling, heating and lighting as shown in Table 5. The energy consumption shown in the table was based on the assumption that the building operates within ASHRAE-thermal comfort zone during the cooling and heating periods [29]. Most of the buildings incorporate energy efficient passive cooling, solar control, photovoltaic, lighting and day lighting, and integrated energy systems. It is well known that thermal mass with night ventilation can reduce the maximum indoor temperature in buildings in summer [30]. Hence, comfort temperatures may be achieved by proper application of passive cooling systems. However, energy can also be saved if an air conditioning unit is used [30]. The reason for this is that in summer, heavy external walls delay the heat transfer from the outside into the inside spaces. Moreover, if the building has a lot of internal mass the increase in the air temperature is slow. This is because the penetrating heat raises the air temperature as well as the temperature of the heavy thermal mass. The result is a slow heating of the building in summer as the maximal inside temperature is reached only during the late hours when the outside air temperature is already low. The heat flowing from the inside heavy walls can be removed with good ventilation in the evening and night. The capacity to store energy also helps in winter, since energy can be stored in walls from one sunny winter day to the next cloudy one.

Table 5. Energy-saving in buildings

Passive Comfort Measures	Active Comfort Measures	Climatic zones			
		Mediterranean	Subtropical	Tropical	Desert
Natural ventilation		6	7	7	7
	Mechanical ventilation	4	5	6	6
Night ventilation		6	7	7	7
	Artificial cooling	3	5	5	6
Evaporative cooling		3	2	2	7
	Free cooling	5	6	6	7
Heavy-weight construction		6	2	2	6
Light-weight construction		3	5	5	4
	Artificial heating	4	0	0	1
Solar heating		6	0	0	0
	Free heating	5	0	0	0
Incidental heat		4	0	0	0
Insulation/permeability		5	0	0	4
Solar control/shading		6	6	6	7
	Daytime artificial lighting	3	3	3	2
Day lighting features		6	5	5	4

* 0 = not important, 4 = important, and 7 = very important (importance is rated from 0 to 7).

One can define four levels of thermal mass as follows:

- Light building: no thermal mass, e.g., a mobile home.
- Medium-light building: light walls, but heavy floor, e.g., cement tiles on concrete floor, and concrete ceiling.
- Semi-heavy building: heavy floor, ceiling and external walls (20 cm concrete blocks) but light internal partitions (Gypsums boards).
- Heavy building: heavy floor, ceiling, external and internal walls (10 cm concrete blocks, with plaster on both sides).
- The exact reduction in the maximum indoor temperature depends on the amount of thermal mass, the rate of night ventilation, and the temperature swing between day and night.

11.1. Energy Efficiency and Architectural Expression

The focus of the world's attention on environmental issues in recent years has stimulated response in many countries, which have led to a closer examination of energy conservation strategies for conventional fossil fuels. Buildings are important consumers of energy and thus important contributors to emissions of greenhouse gases into the global atmosphere. The development and adoption of suitable renewable energy technology in buildings has an important role to play. A review of options indicates benefits and some problems [30]. There are two key elements to the fulfilling of renewable energy technology potential within the field of building design; first the installation of appropriate skills and attitudes in building design professionals and second the provision of the opportunity for such people to demonstrate their skills. This second element may only be created when the population at large and clients commissioning building design in particular, become more aware of what can be achieved and what resources are required. Terms like passive cooling or passive solar use mean that the cooling of a building or the exploitation of the energy of the sun is achieved not by machines but by the building's particular morphological organisation. Hence, the passive approach to themes of energy savings is essentially based on the morphological articulations of the constructions. Passive solar design, in particular, can realise significant energy and cost savings. For a design to be successful, it is crucial for the designer to have a good understanding of the use of the building. Few of the buildings had performed as expected by their designers. To be more precise, their performance had been compromised by a variety of influences related to their design, construction and operation. However, there is no doubt that the passive energy approach is certainly the one that, being supported by the material shape of the buildings has a direct influence on architectural language and most greatly influences architectural expressiveness [31]. Furthermore, form is a main tool in architectural expression. To give form to the material things that one produces is an ineluctable necessity. In architecture, form, in fact, summarises and gives concreteness to its every value in terms of economy, aesthetics, functionality and, consequently, energy efficiency [32]. The target is to enrich the expressive message with forms producing an advantage energy-wise. Hence, form, in its geometric and material sense, conditions the energy efficiency of a building in its interaction with the environment. It is, then, very hard to extract and separate the parameters and the elements relative to this efficiency from the expressive unit to which they belong. By analysing energy issues and strategies by means of the designs, of which they are an integral part, one will, more easily, focus the attention on the relationship between these themes, their specific context and their architectural expressiveness. Many concrete examples and a whole literature have recently grown up around these subjects and the wisdom of forms and expedients that belong to millennia-old traditions has been rediscovered. Such a revisiting, however, is only, or most especially, conceptual, since it must be filtered through today's technology and needs; both being almost irreconcilable with those of the past. Two among the historical concepts are of special importance. One is rooted in the effort to establish rational and friendly strategic relations with the physical environment, while the other recognises the interactions between the psyche and physical perceptions in the creation of the feeling of comfort. The former, which may be defined as an alliance with the environment deals with the physical parameters involving a mixture of natural and artificial ingredients such as soil and vegetation, urban fabrics and pollution [33]. The most dominant outside parameter is, of course, the sun's irradiation, our planet's primary energy source. All these elements can be measured in physical terms and are therefore the subject of science. Within the second concept, however, one considers the emotional and intellectual energies, which are the prime inexhaustible source of renewable power [34]. In this case, cultural parameters, which are not exactly measurable, are involved. However, they represent the very essence of the architectural quality. Objective scientific measurement parameters tell us very little about the emotional way of perceiving, which influences the messages of human are physical sensorial organs. The perceptual reality arises from a multitude of sensorial components; visual, thermal, acoustic, olfactory and kinaesthetics. It can, also, arise from the organisational quality of the space in which different parameters come together, like the sense of order or of serenity. Likewise, practical evaluations, such as usefulness, can be involved too. The evaluation is a wholly subjective matter, but can be shared by a set of experiencing persons [34]. Therefore, these cultural parameters could be different in different contexts in spite of the inexorable levelling on a planet- wide scale. However, the parameters change in the anthropological sense, not only with the cultural environment, but also in relation to function. The scientifically measurable parameters can, thus, have their meanings very profoundly altered by the non-measurable, but describable, cultural parameters.

However, the low energy target also means to eliminate any excess in the quantities of material and in the manufacturing process necessary for the construction of our built environment. This claims for a more sober, elegant and essential expression, which is not jeopardising at all, but instead enhancing, the richness and preciousness of architecture, while contributing to a better environment from an aesthetic viewpoint [35]. Arguably, the most successful designs were in fact the simplest. Paying attention to orientation, plan and form can have far greater impact on energy performance than opting for elaborate solutions [36]. However, a design strategy can fail when those responsible for specifying materials for example, do not implement the passive solar strategy correctly. Similarly, cost-cutting exercises can seriously upset the effectiveness of a design strategy. Therefore, it is imperative that a designer fully informs key personnel, such as the quantity surveyor and client, about their design and be prepared to defend it. Therefore, the designer should have an adequate understanding of how the occupants or processes, such as ventilation, would function within the building. Thinking through such processes in isolation without reference to others can lead to conflicting strategies, which can have a detrimental impact upon performance. Likewise, if the design intent of the building is not communicated to its occupants, there is a risk that they will use it inappropriately, thus, compromising its performance. Hence, the designer should communicate in simple terms the actions expected of the occupant to control the building. For example, occupants should be well informed about how to guard against summer overheating. If the designer opted for a simple, seasonally adjusted control; say, insulated sliding doors were to be used between the mass wall and the internal space. The lesson here is that designers must be prepared to defend their design such that others appreciate the importance and interrelationship of each component. A strategy will only work if each individual component is considered as part of the bigger picture. Failure to implement a component or incorrect installation, for example, can lead to failure of the strategy and consequently, in some instances, the building may not liked by its occupants due to its poor performance.

11.2. Sustainable Practices

Within the last decade sustainable development and building practices have acquired great importance due to the negative impact of various development projects on the environment. In line with a sustainable development approach, it is critical for practitioners to create a healthy, sustainable built environment [34, 35, 36 and 37]. In Europe, 50% of material resources taken from nature are building-related, over 50% of national waste production comes from the building sector and 40% of energy consumption is building-related. Therefore, more attention should be directed towards establishing sustainable guidelines for practitioners. Furthermore, the rapid growth in population has led to active construction that, in some instances, neglected the impact on the environment and human activities. At the same time, the impact on the traditional heritage, an often-neglected issue of sustainability, has not been taken into consideration, despite representing a rich resource for sustainable building practices.

Sustainability has been defined as the extent to which progress and development should meet the need of the present without compromising the ability of the future generations to meet their own needs [38]. This encompasses a variety of levels and scales ranging from economic development and agriculture, to the management of human settlements and building practices. This general definition was further developed to include sustainable building practices and management of human settlements. The following issues were addressed during the Rio Earth Summit in 1992 [41]:

- The use of local materials and indigenous building sources.
- Incentive to promote the continuation of traditional techniques, with regional resources and self-help strategies.
- Regulation of energy-efficient design principles.
- International information exchange on all aspects of construction related to the environment, among architects and contractors, particularly non-conventional resources.
- Exploration of methods to encourage and facilitate the recycling and reuse of building materials, especially those requiring intensive energy use during manufacturing, and the use of clean technologies.

The objectives of the sustainable building practices aim to:

- Develop a comprehensive definition of sustainability that includes socio-cultural, bio-climate, and technological aspects.
- Establish guidelines for future sustainable architecture.
- Predict the CO_2 emissions in buildings.
- The proper architectural measure for sustainability is efficient, energy use, waste control, population growth, carrying capacity, and resource efficiency.
- Establish methods of design that conserve energy and natural resources.

A building inevitably consumes materials and energy resources. The technology is available to use methods and materials that reduce the environmental impacts, increase operating efficiency, and increase durability of buildings (Table 6).

Table 6. Design, construction and environmental control description of traditional and new houses

Design characteristics	Traditional houses	New houses
Form	Courtyard (height twice its width) - open to sky	Rectangle-closed
Construction	Brick walls 50 cm thick, brick roof with no insulation in either	Brick walls 25 cm thick, concrete roof-no insulation
Environmental control	Evaporative air coolers	Evaporative air coolers
Ease of climatic control	Difficult-rooms open into the climatically uncontrolled open courtyard	Moderate-rooms open into the enclosed internal corridor
Maintenance	Well conserved and maintained	New construction
Windows	Vertical and single glazing	Horizontal-single glazing
Urban morphology	Each house attached from 3 sides	Row houses attached from 2 sides
Orientation	Varies-irregular shapes and winding alleyways	North-south (row houses)
Orientation and solar gain	Solar gains less affected by orientation due to shading provided by the deep courtyard	Solar gain determined by orientation-no obstruction (shading)
Sharing solar gain	Significant - due to long-wave exchange or convective exchange between the 4 vertical walls surrounding the courtyard	Minimal- S-wall receives much more solar radiation than the N-wall due to the absence of any interreflection and long-wave exchange
Occupant's social status	Low income families	Low and middle income families

Literature on green buildings reveals a number of principles that can be synthesised in the creation of the built environment that is sustainable. According to Lobo [42], these are: land development, building design and construction, occupant considerations, life cycle assessment, volunteer incentives and marketing programmes, facilitate reuse and remodelling, and final disposition of the structure. These parameters and many more are essential for analysis, making them an important element of the design decision-making process. Today, architects should prepare for this as well as dealing with existing buildings with many unfavourable urban environmental factors, such as many spaces have no choice of orientation, and, often, set in noisy streets with their windows opening into dusty and polluted air and surrounding buildings overshadowing them.

11.3. Buildings and CO_2 Emission

To achieve carbon dioxide, CO_2, emission targets, more fundamental changes to building designs have been suggested [39]. The actual performance of buildings must also be improved to meet the emission targets. To this end, it has been suggested that the performance assessment should be introduced to ensure that the quality of construction, installation and commissioning achieve the design intent. Air-tightness and the commissioning of plant and controls are the main two elements of assessing CO_2 emission. Air-tightness is important as uncontrolled air leakage wastes energy. Uncertainties over infiltration rates are often the reason for excessive design margins that result in oversized and inefficient plants. On the other hand, commissioning to accept procedures would significantly improve energy efficiency. The slow turnover in the building stock means that improved performance of new buildings will only cut CO_2 emissions significantly in the long term. Consequently, the performance of existing buildings must be improved. For example, improving 3% of existing buildings would be more effective in cutting emissions than, say, improving the fabric standards for new non-domestic buildings and improving the efficiency of new air conditioning and ventilation systems [40]. A reduction in emissions arising from urban activities can, however, only be achieved by a combination of energy efficiency measures and a move away from fossil fuels.

11.4. Low Energy Buildings

There is no single, simple formula for achieving low energy buildings. The basic principle is to minimise energy demand and to optimise energy supply through a greater reliance on local and renewable resources. Cities need to take a close look at how to make more efficient use of resources while fulfilling the needs of the people. An energy dimension should be included in the development process to measure the sustainability of urban and building design and growth planning models. Previous experience in public transport systems indicates that density is conductive to profitability and efficiency [43]. A compact urban form with vertical zoning through multi-level and multi-functional urban clusters may be an efficient option for high-density living. There are, however, opportunities for high-density cities to explore and develop effective energy technologies, which can take full advantage of the concentrated loads and high-rise context, such as using district energy systems and vertical landscapes. Designing and constructing low energy buildings require the design team to follow an energy design process that considers how the building envelope and systems work together [45].

As low energy design is becoming more and more complicated, there is a need to develop analytical methods and skills, such as simulation and modelling techniques, for the evaluation of energy performance of buildings and the analysis of design options and approaches [33]. Kausch [43] pointed out that low energy building design is compatible with a wide range of architectural styles. Studio Nicoletti [34] also illustrated the methods of architectural expression for low energy buildings in their projects. For high-density conditions, some of their methods are still valid but adaptation or modification may be needed to satisfy the local requirements. Climate consideration is a key element and starting point for formulating building and urban design principles that aim at minimising the use of energy for environmental control. In densely populated areas, analysis of the climatic and solar conditions is critical for the design optimisation. It should be noted that in urban areas, the group of buildings would in fact modify the climatic conditions surrounding it.

Measures to maximise the use of high-efficiency generation plants and on-site renewable energy resources are important for raising the overall level of energy efficiency. For renewable energy systems, energy storage is still the major technical constraint to their applications [44]. Loads concentration in high-density cities might provide opportunities for better utilisation of renewable energy systems. At present, lack of incentives and shortage of land and space are the key factors limiting the deployment of renewable energy systems. High-rise buildings and high population density make it difficult to find suitable locations for solar collectors and equipment. As the demand for heating energy is relatively low in many buildings because of the warm climate throughout the year, the economic advantage of directly using solar heat is weakened. To promote renewables, it is necessary to create new development patterns and shift from a centralised view of energy sector to a regional perspective [45].

One important aspect often being overlooked is the raising of awareness and the education about low energy design. More efforts are needed to educate the people and establish the culture so that more people would accept and consider low energy buildings an important element of their living and working environment. It is important to recognise that solutions to the energy problems are not simply a matter of applying technology and enforcement through legislation [46]. It requires public awareness and participation as well. Therefore, measures to promote public awareness and education are crucial for the implementation of energy efficiency and renewable energy policies.

In summary, achieving low energy building requires comprehensive strategy that covers; not only building designs, but also considers the environment around them in an integral manner. Major elements for implementing such a strategy are as follows:

11.4.1. Efficiency use of energy

- Climate responsiveness of buildings.
- Good urban planning and architectural design.
- Good house keeping and design practices.
- Passive design and natural ventilation.
- Use landscape as a means of thermal control.
- Energy efficiency lighting.
- Energy efficiency air conditioning.
- Energy efficiency household and office appliances.
- Heat pumps and energy recovery equipment.
- Combined cooling systems.
- Fuel cells development.

11.4.2. Utilise renewable energy

- Photovoltaics.
- Wind energy.

- Small hydros.
- Waste-to-energy.
- Landfill gas.
- Biomass energy.
- Biofuels.

11.4.3. Reduce transport energy

- Reduce the need to travel.
- Reduce the level of car reliance.
- Promote walking and cycling.
- Use efficient public mass transport.
- Alternative sources of energy and fuels.

11.4.4. Increase awareness

- Promote awareness and education.
- Encourage good practices and environmentally sound technologies.
- Overcome institutional and economic barriers.
- Stimulate energy efficiency and renewable energy markets.

12. CONCLUSIONS

Newspapers, TV, schools, universities and politicians rant and rave about being 'green' and doing our bit for the environment, but can we as individuals change things? Energy efficiency brings health, productivity, safety, comfort and savings to homeowner, as well as local and global environmental benefits. The use of renewable energy resources could play an important role in this context, especially with regard to responsible and sustainable development. It represents an excellent opportunity to offer a higher standard of living to local people and will save local and regional resources. Implementation of greenhouses offers a chance for maintenance and repair services. It is expected that the pace of implementation will increase and the quality of work to improve in addition to building the capacity of the private and district staff in contracting procedures. The financial accountability is important and more transparent. Various passive techniques have been put in perspective, and energy saving passive strategies can be seen to reduce interior temperature, increase thermal comfort, and reducing air conditioning loads. The scheme can also be employed to analyse the marginal contribution of each specific passive measure working under realistic conditions in combination with the other housing elements. In regions where heating is important during winter months, the use of top-light solar passive strategies for spaces without an equator-facing façade can efficiently reduce energy consumption for heating, lighting and ventilation. Many cities around the world are facing the problem of increasing urban density and energy demand. As cities represent a significant source of growth in global energy demand, their energy use, associated environmental impacts, and demand for transport services create great pressure to global energy resources. Low energy design of urban environment and buildings in densely populated areas requires consideration of a wide range of factors, including urban setting, transport planning, energy system design, and architectural and engineering details. It is found that densification of towns could have both positive and negative effects on the total energy demand. With suitable urban and building design details, population should and could be accommodated with minimum worsening of the environmental quality.

REFERENCES

[1] Reddy, A; Williams, R; and Johansson, T. *Energy after Rio: prospects and challenges*. United Nations Development Programme (UNDP). http://www.undp.org/seed/energy/-exec-en.html. 2007.

[2] Cavallo, AJ; and Grubb, MJ. Renewable energy sources for fuels and electricity. *London: Earthscan Publications*, 1993.

[3] REN21. Renewables 2007 global status report. www.ren21.net. 2008.

[4] Gilman, K. Water of wetland areas. In: Proceedings of a Conference on the Balance of Water – Present and Future. Dublin: Ireland, p. 123-142. 7-9 September 1994.

[5] John, W. The glasshouse garden. *The Royal Horticultural Society Collection*, UK. 1993.

[6] United Nations. World Urbanisation Prospect: The 1999 Revision. New York. *The United Nations Population Division*, 2001.

[7] WCED. *Our common future*. New York. Oxford University Press. 1987.

[8] Herath, G. The Green Revolution in Asia: productivity, employment and the role of policies. *Oxford Agrarian Studies*, 1985, 14, 52-71.

[9] BRECSU. Energy use in offices. Energy Consumption Guide 19. Watford: *Building Research Energy Conservation Support* Unit, 1998.

[10] Farm Energy Centre. Helping agriculture and horticulture through technology, energy efficiency and environmental protection. *Warwickshire*, 2000.

[11] Randall, M. *Environmental Science in Building*. Third Edition, 1992.

[12] Tiwari, GN; and Goyal, RK. *Greenhouse technology*. New Delhi: *Narosa Publishing House*, 1998.

[13] Santamouris, M; Balaras, CA; Dascalaki, E; and Vallindras, M. Passive solar agricultural greenhouse; a worldwide classification evaluation of technologies and systems used for heating purpose. *Solar Energy*, 53(5), 411-26. 1994.

[14] Santamouris, M; Arigirious, A. and Vallindras, M. Design and operation of a low energy consumption passive greenhouse. *Solar Energy*, 52(5), 371-8. 1993.

[15] Mercier, I. Design and operation of a solar passive greenhouse in the South West France. In: Proceedings of the International Congress on Energy Conservation of Agriculture and Fishculture. London, 1982.

[16] Grafiadellis, M. *Greenhouse heating with solar energy*. In: Von Zabettitz C (Editor), FAO, Rome, 1987.

[17] Fotiades, I. Energy conservation and renewable energies. In: Von Zabettitz C (Editor), FAO, Rome, 1987.

[18] Pacheco, M; Marreivos, S; and Rosa, M. Energy conservation and renewable energies for greenhouse heating. In: Von Zabettitz C (Editor), FAO, Rome, 1987.

[19] Santamouris, M; Mihalakakow, G; Belaras, CA; Lewis, JO; Vallindras, M; and Argiriou, A. Energy conservation in greenhouses with buried pipes. *Energy*, 21(5), 353-60. 1996.

[20] Garzoli, KV; and Blackwell, Jan. Analysis of the nocturnal heat loss from a single skim plastic greenhouse. *J. Agric. Engg. Res*, 26, 203-14. 1981.

[21] Chandra, P; and Albright, LD. Analytical determination of the effect on greenhouse heating requirements of using night curtains. *Trans ASAE*, 23(4), 9994-1000. 1980.

[22] Anne, SA. *Handbook of greenhouse and conservatory plants*. Paston Press. London. 1989.

[23] Lynn, B. *The pleasure of gardening*. ANAYA Publishers Limited. London, 1993.

[24] Paul, F. Indoor hydroponics: *A guide to understanding and maintaining a hydroponic nutrient solution*. UK. 2001.

[25] Cengel, Y. *Heat Transfer-A Practical Approach*. First ed. McGraw-Hill, Inc. 1998.

[26] Heinonen, EW; Tapscott, RE; Wildin, MW; and Beall, AN. *Assessment of anti-freeze solutions for ground-source heat pumps systems*. New Mexico Engineering Research Institute NMERI 96/15/32580, p. 156. 1996.

[27] Huttrer, G. *The status of world geothermal power generation*, 1995-2000. Geothermics 30, 1-27. 2001.

[28] Givoni, B. *Climate consideration in building and urban design*. New York: Van Nostrand Reinhold. 1998.

[29] ASHRAE. Energy efficient design of new building except new low-rise residential buildings. BSRIASHRAE proposed standards 90-2P-1993, alternative GA. American Society of Heating, *Refrigerating, and Air Conditioning Engineers* Inc, USA. 1993.

[30] Kammerud, R; Ceballos, E; Curtis, B; Place, W; and Anderson, B. *Ventilation cooling of residential buildings*. ASHRAE Trans, 90 Part 1B. 1984.

[31] Shaviv, E. The influence of the thermal mass on the thermal performance of buildings in summer and winter. In TC. Steemers, W. Palz, (Eds.), *Science and Technology at the service of architecture*. Kluwer Academic Publishers, p. 470-2. 1989.

[32] Anink, D; Boostra, C; and Mark, J. Handbook of sustainable buildings: *an environmental performance method for selection of materials for use in construction and refurbishment*. UK: James and James, 1996.

[33] Clarke, JA; Grant, AD; Johhnstone, CM; and Maccdonald, I. Integrated modelling of low energy buildings. *Renewable Energy*, 1998, 15, 151-6.

[34] Studio Nicoletti. Architectural expression and low energy design. *Renewable Energy*, 1998, 15, 32-41.

[35] Lund, PD. The energy storage problem in low energy buildings. *Solar Energy*, 1994, 52(1), 67-74.

[36] Naga, AM; Alsallal, K; and Eloliasty, R. The impact of city urban patterns on building energy consumption in hot climates: Alain city as a case study. *In: Proceedings of ISES Solar World Congress*: Taejon. South Korea, p. 170-81. 1997.

[37] EIBI (Energy in Building and Industry). Constructive thoughts on efficiency, building regulations, inside committee limited, InsidEnergy: *magazine for energy professional*. UK: KOPASS, p.13-14. 1999.

[38] Celik, Z. Urban preservation as theme park: the case of Sogukcesme Street. In: Z. Celik, Z. Favro, R. Ingersoll, (Eds.), *Streets: critical perspectives on the public space*. Berkeley (CA): University of California Press, p. 83-94. 1994.

[39] Naga, AM; and Amin, M. Towards a healthy urban environment in hot-humid zones-information systems as an effective evaluation tool for urban conservation techniques. In: *Proceedings of XXIV IAHS World Congress*, Ankara. Turkey, 1996.
[40] Steele, J. Sustainable architecture: principles, paradigms, and case studies. New York: *McGraw-Hill Inc*, 1997.
[41] Sitarz, D. Editor. Agenda 21: The Earth Summit Strategy to save our planet. Boulder (CO): *Earth Press*, 1992.
[42] Lobo, C. Defining a sustainable building. In: Proceedings of 23rd National Passive Conference, ASES'98. Albuquerque (USA): *American Solar Energy Society*, 1998.
[43] Kausch, A. Is low-energy building compatible with any architectural style? *In: Environmental friendly cities*. Proceedings of PLEA 98, Lisbon. Portugal. June 1998, p.217-220.
[44] Sarafiddis, Y; Diakoulaki, D; Papayannabis, L; and Zervos, AA. Regional planning approach for the promotion of renewable energies. *Renewable Energy*, 1999, 18(3), 317-30.
[45] Tong, CO; and Wong, SC. Advantages of a high density, mixed land use, linear urban development. *Transportation*, 1997, 24(3), 295-307.
[46] Lam, JC. Shading effects due to nearby buildings and energy implications. *Energy Conversion and Management*, 2000, 47(7), 647-59.

NOMENCLATURE

E_y	Annual energy yield in kWh
W	The wind speed in m/s
n	The number of data bins converting the wind speed range of the turbine (0.5 or 1 m/s intervals)
f_{wi}	The number of hours per year for which wind speed is w m/s
P_{wi}	Is the power resulting from a wind speed of w m/s
Q^{n}_{conv}	Rate of convection heat transfer per unit area, W/m^2
Q^{n}_{evap}	Rate of heat transfer per unit area due to evaporation, W/m^2
Q^{n}_{rad}	Rate of radiation heat transfer per unit area, W/m^2
Q^{n}_{total}	Total rate of heat transfer per unit area, W/m^2
T_s	Soil surface temperature, °C
T_{surr}	Surrounding temperature, °C
T_∞	Airflow temperature far from the surface, °C
G_r	Grashof number, dimensionless
R_e	Reynolds number, dimensionless
DC	Direct current
HSPF	Heating season performance factor
SEER	Seasonal energy efficiency ratio
Btu	British thermal unit

EER Energy efficiency rating
DX Direct expansion
GS Ground source
EPA Environmental Protection Agency
HVAC Heating, ventilating and air conditioning

Greek Letters

ε Surface emissivity, dimensionless
φ Relative humidity, dimensionless
ρ Density, kg/m^3
δ Stefan-Boltzmann constant, 5.67×10^{-8} W/m^2K^4

Subscripts

a Air
v Vapour
sat Saturated
∞ Far away from the surface
s Surface

In: Buildings and the Environment
Editors: Jonas Nemecek and Patrik Schulz ISBN: 978-1-60876-128-9

Chapter 3

MANAGING PROCESSES RELATED TO PEDESTRIAN MOBILITY THROUGH THE DISTRIMOBS DECISION SUPPSORT SYSTEM

Marco Brambilla*[*] *and Luca Cattelani
Physics of the City Laboratory Alma Mater Studiorum-University of Bologna via Zamboni 72, 40126 Bologna

INTRODUCTION

Processes related to planning, redesign and services assignment inside buildings, squares, parks, sidewalks or more generally indoor and open environments are usually realized without the aid of a decision support system able to consider pedestrian mobility. Usually engineers, architects and planners handle pedestrian mobility in these environments through design patterns, experience and rough estimations. A decision support system able to manage pedestrian mobility, in all its aspects and consequences, is a precious instrument for these professional Figures, that will rapidly become more and more important in the next years. This importance is due to the ability pedestrian mobility simulation gives to these professionals: the power to predict consequences of their choices. It can be expected that pedestrian mobility simulation will help in the evaluation of key parameters of areas including: level of usage ("How many persons pass through this alley?", "How long people stays in front of this showcase?"), kind of population ("What are the kinds of people staying in this waiting room?"), security ("Are these streets sufficiently lighted?") or safety ("Is this alley sufficiently wide for the fluxes passing in?") etcetera. The importance of these kinds of questions (and obviously of their answers) will reasonably become more important due to the evolution of economic and environmental world conditions:

- private cars will reduce their number (due to costs and pollution);

[*] Corresponding author: {brambill, cattelan}@cs.unibo.it

- collective solutions will augment their numbers (busses, undergrounds, trains, airplanes).

These two simple assertions will cause an augment of pedestrian mobility impact in many environments (parks, roads, buildings, stations etcetera). Holding a suitable instrument for managing pedestrian mobility before the definitive point-break (i.e. the number of existing private car starts to decrease) is important key in order to avoid having in the future additional problems caused by raised density of pedestrians.

This chapter describes, after discussing the evolution of pedestrian mobility modeling, several potentialities of this research field through the Distrimobs simulator. A highlight of the most prominent components of the Distrimobs software architecture is given. It includes contents about environment representation, interactions with other existing software and data sources, behaviors of simulated pedestrians (artificial intelligence) and handling of high computational costs (parallelization).

Finally, some applications of the simulator are presented, including the historic Palazzo Poggi building, the pedestrianization of the medieval center of Bologna, the changes brought by Calatrava's Bridge in Venice and the requalification of the Scalo Farini district in Milano.

Decision Support System and Computer Aided Design

Design and Computer Science, a Lasting Synergy

Initial applications of computer science (or, more generally, science) in the design processes were carried out in the 1960s within the aircraft and automotive industries. As usually happen when new researches are leaded by several industries, each design company started independently and often not publicly published until much later. The most important works on polynomial curves and sculptured surfaces (used in the area of tridimensional surface construction) were done by Pierre Bezier (Renault), Paul de Casteljau (Citroen), Steven Anson Coons (Ford), Carl de Boor, Garrett Birkhoff and H. L. Garabedian (GM).

Bezier worked on Bezier's curve, while the original algorithm for creating these curves was by Casteljau (the Casteljau's algorithm). Coons worked on conic curves while, finally, Birkhoff, Garabedian and de Boor worked on splines and cubic splines. Although their works are famous due to the great consequences they had in the computer graphics their target was to aid design of surfaces. They are the pioneers of the computer aided design.

The first milestone in the CAD history is the work by Ivan Sutherland, Sketchpad that allowed the designer to interact graphically with the computer: the design can be fed into the computer by drawing on a CRT monitor with a light pen. It was a prototype of graphical user interface, one of the most important features of modern CAD.

The second milestone event in the history of CAD was the foundation of Manufacturing and Consulting Services by Patrick Hanratty, who created, in 1972, ADAM (Automated Drafting and Machining). CAD systems quickly evolved starting from the 1970's with bidimensional drawings, reaching, with the advances of computer science and hardware, to solid modeling.

The third milestone was the wide adoption of CAD systems. Finally, in the 1980s, several famous solid modeling packages become widely used and started to be taught in schools (AutoCAD, Unisolid, Catia etcetera).

The final milestone was the development of solid modeling kernels that allowed representing, in addiction to surface, materials and their properties (Parasolid, ACIS, SolidWorks etcetera). These were the first years of 1990s and computer science, with CAD, totally changed the design processes. Most important, these advances, made designers and software houses able to understand that every design process can be aided by computer science.

Simulations Applied to Computer Aided Design

Recent CAD evolutions are strictly bounded to simulation because this computer science branch allows for the prediction of effects of planners' choices during design processes. These processes of designing indoor and urban environment simulations have been widely used for daylight, energy, warm/cool and so on. The review work by Tianzhen Hong et al. [1] is an important milestone for the application of simulation to computer aided design. Some paragraphs are provided here:

- "The application of computer-based tools in building design can be broadly divided into two groups, namely, computer-aided documentation, design, and drafting; and computer-based simulation. Today, the first application, which often uses personal computers to produce technical documents and drawings, is already popular with building designers. The latter application often requires the use of engineering tools to calculate envelope heat gains and space heat loads, predict the energy performance of the building, and provide diagnostics to enable automatic control of system and plant operation. While the first application can help improve the productivity of building designers, it has little impact on efficient building performance. Only computer-aided simulation holds the key to improve building energy efficiency."
- "Before the advent of computer-aided building simulation, architects and building services engineers relied heavily on manual calculations using pre-selected design conditions and often resorted to the `rule-of-thumb' method and extrapolations in extending beyond conventional design concepts. This approach had frequently led to oversized plant and system capacities and poor energy performance due to excessive part-load operations. For large or complex buildings, it would be unrealistic to expect that energy-efficient designs could be attempted without resorting to computer-aided detailed building simulation programs (BSPs)."

Although the authors recognize the importance of simulation, they still consider it only for well-described physical aspects, such as energy.

- "Building simulation programs can be grouped into two categories, namely, design tools (DTs) and detailed simulation programs (DSPs). DTs are more purpose-specific and are often used at the early design phases because they require less and simpler input data. For example, DTs are very useful in the compliance checking of prescriptive building standards. Because DTs are easy to develop and test, they have proliferated. Many of them are developed for in-house use, while some can be found in the public domain. On the other hand, DSPs often incorporate computational techniques such as finite difference, finite element, state space, and transfer function for building load and energy calculations"

Another important consideration of the authors is about a distinction between types of simulation software for building. There are design tools and detailed simulation programs.

- "The potential user is faced with the difficulty of choosing a suitable program from those available. Which BSP should one select? The answer to this question, unfortunately, is not straightforward. The choice should be made after carefully assessing the requirements of the user and matching them with the capabilities of the BSP. There are three vital factors to consider from the user's side. The first concerns need or purpose. Understanding the nature of the problem that the user expects to solve with the use of a BSP is an important criterion. [...] The second relates to budget. The budget to purchase and use a BSP includes software cost, maintenance, if necessary, and the cost of the computer platform to run the BSP. In addition, provision should be made for user training. The third is the availability of facilities. The user should select a BSP that can be run on existing computer facilities, or when anticipated investment in new computer resource is bearable."

The authors finally recognize the importance of choosing the right, or at least the better, simulation program in order to get realistic results and, consequently, to be aided by simulation programs. This fact has been divulged by the work presented by Hensen and Clarke [2] at the international conference on design and decision support systems in architecture and urban planning. Here we give two important sentences extracted from the proceedings of the same conference:

- "Simulations can thus be used for building analysis and design in order to achieve a good indoor environment that is sustainable to care for people now and in the future".
- "Although most practitioners will be aware of the emerging building simulation technologies - and their potential for building design -, few as yet are able to claim expertise in its application. This situation is poised to change with the advent of:
  performance based standards;
  societies dedicated to the effective deployment of simulation - such as the International Building Performance Simulation Association;
  appropriate training and continuing education;
  the growth in small-to-medium sized practices offering simulation-based services."

These works testify to how simulations are highly-regarded by computer aided design producers. Nevertheless, developers of CAD software consider simulation only for well-described physical phenomena. Pedestrian mobility is very important for the evaluation of indoor and outdoor environments and, although it isn't a well known phenomenon, its simulation started to be undertaken some years ago.

Pedestrian Mobility Simulation

Research and development in the field of pedestrian mobility simulation interests both the academic and business worlds as confirmed by the attempt to model pedestrian mobility using different techniques (least effort algorithms [3], gas kinetics [4], social forces [5], chronotopical [6], cellular automata [7-8] and multi-agent [9-10-11]) and by the works of some software companies (such as Simwalk [12] and Legion [13]).

Multi-agent based simulation is considered a very powerful approach for modeling pedestrian mobility; the decision support system presented in this chapter, the Distrimobs simulator, is based on a multi-agent model.

Passing over the names of the computer science approaches, the characteristics of the real world that the simulators are able to handle are more important.

These are:

1. Spatial representation: if the space is continuous or discrete.
2. Time representation: if the space is continuous or discrete.
3. Finality: if the simulator focuses an ad-hoc scenario or is general-purpose.
4. Environment complexity: how the model is able to manage complex and structured environment.
5. Physics of agents: how the model can represent physics aspects of the agents such as encumbrances, speeds and so on.
6. Behavior of agents: how the model can represent behavioral aspects of the agents such as paths, times and so on.
7. Large–scale: the ability to represent a large number of pedestrians.
8. Realism: how realistic the model is and whether it has been validated.
9. User friendliness: how the input, the simulations, the analysis are understandable to users (engineer, architects but also other people).

From the point of view of engineers, architects and so on, it would not be trivial to Figure out, considering the above parameters, which support system is the best choice for their use. Hensel et al. [2] suggest three vital factors to consider about simulator:

1. "The first concerns need or purpose. Understanding the nature of the problem that the user expects to solve with the use of a BSP is an important criterion. Choosing an `overpowered' BSP is not only unnecessary and expensive but can be costly when mistakes are made due to the complexity of the software".
2. "The second relates to budget. The budget to purchase and use a BSP includes software cost, maintenance, if necessary, and the cost of the computer platform to run the BSP. In addition, provision should be made for user training".
3. "The third is the availability of facilities. The user should select a BSP that can be run on existing computer facilities, or when anticipated investment in new computer resource is bearable".

In Table 1 a brief comparison table of the most famous applied simulators, existing at the current time, is showed.

Table 1: Comparison table between applied simulators.

	Distrimobs	Legion	SimWalk	CAST	UAF
Paradigm	Agent-based	Unknown	Physical forces	Unknown	Agent-based
Space	Continuous	Unknown	Continuous	Unknown	Unknown
Time	Continuous	Unknown	Discrete	Unknown	Unknown
Scalability	High	Unknown	Unknown	Unknown	High
HL behavior	Motivation	Unknown	Manual	Unknown	Unknown
ML behavior	Graph	Unknown	None	Unknown	Unknown
LL behavior	Steering behaviors	Unknown	Social Force Model	Unknown	Least effort algorithm
Input	DXF, DWG	Unknown	CAD	DXF, 3DS	CAD
Output	3D, 2D	3D, 2D	3D, 2D	3D, 2D	3D, 2D

Legion is one of the most used pedestrian simulation software but nothing about its model is known. SimWalk is based upon the Social Force Model. Although it claims to have a continuous space, videos seems to be generated from a cell-based model. CAST models pedestrian inside airports. Its engine is unknown. UAF, Urban Analytics Framework, is based upon the least effort algorithm. The real implementation of this method is unknown. Finally, UAF seems to be highly scalable.

The natural conclusions to this brief introduction to the application of pedestrian mobility seem to be just positive sentences; it may seem that a professional just needs to find the simulator that best fits his needs and use it. Unfortunately, it's not so simple. In order to be a good design support system, a simulator needs to be predictive; this means that the simulations produced by this simulator will be realistic. Regrettably, the realism of a simulator cannot be easily judged due the following main causes.

- The almost total lack of scientific data about pedestrian behaviors (both in the academic and commercial literature).
- The large number of theoretical-only pedestrian mobility models in the academic literature.
- The interest authors of academic and commercial mobility models have in avoiding having to validate their models.

The professionals interested in some kind of simulator will need to check the accuracy of the simulator unless the simulator itself provides a certain kind of results validation.

A complex work of validation of the Distrimobs simulator has been started using the few papers in the literature that use a scientific approach to real pedestrian mobility analysis [14-17]. The early results are encouraging and a full publication about the quality of the realism of the Distrimobs simulator is already planned.

The Life-Cycle of a Process Aided by the Distrimobs DSS

From Ideas to Proposals

The Distrimobs decision support system may interact with several life periods of an environment: from the early design, through the space assignment, the space valorization and so on, to the requalification. The early design of an environment includes the sketches of the environment and their realization through software able to manage 2D or 3D representations. These preliminary operations are led by one or more professional Figures such as engineers, architects or city planners. These sketches are typically analyzed by a team of experts and, after some modifications; the sketch became a candidate project. In this phase Distrimobs become useful for professionals by allowing for quick analysis of pedestrian mobility in the environment. Indeed Distrimobs can load and manage all the standard CAD documents (as described in a following subsection), then, after loading the sketch, it gives the opportunity to randomly generate mobility inside the environment and consequently to view in a few seconds how pedestrians will be able to behave in their sketches.

Agencies usually produce more candidate projects (also named proposals) and for each of them many demonstrative videos are produced. Demonstrative videos show customers, sponsors, holdings and so on the potentialities of the project. Distrimobs is also really useful in this phase. In fact, the Distrimobs decision support system includes a tridimensional viewer of the simulated environment that allows for the exploration of the simulation both spatially and temporally.

Defining the Environment, Persons and their Behaviors

The phase of characterization of the environment, that typically follows the above one, usually suggests to analyze and simulate the most important parameters of the designed environments (related to one or more candidate project). Here the classical CAD software perform analysis and simulation such as finite element analysis, stress analysis on components and assemblies, thermal and fluid flow analysis (computational fluid dynamics), warm/cool analysis, kinematics, mechanical event simulation, simulation of casting, molding, and die press forming...

The Distrimobs DSS is able to simulate, and thus to predict with a tolerance dependent on many parameters, pedestrian mobility inside environments. As other simulators used in computer aided processes, Distrimobs needs data input in order to provide realistic results. This input data mix geometrical (the map of the environment), numerical (the fluxes of pedestrians with times and sizes) and behavioral (the interactions of pedestrians with the environment) information. As other simulators do, the more accurate these data are, the more realistic the simulation performed by Distrimobs is.

The map describes all the geometrical information of the projects. Typically, the document is produced by the professional through the drawing (CAD) software. When the drawings are loaded in the Distrimobs simulator, some simple operations should be undertaken in order to define the geometrical information and then obtain a more realistic simulation. These operations are:

1. define an orthogonal rectangle bordering the map;
2. define which shapes are accessible by pedestrians and which are inaccessible;
3. define the height of the shapes (if the drawing is bidimensional);
4. define the types of accessible and inaccessible shapes (i.e. sidewalk, waiting room, fence, fountain, etcetera).

Note that the Distrimobs suite allows one to perform the last three operations using the layers from the original drawing document. This means that, in the case of a project of a historical center, if the document contains a layer for the sidewalks, a layer for the buildings and so on, the import operation is quick and straightforward.

The fluxes describe how people move around the obstacles. A good description of a flux includes (at least):

1. times (at what time does the flux start/end?);
2. magnitude (how many people does the flux include?);
3. physical characteristics (what is the speed of the flux?);

Note that the Distrimobs suite can describe the fluxes (and their parameters) in different manners: from an intuitive graphical user interface to a code-based approach (see the subsection about the artificial intelligence of the agents). Note also that fluxes are used only to describe the input for the simulator. As explained in a following subsection Distrimobs simulates pedestrians individually.

The behaviors describe the interactions pedestrians have with themselves and with the environment. For example these can be: pedestrians looking at showcases, going to the toilet, waiting for a bus or passing through a particular passage. As showed in a later subsection, there are many ways to describe behaviors of agents. The Distrimobs suite is designed to offer a usable graphical user interface allowing for the description of all the parameters of the motivational model. Distrimobs embraces the motivational model but leaves the way open to many other models, giving the possibility of directly writing the artificial intelligence of agent (even of every single agent if necessary).

Now the Distrimobs suite lets the user save these input data into an xml document (called EDML) that represents the real input for the Distrimobs simulator; it contains all the data useful for a pedestrian mobility simulator: geometries, fluxes and behaviors.

Simulations and Choices

Starting from one or more EDML documents, Distrimobs can perform the simulation(s) and give the results. Distrimobs will give as output special xml documents (RDML) that can be used to view, analyze and explore the simulated environment. So, starting from each proposal, one or more EDML documents is created and many RDML documents are obtained.

The designers, the planners, the people responsible for the projects have, at this point in its life cycle, a powerful tool: a predictive projection of their design. Furthermore, interacting with the graphical user interface of the Distrimobs simulator, professionals can make modifications to the EDML documents in their components.

- Geometry: walls can be removed, doors can be enlarged, bridges can be added and so on. Professionals can try different proposals and solutions and determine the best geometrical configuration. The applications of these kinds of experiments are uncountable. Some examples: testing where to place a security exit, testing if the position for a fountain is right, understanding the best position for an advertising box...
- Fluxes: the magnitude of fluxes can be varied in order to test the suitableness of the designed environment. The graphical user interface also lets the user change the starting and arriving times in addition to the physical characteristics of the agents. These features allow for testing, for example, how the designed environments react to tripled amount of people or to an unexpected flux.
- Behavior: the actions pedestrians perform in the environment can be changed in all aspects. These allow for changing many aspects of the pedestrian mobility in indoor and outdoor environments. Some examples are: the preferred restaurant of people, the preferred stairway, the waiting time for a service, the place where people meet and so on. It's easy to imagine how these parameters are very important for the mobility of the whole system, in particular due to the complex interactions existing between them. For example, consider that the waiting time for a service is strictly related to the length of the queue and consequently, the accessibility of the places near the queue.

Some applications of these experiments have been used in the recent studies we are leading about the pedestrian mobility inside a complex and ancient environment (see the section about the applications of Distrimobs).

Experimenting and testing the different proposals and hypotheses, evaluating the different simulations and deciding the best choice for every project: these are the remaining phases of the life-cycles of projects. Professionals can repeat these more and more times, choosing at every phase the, hopefully, best choice. For example, consider the life-cycle of an exhibition hall. It is expectable that Distrimobs will be used when the building is under design by civil engineers and architects (the more important tests will be related to the geometry of the building). Nevertheless, after its realization, the exhibition hall usually hosts conferences, exhibitions and so on; at this point Distrimobs will be used to indicate the best placement of stands, coffee bars, advertisements, desks of cashiers and so on. In this case the most important tests will be related to the fluxes. Finally, during the planning of an exhibition, Distrimobs is useful for testing particular happenings too. Examples are: what happens if the waiting time at the cashier becomes too long? What happens if the people decide to use the bathroom very frequently? These tests are realizable through the configuration of the behavior of the simulated pedestrians.

Letting the professionals make the best choice: this is the main target of the Distrimobs simulator acting as a decision support system during the whole life-cycle of professionals' projects.

Showing Innovations and Motivating Choices through Simulations

In order to have a more direct interpretation about simulations, professionals may view these documents (RDMLs) through many tools included in the Distrimobs suite. The currently developed tools can be grouped in two main categories. These categories derive from two applications belonging to the Distrimobs suite:

Figure 1. Snapshot of DiVE during the exploration of a simulation.

- DiVE (see Figure 1), a tridimensional explorer able to represent the whole simulated environment (with buildings, walls, roads, etcetera) and the agents, representing the latter as spheres, cylinders or humans. DiVE lets the user move freely in the simulated space and time. The user can consequently observe at every point of the environment in every time, acting, for the time-related behavior, similarly to a movie player. Furthermore, similarly to a movie player, the user can decide to increase or reduce the showing speed. Through DiVE, movie files can be produced and then shared with customers, sponsors, media and so on. These movies are typically the best available brochure of professionals' projects.
- Distrimobs Analyzer (see Figure 2), an analyzer that allows us to aggregate different parts of the simulated environment and creates a particular analysis for each part. These analyses are then represented in different manners. Some examples of the analyses that can be performed by Distrimobs Analyzer are: density (see Figure 3), speed, direction (see Figure 4), attractive points, etcetera.

These tools allow the Distrimobs simulator to be a useful decision support system in the whole life-cycle of a design process. One of the most important uses of the several simulation representations Distrimobs has is the simplifying of the distribution and better divulgation of the projects. Having the support of this kind of software will become more and more important in the near future. When professionals start to use pedestrian simulators a little more widely during their processes, it can be expected that this will affect all other professionals: a better project will be better paid and more-highly regarded.

A Short Description of the Distrimobs Simulator

Main Potentialities

One of the most prominent strong points of the Distrimobs simulator is its underlying dynamic model, which we deem more realistic when compared to other models developed for the same task. Distrimobs runs environments having continuous space and time, where pedestrians have personal encumbrance, speed and sight range. We think it is important to delve in more detail about the characteristics of the model to fully understand such fundamental design choices.

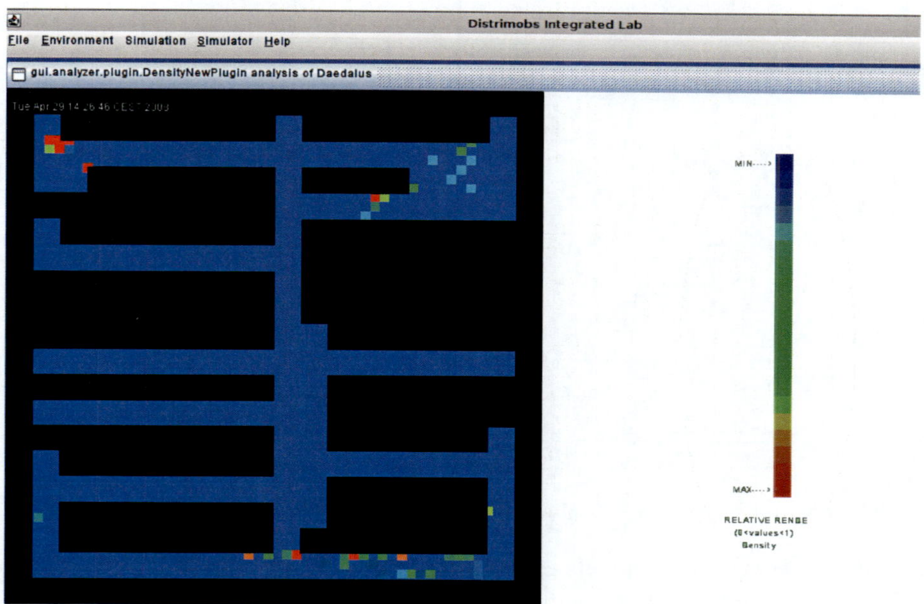

Figure 2. Snapshot of the Distrimobs Analyzer.

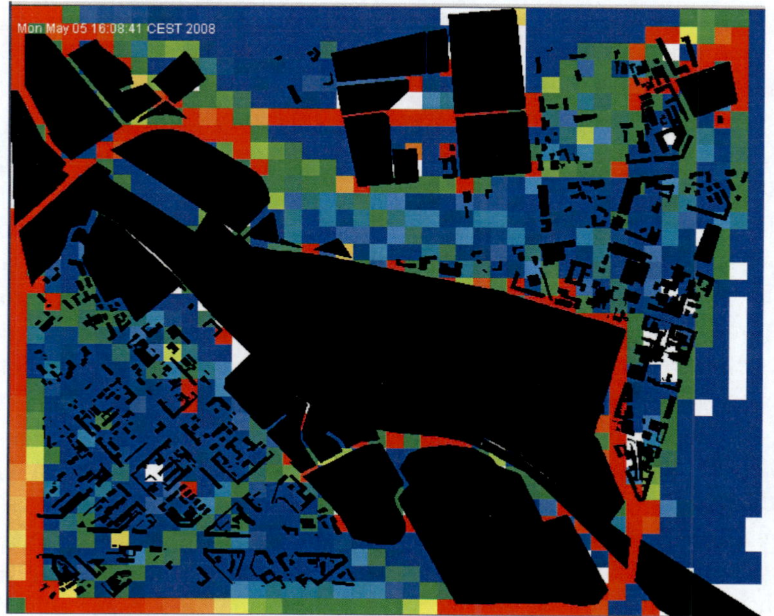

Figure 3. Snapshot of the Distrimobs Analyzer during an analysis of the pedestrian density.

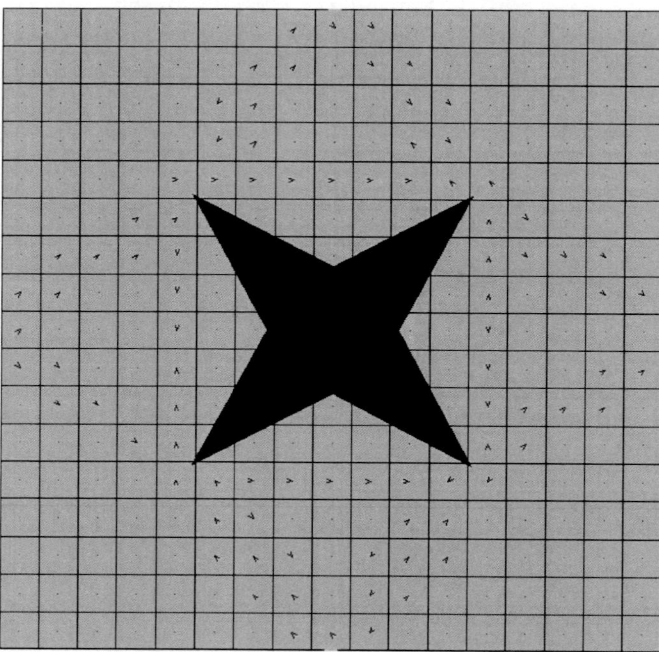

Figure 4. Snapshot of the Distrimobs Analyzer during an analysis of the average direction.

Some pedestrian mobility models use a discretization of the space in the form of a grid of positions (with tiled squares or hexagons) or in the form of a graph of interconnected, situated nodes [8]. This limitation causes pedestrians to "teleport" between a position and another and does not allow positions not included in the list of predefined ones (all intermediate positions for instance). A continuous representation of space, instead, where every reachable position <x, y, z> is possible, permits a more fluid and realistic movement, and an accurate representation of environment elements (buildings…) in their actual shape, without the need to discretize them on a grid or with other simplifications. It can also represent different sight ranges of different persons, and lets them choose, while selecting direction, among a full spectrum of 360°.

As with space continuity, time continuity also is an important element of the Distrimobs model. Probably the most relevant aspect that this feature implies is that collisions between pedestrian pairs and between pedestrians and environment obstacles are taken into account for the entire movement path, and not just at discrete steps. This improves the realistic avoidance of movements not feasible in practice (because they implicate body overlapping for a time interval), or, on the other hand, movements that are too conservative if the model compensates for this through a sampling-based collision detection mechanism with excessively large distances between bodies. It also lets different people walk with different speeds, as in the real world.

In Distrimobs, pedestrians have a space encumbrance: they are represented by moving disks. Other models use point-like agents instead. This choice may be motivated by the necessity to have faster computations, because collision detection and path planning are simpler when space occupancy is not taken into account. Our choice is to accept the extra computational cost and use a parallel and distributed architecture, if needed, to reduce real-time requirements. Simulated pedestrians with encumbrance lets us, among other things, represent people with different sizes; we think this non-homogeneity has a good probability of being crucial to mobility.

The Distrimobs method for parallelization of computational costs, as described in a following subsection, requires the simulation model to be step based. This means that there is a simulation parameter telling which time length a step must be, and all peers do an optimized synchronization at the end of every time step. Artificial intelligence of a single mobber resides on a single peer (though the opposite is not true: usually a single peer will contain AIs of hundreds/thousands of mobbers) and makes decisions at the start of every time step. This desired action is propagated, if needed, to other peers and action results are computed consistently across the parallel architecture.

After the decisions of the mobber AIs and their distribution to every interested peer (only one peer, the same where the decision was taken in most cases) their movement is executed by applying clear rules, stating that a mobber that does not encounter obstacles moves as the AI requires, while there are deterministic rules to handle mobbers that are about to collide.

The rule that mobber AIs make decisions only one time per step is not a big drawback of the parallelization, because with a sufficiently small step duration (we have found a quarter of a second to appear as a good trade off) the steps are unperceivable for an observer.

It is important to underline that having synchronizations and AI computations only once every time step does not mean that the simulation time is not on the continuum. Mobber movements are effectively computed with continuous time, only directional choices are updated with a step based method.

Interactions with Other Software Solutions

The interactions a software has with the real environment are at the base of productivity and good usability of a new kind of software. Hence, as already said, Distrimobs tries to be able to load a large variety of drawings documents. A special tridimensional cad editor ("Distrimobs CAD") belongs to the suite of Distrimobs and through a plug-in based engine allows loading, at the moment, the following documents:

- SHAPE: a format widely adopted by urban and city planners;
- DWG: the most used format for drawing documents. This format is a closed-source format owned by Autodesk. Distrimobs is able to load these kinds of document thanks to the Open Alliance efforts.
- DXF: the standard and open format for drawing documents.

Finally, the plug-in based engine allows importing into the Distrimobs CAD all the kinds of geometry based documents (and some of them are already planned, such as GML).

More interactions between Distrimobs and other software exist in the opposite direction, the exiting one. Distrimobs produces an xml document (RDML) as output of the simulator. Then this file is elaborated by DiVE and Distrimobs Analyzer and many kinds of documents are produced. Some of them are movies, images and spreadsheets.

Environment Representation

Environment is the main input of a simulation. It may be subdivided into two main parts: description of a map and its features, and description of mobbers and their behaviors.

A map may be subdivided again in two aspects: the geometrical properties and the motivational properties.

The geometrical properties of the map are imported into the Distrimobs format for environment representation starting from a CAD file (more specifically from a Shapefile, DWG or DXF). This import is done automatically by a Distrimobs tool, requiring just some seconds of processing for a map of a city district. It may cause the user some additional work if the original CAD is not well-formed enough to be directly imported, for example if building polygons are not closed or are overlapping. In this case, the process becomes semi-automated with the user directing some helper tools that try to solve various problems (such as closing open polygons while preserving the polygon's original shape as much as possible). The Distrimobs tool realized to import and manage maps allows it to import tridimensional environments too.

The possibility of using previously existing CAD is a very important feature because, taking into account the spread of CAD applications between designers and city planners, it is normal to have an electronic representation of the geometry ready to use.

In Distrimobs, geometrical properties of the map include a rectangular map border and closed, non overlapping simple polygons (simple polygons are polygons without self intersections) like movement obstacles (buildings, fences, canals…) and walkable areas (roads, building insides).

Motivational properties of the environment are bound to the Distrimobs motivational model, and need to be filled only if the motivational layer of the artificial intelligence is used. If the motivational layer is not in use, other ways of representing non-geometric features of the environment are needed, such as directly inserting activities into the agenda.

Motivational properties are expressed for every obstacle and every walkable area. They express what motivations they may satisfy in mobbers. They are in the form of pairs <name, value>, where the name is the name of the motivation, and the value is the intensity by which the motivation is satisfied by the environment feature. Values are (theoretically) real numbers (floating points in practice). A zero value means that the environment feature is not relevant regarding that motivation. A positive value means that mobbers interested in that motivation may be satisfied by reaching that place. A negative value means that sensitive mobbers prefer to stay away.

There is no limit to the number of pairs that a feature may have. This number depends on the degree of detail the user chooses to model. Another important aspect is that motivational properties may change as simulated time advances.

For instance, a restaurant may have a couple <eat, 50> during opening hours, and a couple <eat, 0> while closed. A smog-producing factory may have a couple <nature, -100> to represent dislike by pedestrians who are interested in natural landscapes.

An all-around description of the proposed motivational model, its strong points and its drawbacks is outside the scope of this paper. It has been chosen as the main way to express high-level reasoning in the Distrimobs simulator because it allows for a map of all the main human activities and is relatively easy to understand and fast to configure, dramatically reducing the time needed for the user to configure the simulations, for instance with respect to setting individual paths, because it leaves a lot of decisions in the hands of the artificial intelligences.

Configuring mobbers includes defining the types of mobbers present in the simulation, with their entering fluxes, body characteristics and "psychological" motivations.

Mobbers are partitioned in types; this may appear as an undesirable limitation, but it is motivated by the necessity for the user to characterize a good percentage of the population at a time, when defining their bodies and motivations. Characterizing every single mobber is not prohibited, it is sufficient to define a different type for every mobber, but is very tedious and unpractical for big environments, including, for instance, hundreds of thousands of mobbers.

Mobbers enter the simulation through specified areas, called generative areas. These may represent roads starting from map borders, building exits (if the building inside is not simulated too), public transport arrival places, car parking lots and so on.

For every generative area it is necessary to specify which kind of pedestrian arrives and with which frequency distribution.

The simulator takes care, in a pseudorandom and reproducible way, of generating mobbers according to the distributions specified. Every mobber is an instance of a mobber type, but for each of them some parameters are randomized, again with a distribution specified by the user, to represent personal physical and behavioral characteristics. Behavioral differences are produced by differences in motivations, so that mobbers of the same type may make different choices.

Body parameters include size, maximum speed and sight range. Behavioral configuration is more complex and includes an initial motivational state, as a set of couples <motivation, value>, and a dynamic that specifies motivation evolution based on previous motivational states, passed time and some selected environment observables, as perceived by the mobber. For example if at time 1 a mobber has a motivation <coffee, 100> and at time 2 he has drunk his coffee (environment observable): at time 2 its motivation may become <coffee, 0>.

All simulation configurations described in this subsection (map geometry, map parts motivations, mobber types and generative zones), are memorized in an XML document called EDML, which stands for Environment Distrimobs Markup Language [9-10]. The choice to use an XML document type is justified by the fact that XML is simple to read by a plethora of software tools and APIs, and that by defining it through an XML-Schema, the EDML standardization is formal and clear. Information included in EDML is sufficiently general to be useful also for other mobility tools, it being a map representation enriched by metadata about geometrical entities meaning and pedestrian population.

Architectural Elements

In this subsection, Distrimobs software's architectural elements are presented. The description follows a computer science approach, but we have also tried to approach most of the arguments in a language familiar to other audiences.

Distrimobs' program code is partitioned in various modules that collectively form its software architecture. Each of these modules has a responsibility to handle some of its functional aspects or to give services for other modules. The coarse-grained subdivision presented here has the following main actors: computational geometry/dynamics, graph abstraction, parallelism/distribution, simulation model and artificial intelligence. We will briefly describe their responsibilities and their interactions.

The geometrical nature of Distrimobs environment requires a software layer capable of executing needed operations in a fast and reliable way. One of the most time-consuming tasks is the collision detection and avoidance. In Distrimobs, mobbers overlap is forbidden, predicted and avoided in advance. Collision detection and handling is a complex field with a vast literature (see [18-19-20]) and requires dynamic computations that, for good simulation results, must be accurate and, when the simulator has lots of moving entities like Distrimobs, fast.

As known, when doing floating point computations there is always an unavoidable precision loss, but a careful design of the algorithms may give results with orders of magnitude that are more accurate than a naïve one. More precision often means more CPU time, and a good trade off must be researched.

Another important aspect to improve performance is choosing fast data structures for most-used spatial queries. We will not delve into this chapter in the technical details of the data structures chosen to contain information about mobber and obstacle positions; for some inspiring literature see [21-22].

The geometrical/dynamical module is an API without dependencies from any other of the presented modules, and is used by the simulation model to update the environment state, and by mobber AIs to analyze perceptions and make decisions involving physical dynamics.

For high-level planning regarding the map, a graph abstraction is used to improve performance. To use every detail of the geometry even when reasoning about high level planning would be very computationally intensive, and is not even what real people do. Using a graph improves performance by orders of magnitude and, with the choice of a good abstraction, has little slack from decisions optimality.

Mobbers' AIs are featured with memory, containing a selection of what they have previously perceived, computed and chosen. It is constantly accessed to drive new decisions. This memory requirement is an important bottleneck, because, with many mobbers, the simulator may easily fill the primary memory of a host, and, as it is known, working with secondary memory would greatly reduce performance.

The chosen solution is a parallelization and distribution method where mobbers are spread across more peers to have access to their combined primary memory. Mobbers may migrate between peers and there is a need to synchronize their movements in a consistent global state too. These and other communication needs are taken care of by the parallelization and distribution module, better explained in the proper subsection. This module communicates with the model module, its responsibility being to mask communications between peers and let single models export their local state and synchronize on a common, consistent global state.

Environment state is managed by a module called simulation model. Every peer has a model. It communicates with the parallelization and distribution module to share information with other models residing on other peers, in order to maintain a common global state. It gives the AIs their perceptions, based on mobber positions and sight ranges, and asks them for actions. Using mobbers' states and desired actions, the model advances simulation time, handling environment rules, such as non-overlapping of bodies.

The model module interacts with the parallelization/distribution module to synchronize its state with models of other peers, and with mobber AIs to give them perceptions and ask their desired actions. It also uses the geometrical/dynamical module to advance environment state with respect to the physical rules of the Distrimobs simulator.

Distrimobs' artificial intelligence follows the agent paradigm. As such it receives perceptions from the outside (the model passes them) and answers outside requests to choose the next action, if any. This request comes from the model too.

Agent architecture means that artificial intelligence is well encapsulated; every mobber has its personal one and communicates with the outside only through the model.

Artificial Intelligence

Every mobber is featured with an individual artificial intelligence, based on the agent paradigm. This means that mobbers do not communicate between themselves directly through their AIs, but only indirectly through the environment. They receive perceptions from the environment and may take actions that affect it.

Perceptions are in the form of sight information. This includes relevant information about visible mobbers (size, speed...) and static features (roads, buildings...). Actions are mostly movements: a mobber specifies the desired speed and direction. Perceptions received and selected actions are continuously updated during the lifetime of a mobber.

In a simulation, artificial intelligences may be partitioned by the kind of mobber (every environment may include particular mobber kinds, such as tourist, employee, student...), and, within specific kinds, every single agent may have different parameters to map different simulated psychologies. Therefore, by having a computational model able to support a personal AI for every agent, it is possible to represent the wide range of different personal behaviors found in the real world.

The artificial intelligence for a kind of mobbers may be specified mainly in three different ways.

- Programming artificial intelligence code directly. This is the most configurable but also the most time-consuming alternative. Usually it is not necessary to exploit such complete control, but if it is the case the programmer may use a powerful API so that it is not necessary to reinvent the wheel (such as manually place the points of the paths). This API includes different representations of the map, including a graph abstraction of it, a fast, highly customizable pathfinder, a modular framework for composing simple behaviors to react to the viewable environment state (avoiding other pedestrians, etc.) and a geometric/dynamics library. By using this direct programming alternative the coder may use the Distrimobs simulator with pedestrians featured with any kind of decision mechanism, such as mechanisms proposed by other authors as seen before in this chapter, or with a modeling of a proposed sociological/psychological theory about pedestrian choices.
- Filling agendas. Distrimobs features a point and click interface to program mobber agendas (see Figure 5). Every type of mobber may be programmed with a different agenda, and by giving parameter distributions and randomizing the agendas every mobber will behave differently, even from other mobbers of the same kind. At agenda level, as next explained, the user chooses activities that a kind of mobber wants to do,

and organizes them in a structure, the agenda, which takes into account temporal and logical relationships between them, such as priorities and time constraints. This method is the fastest and simplest of the three when the simulated environment the user wants to model contains only a small number of different kinds of pedestrians, and when the activities they must do are not too complex. The process is simple and does not require information technology skills.

- Using motivations. This is the most complete built-in method that Distrimobs features to produce pedestrian behaviors. The user specifies the behavior of a source of motivations for every kind of mobber. The source of motivations creates initial motivations and handles their dynamics, their evolving in time. Starting from motivations the AI creates and updates activities on the fly, thus placing the motivational level higher than the agenda's one. Using motivations is the most complete and realistic way, excluding direct coding, that a user may exploit to define mobber AIs. Using motivations does not require programming skills, but needs good data (or guessing) about pedestrians' real motivations. The work is repaid by greater realism, especially in the case of unsystematic mobility, and by the possibility to let the AIs automatically manage very complex agendas and therefore a high spectrum of different behaviors.

A rtificial intelligence architecture is organized in layers. Lower layers handle simple reactions to simple stimuli, and are used to make decisions with a short time-span (avoiding a collision with a mobber or a wall, pass near streetlights…). Upper layers are, on the opposite, devoted to making decisions using more deep knowledge and influencing further in the future, such as choosing long-term goals.

As in other areas of computer science, the paradigm of subdividing a complex problem into smaller levels, where each level interacts only with the immediate higher and lower ones, has also spread widely in artificial intelligence models. If correctly applied, this may improve understandability, encapsulation, extendibility, maintenance and reusability of the framework and of the single parts. For examples of layered architectures in artificial intelligence, see [23].

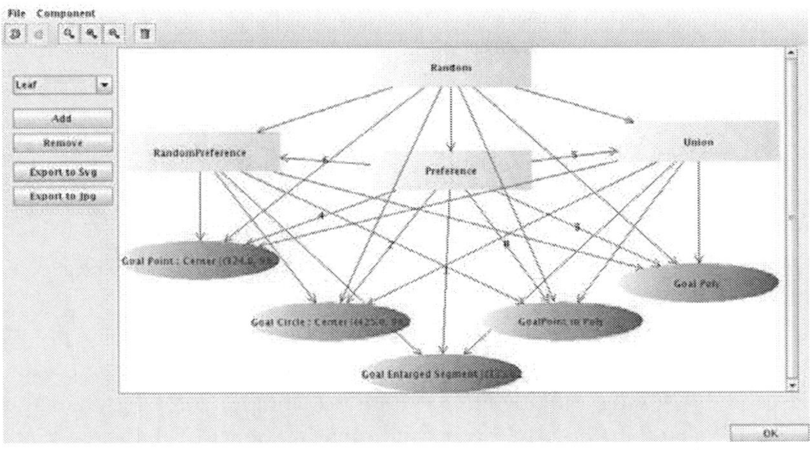

Figure 5. Snapshot of the configuration tool for the agenda of the agents.

The higher level is the motivations source, which handles motivations dynamics, taking into account mobber characteristics, previous motivational states and perceived environment states.

Motivations are used by the activity generation layer to produce activities that the mobber needs to execute to satisfy its motivations. This maps abstract objectives (eat) into concrete ones (go to restaurant X), taking into account services offered by the environment.

The agenda level has activities inserted from the level above and acts as a scheduler for these activities, selecting the next one to be executed, taking into account their priorities, spatial and temporal constraints, causal constraints between activities and contingent state of the environment. We may see the highest three levels of Distrimobs' AI as a means of generating, organizing and selecting goals, so Distrimobs AI architecture may be seen as a goal-oriented one.

When an activity requires a movement in the space (as is usual in a pedestrian mobility simulation), the global pathfinder layer is used to produce a path on a precompiled graph. This graph, which is an improvement on the visibility graph that takes into account mobile agents size, which we call the disk-visibility graph, is used at this level to abstract the geometric characteristics of the space, simplifying the possibly long task of producing long-range planning of the route to take. The result of the plan is a sequence of situated nodes.

The lowest of these levels is the local steering one, where local planning is done by using a composition of steering behaviors (see for example [24-25-26]). This layer uses view perceptions and directions from the global path to avoid mobbers and other obstacles while going for the next node and trying to satisfy motivations influencing short-range decisions, such as liking or disliking having mobbers at a short distance or of looking at buildings of touristic interest. This level computes fast and relatively simple decisions that are reevaluated continuously to guarantee responsiveness to stimuli.

Parallelization of the Computations

A very important feature of a simulator that aims to reproduce pedestrian mobility inside real environments (such as railway stations, malls, theme parks, historical centers of cities and so on), is the ability of simulating hundreds of thousands of simultaneous pedestrians. For example, it is supposed that during the carnival of Venice a maximum of about three hundred thousand pedestrians simultaneously populate the historical center and during the three days of the carnival a total of one million people participate in it [27]. In order to achieve such a large number of pedestrians we retain the parallelization of the models as the main, and probably the only feasible solution.

Parallelization of simulation is a well-known field of computer science and it has a vast literature (see as example [28-29-30]). There are two main approaches to parallelism (and or distribution) of simulators: the optimistic one and the pessimistic (or conservative) one. Distrimobs embraces the pessimistic one using a time-step based synchronization. A full description of the motivations behind this choice are outside the scope of this chapter: the technical questions aren't so important to understand how Distrimobs simulator is a valid decision support system. The Distrimobs simulator, keeping continuous time and space, synchronizes at every time the processing units (processors or calculators).

Artificial intelligence typically requires a lot of computations; furthermore it is featured with memory allowing mobbers to store data in a personal, persistent and potentially unbounded object. This capability allows mobbers to store data, deduce and finally learn. Please note that without memory, mobbers with the same state of the AI (including motivations, pseudo-random number generator values, etcetera), if placed in the same position many times with equal perceptions, behave exactly in the same manner. Similar pedestrians are something like silly men that continue to take the same congested route.

Mobbers use lots of resources of both primary memory (IO bound) and computations (CPU bound). Being featured with an unlimited individual memory, using it for storing information and computing decisions, mobbers require a lot of space in primary memory. Furthermore each mobber has a powerful layered artificial intelligence that uses perceptions and memory in order to compute decisions. These AIs require a lot of computational time. Memory and computations requirements are highly related to the number of mobbers existing in a simulation: the more the simulated mobbers, the more the requirements (CPU and memory). Distrimobs handles these requirements through parallelization (and distribution), subdividing the simulated map in many zones and assigning each processing unit with some of these zones. Each zone contains mobbers and these are managed by the processing unit. The Distrimobs parallelization system includes a load-balancing subsystem.

A short analysis of the performances of the Distrimobs parallelization system has been presented at the Italian Conference of Science Computation [31]. The results of this analysis showed that the parallelization system appears to be able to manage the entire Carnival of Venice.

Application of Distrimobs to Real Environments

The Case of Palazzo Poggi

Palazzo Poggi is the headquarters of the Alma Mater Studiorum – University of Bologna and contains secretariats, historical rooms used by the council of the senate and other university governing bodies, several rooms belonging to Museum of Palazzo Poggi, the historical library of the University and finally, the offices of the whole administration. The palace is located in the historical centre of Bologna and has a bounding rectangle 120 m wide and 110 m tall. The application of the Distrimobs simulator to Palazzo Poggi (see Figure 6), has been recently published by a scientific journal [32]. The full description of this application is outside the scope of this paragraph, but some key points of this application are duly reported:

- Distrimobs has been used to test the capacity of the secretariats to handle fluxes with a magnitude of about 200%, 250%, 350% of the average.
- Distrimobs was used to evaluate optimization of the schedule of visitors to the museum, allowing it to bring the number of simultaneous visiting groups to 200% of the average.
- Distrimobs has been used to understand the real usefulness of the opening of a new door near the hub of pedestrian mobility in Palazzo Poggi.

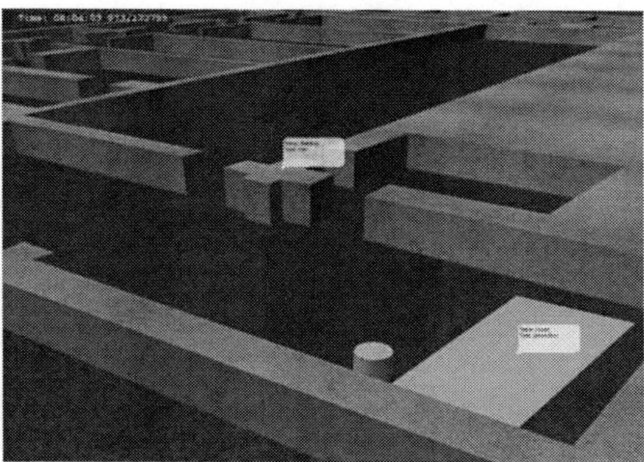

Figure 6. Snapshot of DiVE during the exploration of a Palazzo Poggi simulation. In particular, the entrance of a mobber is showed.

This application is the first, to the best of our knowledge, real application of a pedestrian mobility simulator to a real complex environment.

The Historical Center of Bologna

In recent years, the pollution and the lack of suitable parking sites caused by the pervasive utilization of private cars, brought the administration of the municipality of Bologna to start a process of pedestrianization. This process includes the part of the historical center named "Cittadella Universitaria". Situated on the east-side of the historical center, the Cittadella contains almost all the departments and faculties of the University, a lot of pubs, restaurants and so on. In other words this part of the city gravitates around the students (shortly: studying, eating, entertaining, sleeping) and they usually don't use private cars in order to move.

The pedestrianization process started during 2008; it started with a small portion of the Cittadella (see Figure 7). The only vehicles allowed to pass through the pedestrian-only roads are: emergency vehicles (police, firefighter, ambulance, etcetera), public transport vehicles (bus, tram, etcetera), electric-powered and residential vehicles. It's easy to imagine how this process brought a lot of supporters but also a lot of detractors. The positive consequences are uncountable but some negative consequences exist and they are all related to economical interests. Some of the raised questions follow:

- Will my shop continue to sell its products or will people go outside the Cittadella in order to buy items with the comfort of a car?
- Will the value of my flat grow or decline after this process?
- Will students rent my flat?

Understanding whether pedestrianization is a good administrative choice is beyond the purpose of this paragraph - we're talking about simulation but, obviously, a simulator cannot make this choice in place of a city planner. Despite this, a simulator can aid the planner in making this decision, and the study of the Cittadella through Distrimobs is under development with this aim (see Figure 8). The usefulness of this simulation is to give an evaluation tool for citizens, administrators, tourists, students and so on that is able to answer, for example, the following questions:

- Where are the most useful places for placing bus stops?
- Which are the most frequented sites?
- In front of which showcase will the great fluxes of pedestrians pass?
- Which sites will be most isolated and unfrequented?

This study is under development and we hope that this study will be a good example of the usefulness of pedestrian mobility simulation.

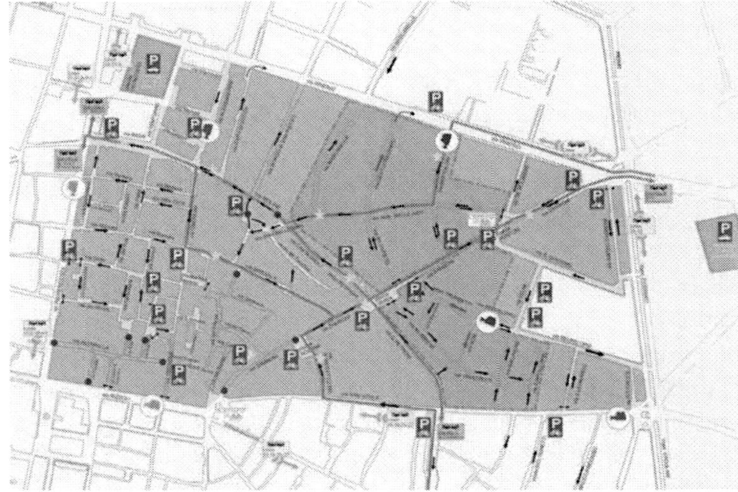

Figure 7. The process of pedestrianization of the "Cittadella Universitaria". The orange zone represents the first step, the grey one represents the remaining ones.

Figure 8. Snapshot of DiVE exploring the simulation of the pedestrianization of the Cittadella.

Ponte Della Costituzione

A study about the building of the Ponte della Costituzione in Venice was done, before its actual opening (September 2008), as a way to test the Distrimobs simulator with a real case of an infrastructure that existed only on paper.

This application was not done with the intent of thoroughly analyzing all aspects of observed and forecasted pedestrian mobility, but just to know if it is possible to do it. Therefore, the investigation was stopped at a relatively shallow level.

This study required the configuration of two environments with a small but significant difference: the bridge itself. One environment modeled the area without the upcoming bridge, while the other included it. The interest was in analyzing forecasted differences between the two scenarios.

The simulated area includes a part of Sestiere Cannaregio surrounding the Santa Lucia station, Ponte Scalzi, part of Sestiere Santa Croce and Sestiere San Polo (see Figures 9 and 10). Overall, more or less a sixth of the historical center of Venice is included, on an area of 1000 x 1500 meters. Having this extension, these simulations are also useful as an intermediate step for the simulation of all of Venice, which is one of our targets.

The simulations showed some important differences caused by the presence of the new bridge. The main ones are probably that pedestrians coming from Sestriere Cannaregio have a more direct route going to Sestriere di Dorsoduro and the west part of Sestriere Santa Croce (where you also find the only parking lot of Venice), crossing Ponte della Costituzione instead of Ponte Scalzi. On the opposite side, pedestrians coming from the west part of Sestriere Santa Croce have a more direct route to Sestriere Cannaregio. Therefore, with Ponte della Costituzione there is less pedestrian traffic on Ponte Scalzi, Fondamenta San Simeon Piccolo and Fondamenta de la Crosa (pedestrian roads south of Canal Grande, from Ponte Scalzi going westward). These differences in traffic mean an easier and faster transit for walking people, better safety in crowded situations (such as during the carnival) but also less work for commercial activities in Fondamenta San Simeon Piccolo and Fondamenta de la Crosa due to a smaller number of pedestrians passing there.

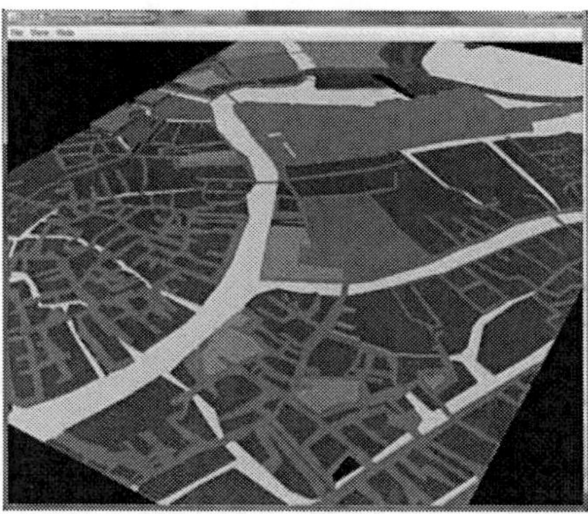

Figure 9. Snapshot of DiVE exploring the simulation of Ponte della Costituzione (planar view).

The simulations of Ponte della Costituzione showed that the placement of new infrastructures influencing pedestrian mobility in big urban areas may be studied with pedestrian mobility simulators.

Figure 10. Snapshot of DiVE exploring the simulation of Ponte della Costituzione (particular on the bridge).

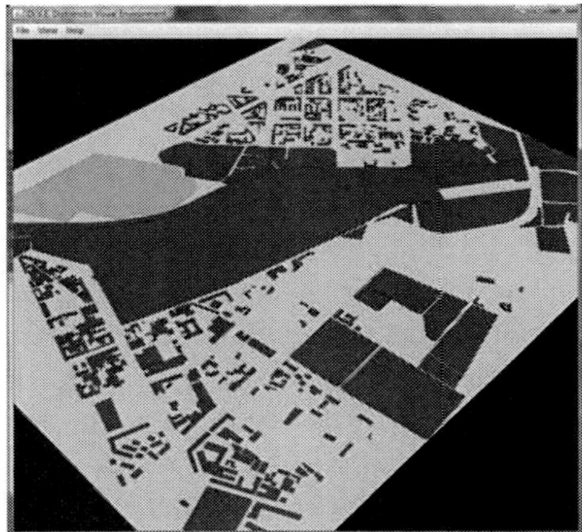

Figure 11. Snapshot of DiVE exploring the simulation of Scalo Farini (planar view).

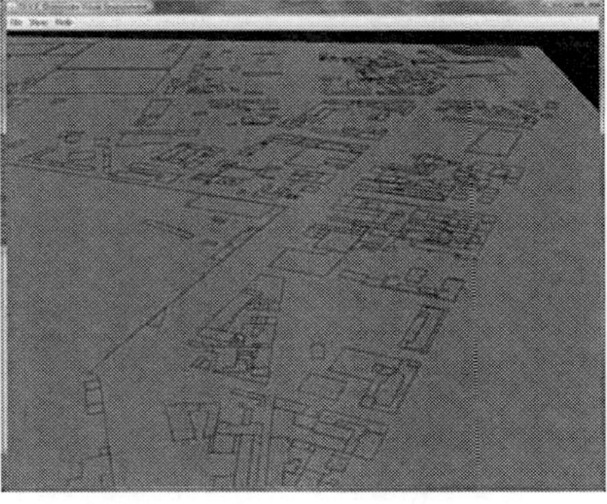

Figure 12. Snapshot of DiVE exploring the simulation of Scalo Farini (volumetric view).

Scalo Farini District

In 2008, the Physics of the City Laboratory started an ongoing collaboration with the Laboratorio di Città di Piacenza of the Polytechnic of Milano. The group from Piacenza is composed of urban planners and architects and works on a project about the requalification of the Scalo Farini district in Milano, spanning about four square kilometers.

This project is a meaningful concrete example of collaboration between urban designers and computer scientists to enrich the design process with new technologies.

The Scalo Farini district is occupied by an old railway station that isn't used anymore (see Figures 11 and 12). The requalification project is about filling the area with a park containing cultural facilities like theaters and auditoriums, with mobility being almost exclusively pedestrian in nature.

While urban planners work on the project, the Physics of the City Laboratory follows its progress by setting up simulations for the proposed environment, adding detail as it becomes available. We think this step-by-step attention may be a relevant addition to the design process, creating a supplementary source of information to the ones planners already have at hand.

The dimension of the environment, four million square meters, is the biggest one studied with the Distrimobs simulator (compared to the one million and half square meters of the Ponte della Costituzione simulations) and is another step toward the scale of the entire city of Venice.

CONCLUSION

Software tools are widely used in urban and building design; we have presented a brief historical excursus about them, reducing the scope first to simulators for various phenomena (light, heat...) then to pedestrian mobility simulators only.

Pedestrian mobility simulation is a relatively new research field that is starting to have real applications. This chapter introduces fields of application already in existence or that we suggest may exist in the near future.

Designing for pedestrian mobility may be seen as a problem-solving activity, with its own recurring issues. How the most common problems are handled by a simulation system may be used as a benchmark for the overall quality of the system's functionality (but usability and integration are important factors too). This consideration leads us to regard as a fundamental step in mobility models research the comparison with rigorously-acquired experimental data about movements of real people. We have seen, unfortunately, that only a scarce quantity of such observations exists.

For simulation techniques to be effective in practice, not only must they provide useful hindsight, but also have acceptable time and resources requirements. We have seen what a typical environment simulations life cycle is like, and, through real case studies, how it interacts with the other, more canonical, aspects of design/decision-making. In particular, we have outlined how simulation output may be analyzed through various tools to meet different needs.

We have included a survey about the Distrimobs simulator, describing its major features, its software architecture and its interactions with other tools. Particularly the availability of tools importing from very diverse CAD format, such as DXF, DWG and Shapefile, permits a relevant speed up in the simulation configuration process. The environment representation, in the novel open file format EDML, a type of XML, simplifies integration with other existing GIS software (see above). Other strong points reside in a modern, agent-based artificial intelligence, ready for a motivational model of unsystematic mobility in continuous space and time, and in a parallelized and distributed architecture for massive simulations.

Considering the fact that pedestrians walk, while moving, between potentially very distant zones, we recognize the importance of simulating a large environment all at once. In the cases of Distrimobs application presented, we have shown a trend in increasing simulated area, going from the single building, like Palazzo Poggi, to city districts and, in the near future, we hope, to a whole city like Venice.

Seeing how communication between different actors of the design process is a very important issue, having various tools to help in the presentation and analysis of simulation results is equally critical. In our opinion, a complete suite must include the possibility of visually exploring the simulation with a 3D tool, and an analyzer that presents more aggregate visual and numerical results.

Due to general economic trends, pollution, and the exhaustion of oil deposits, it may be argued that private transportation will be used less in the future, while more often persons will make use of public transportation. Considering this, in most cases public transportation takes a passenger to a place more distant than where the nearest parking lot would be. Therefore, this shift to public transportation would increase the pedestrian population both in outdoor environments and in indoor public transportation hubs. If this guess is correct, we expect a proportionally increasing interest in pedestrian mobility.

Pedestrian mobility simulators are receiving growing interest from scientific and economic worlds; in response, groups studying and developing simulation software and models are growing too. There are numerous proposed solutions, adopting different techniques but also different data formats. We think that there will be a trend to unify elements upcoming from different working groups as soon as there is large agreement on their efficacy, and this fact will be beneficial in causing an acceleration in the acquisition of new knowledge and in the improvement of simulators' predictive capabilities.

ACKNOWLEDGMENTS

The authors gratefully acknowledge the senior researchers of the Physics of the City Laboratory (A. Bazzani, B. Giorgini and S. Rambaldi) for the possibility to carry out research in this field. The authors also thank the following senior year students for their contributions: A. Bizzarri, R. Bratta, A. Semeraro, M. Nucci, A. Mommo, K. Vukatana, L. Bigliardi, F. Giunchedi, A. Pieropan and E. Agolli.

REFERENCES

[1] Hong, T; Chou, SK; Bong, TY. Building simulation: an overview of developments and information sources. *Building and Environment*, 2000, 35(4), 347–61.

[2] Hensen, JLM; Clarke, JA. Building systems and indoor environment: simulation for design decision support system. In: Proceedings of the international conference on design and decision support systems in architecture and urban planning, 2000. *Architecture*, 177–189.

[3] Keith Still, G. Crowd dynamics. PhD thesis, *University of Warwick*, 2000.

[4] Hoogendoorn, S; Bovy, PHL. A gas-kinetic model for simulating pedestrian flows. *Transportation Research Board*, 2000, 1710, 28–36.

[5] Helbing, D; Molnar, P. Social force model for pedestrian dynamics. *Physical Review E*, 1995, 51(5), 4282–6.

[6] Bazzani, A; Giorgini, B; Servizi, G; Turchetti, G. A chronotopic model of mobility in urban spaces. *Physica A*, 2003, 325(3–4), 517–30.

[7] Batty, M. Cities and complexity understanding cities with cellular automata, agent-based models, and fractals. Cambridge, MA: MIT Press, 2005.

[8] Bandini, S; Manzoni, S; Vizzari, G. Situated cellular agents: a model to simulate crowding dynamics. *IEICE Transactions on Information and Systems*, 2004, E87-D(3), 669–76.

[9] Brambilla, M. Distrimobs, simulatore di mobilità pedonale: studio e progettazione dell'architettura distribuita ed adattiva su cluster di blades. *Master thesis*, Alma Mater Studiorum University of Bologna, 2005.

[10] Cattelani, L. Distrimobs, simulatore di mobilità pedonale: studio e progettazione dell'intelligenza artificiale degli agenti. *Master thesis*, Alma Mater Studiorum University of Bologna, 2005.

[11] Klugl, F; Rindsfuser, G; Large–scale agent–based pedestrian simulation. *Multiagent System Technologies, pages*, 145–156, 2007.

[12] Simwalk Team, Savannah Simulations, AG. http://www.simwalk.com

[13] Legion International Limited, http://www.legion.com

[14] Kretz, T; Grunebohm, A; Kaufman, M; Mazur, F; Schreckenberg, M; Experimental study of pedestrian counterflow in a corridor. Physik von Transport und Verkehr, *Universitat DuisburgEssen*, 2006.

[15] Kretz, T; Grunebohm, A; Schreckenberg, M; Experimental study of pedestrian flow through a bottleneck. Physik von Transport und Verkehr, *Universitat DuisburgEssen*, 2006.

[16] Seyfried, A; Rupprecht, T; Passon, O; Steffen, B; Klingsch, W; Boltes, M; New insights into pedestrian flow through bottlenecks. Central Institute for Applied Mathematics, *Research Centre Julich*, 2007.

[17] Ando, K; Oto, H; Aoki, T; Forcasting the flow of people: a simulation system for the pedestrian movement, 1988.

[18] Fauerby, K; Improved collision detection and response, www.peroxide.dk, 25/07/2003.

[19] Ranta-Eskola, S; Physics in BSP-Trees: Collision Detection, DevMaster.net, 27/09/2003.

[20] Nettle, P; General collision detection for games using ellipsoids, GameDev.net, 19/10/2005.

[21] Samet, H. The Design and Analysis of Spatial Data Structures, Reading, Addison-Wesley, 1990.

[22] Hurtado, F; Noy, M. "Triangulations, visibility graph and reflex vertices of a simple polygon", Computational Geometry: *Theory and Applications*, (6), 1996, 355-369.

[23] Goldberg, D; Cicirello, V; Bernardine Dias M; Simmons R; Smith S; Smith T; Stentz A. A Distributed Layered Architecture for Mobile Robot Coordination: *Application to Space Exploration*. In Proc. 3rd International NASA Workshop on Planning and Scheduling for Space, 2002.

[24] Renoylds, CW. Steering behaviors for autonomous agents, *Game Developers Conference*, 1999.

[25] Tomlinson, SL; The long and short of steering in computer games, *International Journal of Simulation*, 1-2(5).

[26] Fuertey, F; Simulating the collision avoidance behavior for pedestrian, *Master Thesis*, University of Tokio, 2000.

[27] Individuals and groups in movement: sociological tools and new technologies for the study of mobility, touristic events and urban transformations, *PRIN,* 2005-142981, http://www.ricercaitaliana.it

[28] Fujimoto, RM. "Parallel and Distributed Simulation Systems" *in: Winter Simulation Conference*, 2001.

[29] Madisetti, VK; Walrand, JC; Messerschmitt, DG. "Asynchronous Algorithms for the Para Simulation of Event-Driven Dynamical Systems Ie[", *ACM Transactions on Modeling and Computer Simulation*, 1(8), 1991, 244-274.

[30] Marilleau, N; Lang, C; Chatonnay, P; Philippe, L. "An Agent Based Meta-Model For Urban Mobility Modeling" in: *Proceedings of the First Internation Conference on Distributed Frameworks for Multimedia Applications*, 2005.

[31] Brambilla, M; Cattelani, L. The Distrimobs approach for parallelization of pedestrian mobility computations, *Il Nuovo Cimento* (*The European Physical Journal*), to be published.

[32] Brambilla, M; Cattelani, L. Mobility analysis inside buildings using Distrimobs simulator: *A case study, Building and Environment*, Volume 44, Issue 3, March, 2009, Pages 595-604.

In: Buildings and the Environment
Editors: Jonas Nemecek and Patrik Schulz
ISBN: 978-1-60876-128-9

Chapter 4

ADSORPTION AND ADVANCED OXIDATION PROCESS FOR ENVIRONMENTAL CLEANING OF POLLUTED WATER

Yuelian Peng[a] and Shaobin Wang[b*]

[a]College of Environmental and Energy Engineering, Beijing University of Technology, Beijing 100022, P.R. China

[b]Department of Chemical Engineering, Curtin University of Technology, GPO Box U1987, Perth WA 6845, Australia,

ABSTRACT

Soil, water, and air pollution is an important issue for human society. In recent years, water pollution from industrial and household sources is becoming more and more serious and the world is facing clean water crisis. Wastewater treatment and recycling will provide a solution. For water treatment, adsorption and advanced oxidation process are the two important physical and chemical technologies. In this paper, we will report an investigation of using these two methods in treatment of dye containing wastewater.

In adsorption, zeolites synthesized from fly ash (FA) have been tested for methylene blue adsorption. One-step hydrothermal and two-step fusion and hydrothermal syntheses were investigated for zeolite synthesis. The effect of aging process during the two-step synthesis was studied. It was found that zeolite P would be formed during the hydrothermal reaction; however, the conversion efficiency depended on the preparation method. One-step hydrothermal treatment could not convert fly ash to zeolite efficiently while the two-step fusion and hydrothermal treatment with aging process would achieve a complete conversion of fly ash to zeolite. Those synthesized zeolites show higher dye adsorption capacity than fly ash itself and the fusion-hydrothermal synthesized zeolite exhibits the highest adsorption capacity. Fly ash will exhibit adsorption capacity of 5 x 10^{-6} mol/g, while the synthesized zeolite (FA-3) will show adsorption capacity of 9 x 10^{-5} mol/g.

For advanced oxidation process, a nanosized magnetite (Fe_3O_4) has been employed as a heterogeneous Fenton catalyst for decolorization of methylene blue in aqueous

---

* Corresponding author: Email: Shaobin.wang@curtin.edu.au

solution. It is found that the nanosized magnetite is quite effective for dye decolorization at low catalyst loading and H_2O_2 concentration. A 100% decolorization of methylene blue can be achieved within 30 min at the conditions of 2 g/L magnetite and 300 μM H_2O_2. The catalyst exhibits remarkable stability of performance and low metal loss. The maximum concentration of Fe in solution after 3 h oxidation is less than 1.9 mg/L. Life cycle tests also indicate that no decrease in decolorization efficiency was observed after 6 round tests.

INTRODUCTION

Colored wastewater from various industrial processes not only affects the aesthetic merit and water transparency of receiving water bodies but also causes environmental concerns about the possible toxicity and carcinogenicity of some organic dyes. Various physical and chemical methods such as coagulation and flocculation, oxidation or ozonation, membrane separation, ultra chemical filtration, chemical treatments, activated carbon adsorption, etc. have been used for the dye removal but they are either quite expensive or cannot be applied to large volumes of water. Thus in recent time, alternative methods for the removal of dyes from wastewaters with the use of various low cost adsorbents is in high demand. The biggest advantage of the use of waste materials as adsorbents is their low cost, versatility and easy operations. However, physical adsorption is usually non-destructive, and the post-treatment of the solid wastes is necessary and expensive. Chemical treatment using strong oxidants has led to more successful results. Advanced oxidation processes make them a suitable choice for the removal of toxic chemicals from wastewaters in the recent years.

Fly ash (FA) is a by-product of coal and biomass combustion. The disposal of fly ash causes significantly environmental problems. It has been known that the main components of fly ash are aluminosilicate, which makes them a cheap and readily available source of Si and Al for zeolite synthesis. Since the pioneer work of Holler and Wirsching (Holler and Wirsching, 1985), who were the first researchers to synthesize zeolites from fly ash, a lot of investigations have been conducted to convert fly ash to various zeolites in the past decade (Querol et al., 2002). The most common method used for the conversion of fly ash into zeolites involves a hydrothermal process, whereby the fly ash is mixed with an alkali solution at different conditions of temperature, pressure and reaction time. Recently, a fusion method has also been proposed to improve the total conversion of fly ash (Shigemoto et al., 1993) and has been used in several investigations (Chang et al., 1999; Kumar et al., 2001; Molina and Poole, 2004). Chang et al. (1999) and Kumar et al.(2001) reported the application of a fusion step to get supernatant solution from fly ash for synthesis of mesoporous aluminosilicate. Conversion of fly ash into zeolite provides a high-value utilization of fly ash because zeolite usually can be used as adsorbent, ion exchanger and catalyst or catalyst support. However, few investigations have been reported on the effect of two methods on the properties of produced zeolites as well as its applications. Molina and Poole (2004) evaluated the two methods for zeolite conversion at different conditions of temperature, time and proportion of NaOH:FA:H_2O in order to optimize the synthesis and compared the ion exchange capacity of the synthesized zeolites X.

In the past a few years, fly ash has been exploited as a low-cost adsorbents for dye removal from wastewater but the adsorption capacity is generally low compared with other adsorbents (Wang and Wu, 2006; Hsu, 2008). We have reported investigations using chemical pre-treatment of fly ash to increase the adsorption capacity (Wang et al., 2005b) and hydrothermal conversion of fly ash to zeolite for dye adsorption (Wang et al., 2005a). Although Molina and Poole's work provides important information on fusion conditions, the effect of aging product from fusion reaction was not investigated. Also few investigations have been conducted in the application of the synthesized zeolite for wastewater treatment.

Advanced oxidation processes (AOPs) are usually used for treating non biodegradable wastewater. In particular, oxidation with Fenton's reagent, which is based on ferrous ion and hydrogen peroxide, is a proven and effective technology for destruction of a large number of hazardous and organic pollutants (Neyens and Baeyens, 2003; Pignatello et al., 2006). Advantages of Fenton's reagent over other oxidizing treatment methods are numerous, including high efficiency, simplicity in destroying the contaminants, stability to treat a wide range of substances, non-necessity of special equipment, etc. Besides, operating conditions are usually mild, and hydrogen peroxide is easy to handle and the excess decomposes to environmentally safe products. The oxidation products are usually low molecular weight compounds that are often more easily biodegradable or, in some instances, the organic compounds reduced to carbon dioxide and water (Wang, 2008). However, it should be pointed out that the homogeneous Fenton process has a significant disadvantage: homogeneously catalyzed reactions need up to 50–80 ppm of Fe ions in solution, which is well above the European Union directives that allow only 2 ppm of Fe ions in treated water to be dumped directly into the environment. In addition, the removal/treatment of the sludge-containing Fe ions at the end of the wastewater treatment is expensive and needs large amount of chemicals, manpower and energy.

To overcome the disadvantages of the homogeneous Fenton or Fenton-like processes, some solid catalysts containing iron, such as γ-FeOOH, α-FeOOH, Fe_2O_3, and basic oxygen furnace slag (Chou et al., 2001; Feng et al., 2004; Chiou et al., 2006; Wu et al., 2006; Yeh et al., 2008) have been tried in order to avoid the catalyst-recovering step. Indeed, some attempts have also been made to develop Fe ions or Fe oxides into porous supports as heterogeneous catalysts, such as Fe(III) supported on metal oxide, zeolite and clay (Feng et al., 2003; Kuznetsova et al., 2004; Hsueh et al., 2006; Makhotkina et al., 2006; Molina et al., 2006; Ramirez et al., 2007a). It has been observed that depending on the conditions employed these materials can promote the oxidation of different organic compounds, such as aromatic and aliphatic acids, phenols, aromatic hydrocarbons, chlorocompounds and textile dyes with hydrogen peroxide. However, many of these systems showed low activities or strong iron leaching due to low pH, which resulted in the classical homogeneous Fenton mechanism.

Among iron oxides, magnetite (Fe_3O_4), has several features which make it very interesting in the Fenton reaction, such as (i) it contains Fe^{2+} which might play an important role for the initiation of the Fenton reaction according to the classical Haber–Weiss mechanism (Eq. (1)), (ii) the octahedral site in the magnetite structure can easily accommodate both Fe^{2+} and Fe^{3+}, allowing the Fe species to be reversibly oxidized and reduced while keeping the same structure and (iii) Fe ions in the magnetite structure can be substituted by several transition metals with varied redox properties which can be used to tune the catalytic properties of these materials. Moreover, magnetites are magnetic materials and can be easily separated from the reaction medium by a simple magnetic separation procedure (Costa et al., 2006). However, few investigations have been reported using magnetite as Fenton catalyst.

$$Fe^{2+} + H_2O_2 \longrightarrow Fe^{3+} + OH^- + OH\bullet \quad (1)$$

Costa, et al. (2006) prepared magnetites $Fe_{3-x}Mn_xO_4$, $Fe_{3-x}Co_xO_4$ and $Fe_{3-x}O_4$ and the iron oxides Fe_3O_4, γ-Fe_2O_3 and α-Fe_2O_3 and tested for the oxidation of organic molecules in aqueous medium. Maghemite, γ-Fe_2O_3 and hematite, α-Fe_2O_3, showed no activity in decolourization of methylene blue. Fe_3O_4 and $Fe_{2.54}Ni_{0.44}O_4$ showed low activity with only 10% color reduction after 50 min while Mn and Co substituted magnetites show high oxidation activities. Moura et al. (2006) reported an investigation of oxidation of phenol by a Fe^0/Fe_3O_4 composite and H_2O_2. The results suggest that the oxidation proceeded on the surface of the Fe^0/Fe_3O_4 composite via a classical Fenton-like mechanism. Baldrian et al. (2006) also prepared heterogeneous catalysts based on magnetic mixed iron oxides (MO• Fe_2O_3; M: Fe, Co, Cu, Mn) for the decolorization of several synthetic dyes. However, the most effective catalyst, FeO•Fe_2O_3, produced more than 90% decolorization of 50 mg/L Bromophenol Blue, Chicago Sky Blue, Evans Blue and Naphthol Blue Black within 24 h has to be at much higher concentrations (25 mg/mL Fe_3O_4 with 100 mM H_2O_2).

In this paper, we report an investigation employing two methods for fly ash conversion to zeolite considering the aging effect and test their adsorptive behavior in dye adsorption by obtaining adsorption isotherm and capacity. We also report an investigation of using Fe_3O_4 nanoparticles as an iron source in the heterogeneous Fenton reaction to degrade a typical dye, methylene blue (MB), at low catalyst loading and H_2O_2 concentration. The effects of parameters such as pH and reactant concentrations including magnetite loading, H_2O_2, initial dye concentration on Fenton reaction were examined. Moreover, the stability and leaching of the magnetite catalyst were also investigated.

2. Experimental

2.1. Chemicals and Materials

A commercial basic dye, methylene Blue (MB), was used in this study, which was obtained from Unilab. H_2O_2 (30%) was obtained from Chem-Supply, Australia. Fe_3O_4 nanopowder (20 -30 nm particle size, ≥98% metals basis) was obtained from Sigma-Aldrich (USA). Coal fly ash (FA) was first collected by electroprecipitation process from an electric power station in Western Australia. Then the coal fly ash was sieved to get the section with particle size less than 45 μm to remove unburned carbon. It has been found that the unburned carbon in fly ash can be removed by sieving. All sample solutions were prepared using deionized water from the Millipore Milli-Q system. $Fe(NO_3)_3.9H_2O$ and $FeSO_4.7H_2O$ (both from Fluka p.a.) were used as Fe^{3+} and Fe^{2+} cation source.

2.2. Conversion of Fly Ash to Zeolite

For hydrothermal reaction, 4 g fly ash was mixed with 8.4 g NaOH and dissolved in 50 ml water and set in an oven at 100 °C for 24 h (referred to as FA-1). For the fusion method, the fly ash (4 g) was mixed with sodium hydroxide solid (8.4 g) and the mixture was set in an oven at 300 °C for 1 h to obtain a fused mass. The second sample was then obtained by directly mixing the fused fly ash solid with 50 ml water and putting in an autoclave for hydrothermal conversion at 100 °C for 24 h (FA-2). The third sample was obtained by aging the fused sample for 24 h at room temperature and then loaded in an autoclave for 24 h at 100 °C for hydrothermal conversion (FA-3). After different treatments, all the slurry samples were filtered, washed and dried at 110 °C overnight.

2.3. Characterisation of Samples

The phases of fly ash and the derived materials were determined by XRD analyses with an automated Rigaku miniflex diffractometer using Co Kα radiation at 40 kV and 40 mA over the range (2θ) of 5 –80 °.

The surface area, total pore volume and pore size distribution of all samples were determined by N_2 adsorption under –196 °C using Autosorb (Quantachrome Corp.). All samples were degassed at 200 °C for 4 h, prior to the adsorption experiments. The BET surface area was obtained by applying the BET equation to the adsorption data. The total pore volume was calculated using adsorption value at p/p^0= 0.95. The pore size distribution was obtained using BJH method.

2.4. Adsorption Test

The adsorption behaviour of FA and the synthesized products was investigated for methylene blue adsorption. Sorption studies were performed by batch technique to obtain rate and equilibrium data. Adsorbents of 20 mg was added to 100 mL of methylene blue (MB) solution with varying concentrations, which were agitated intermittently for the desired time periods in a shaker at 30 °C. The concentrations of dye were determined at the corresponding λ_{max} = 665 nm using a spectrophotometer.

2.5. Oxidation Test

All experiments were carried out in a 100 ml batch reactor. Typically, the reactor was charged with 100 ml of dye solution at a room temperature and 150 rpm stirring. Then, the Fe_3O_4 powder was added followed by H_2O_2 solution, this being considered the initial instant of reaction (t = 0). For most of the experiments, the initial pH was set at 2.67-2.69. For the experimental runs at different pH, the values were adjusted using sulphuric acid or sodium hydroxide solutions and measured using a pH meter (PHM 250 Ion Analyzer). Aliquots were withdrawn from the reactor at selected intervals for spectrophotometric analysis. At each sampling time, the catalyst was separated by centrifugation (1 min) and the supernatant was immediately used for analysis. The concentration of MB was determined by UV–Vis spectrophotometer (Spectronic 20 Genesis spectrophotometer, USA) and the absorbance measured at the maximum visible absorbance wavelength λ_{max} (663.9 nm) by the UV/Vis spectrophotometer was used to calculate the decolorization efficiency. For sample solutions having higher concentrations than 5 mg/l MB, deionized water was used to dilute the solution to below 5 mg/l.

The effects of pH, dye, H_2O_2 and Fe_3O_4 concentrations on dye decolorization were examined by varying one factor while keeping the other three fixed. The lifetime of the catalyst Fe_3O_4 was also investigated. When determining the lifetime of the Fe_3O_4 catalyst, after the reaction the solid particles were separated by centrifugation and washed with pure water one time for next round reuse.

To evaluate the mineralization of the dye, total organic carbon (TOC) was measured using a Shimadzu 5000A spectrophotometer (TOC-5000 CE). The concentration of iron in solution was determined using atomic absorption spectrometer (Varian AAS 110).

3. Results and Discussion

3.1. Conversion of Fly Ash to Zeolite and Adsorption Test

3.1.1. Characterization of fly ash and zeolite

The XRD patterns of fly ash, hydrothermally and fusion-hydrothermally treated materials are presented in Figure 1. As seen that FA-1, FA-2 and FA-3 present new XRD peaks, corresponding to zeolite P. The major phases of the fly ash are quartz, mullite, with minor hematite and magnetite. Apart from zeolite P, strong peaks of quartz and mullite still appear on FA-1 and FA-2 while no peak occur on FA-3. This suggests that aging process will promote the dissolution of aluminosilicate in fly ash. For the three samples, XRD peaks for zeolite P are becoming stronger and stronger, which suggests high conversion of fly ash to zeolite.

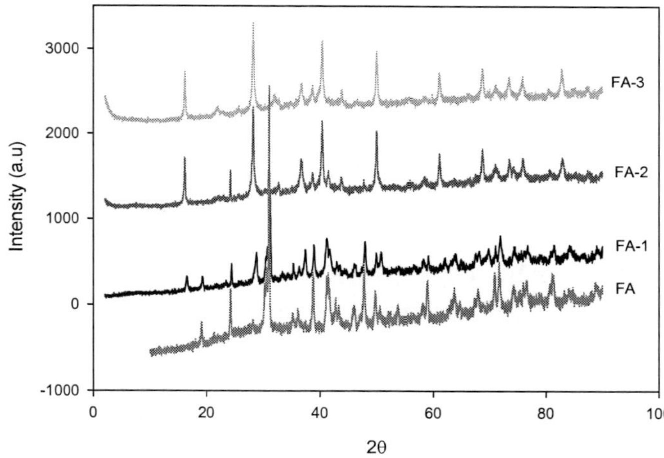

Figure 1. XRD patterns of fly ash and its derived materials.

Previous investigations have shown that fly ash could be converted to various zeolites in hydrothermal conversion depending on fly ash source and synthesis conditions (Querol et al., 2002). The resulting material usually is a mixture of zeolites like hydroxysodalite (HS) or P together with small proportions of zeolite X and A. Molina and Poole (2004) found that zeolite X is the dominated phase in their hydrothermal and fusion conversion processes. Hsu (2008) reported the presence of Na-A zeolite and HS in their hydrothermal conversion of fly ash.

Table 1 shows the BET surface areas and pore volumes of all adsorbents. FA has the lowest surface area and pore volume of 8.4 m^2/g and 0.010 cm^3/g, respectively. After treatment, the samples present much higher surface area and pore volume than the fly ash. FA-1 by one-step hydrothermal conversion exhibits lower surface area and pore volume than those obtained by fusion-hydrothermal conversion. FA-3 gives the highest value of surface area and pore volume of 78 m^2/g and 0.136 cm^3/g. The increase in surface area and pore volume are due to the formation of porous zeolite.

Figure 2 shows the pore size distribution of four samples. It is seen that FA does not show porous structure while other three samples produce well developed pore structure with similar maximum peak position. FA-3 also presents a larger pore around 12 nm, which suggests that it may be good for adsorption of large size molecules and thus showing higher adsorption.

Table 1. Physicochemical properties of fly ash and its derived adsorbents.

Sample	$S_{BET}(m^2/g)$	$V(cm^3/g)$	Sample	$S_{BET}(m^2/g)$	$V(cm^3/g)$
FA	8.4	0.010	FA-2	52.9	0.087
FA-1	40.0	0.083	FA-3	77.9	0.136

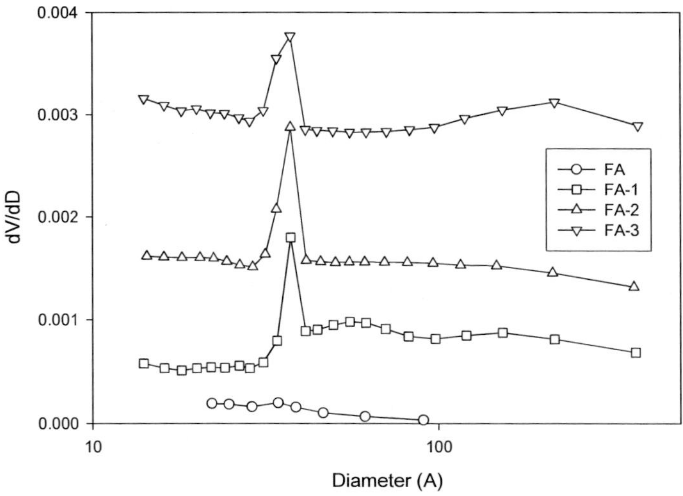

Figure 2. Pore size distribution of FA and treated FAs.

3.1.2. Dynamic adsorption and equilibrium

The dynamic adsorption of MB on fly ash and treated fly ashes is shown in Figure 3. As can be seen that four adsorbents exhibit varying adsorption behavior. FA presents very low equilibrium adsorption around 5 x10^{-6} mol/g and the equilibrium can be reached after 72 h. FA-1 can reach an adsorption equilibirum around 100 h but at high equilibrium adsorption around 6 x 10^{-5} mol/g. FA-2 and FA-3 exhibit higher adsorption than FA-1 and will reach equalibrium at a longer time around 150 h. The equilibrium adsorption for FA-2 and FA-3 is 7.5 x10^{-5} and 8.5 x 10^{-5} mol/g, respectively.

It has been known that adsorption depends on the surface and textural properties of adsorbent. Table 1 has shown that the surface area and pore volume are greatly improved after hydrothermal treatment due to the formation of zeolite P. The order of surface area and pore volume is FA-3 > FA-2 > FA-1 > FA, which is the same order as the MB equilibrium adsorption on the solids, which suggests that the surface area and pore volume control the adsorption of MB on these materials. In addition, zeolites generally exhibit cation exchange property. Molina and Poole (2004) have found the increase in cation exchange of zeolite from the conversion of fly ash by hydrothermal and fusion processes. The XRD results in this work also have shown the intensity of zeolite P in the order of FA-3> FA-2 >FA-1. Therefore, the more conversion of fly ash into zeolite P will result in higher cationic dye, MB, adsorption.

The two important models describing the adsorption kinetics have been used to fit the dynamic adsorption. The two models are the Lagergren first-order model (Ho, 2004) and Ho's pseudo second-order model (Ho and McKay, 1999). The first-order kinetic equation may be expressed as:

$$\frac{dq_t}{dt} = k_1(q_e - q_t) \tag{2}$$

where q_e and q_t are the sorption capacity (mol/g) of dye at equilibrium and at a time t, respectively, and k_1 is the rate constant for pseudo-first order sorption (l/h). After integration and applying boundary conditions, viz. that the initial conditions are $(q_e - q_t) = 0$ at $t = 0$, equation (2) becomes:

$$q_t = q_e(1 - e^{-k_1 t}) \tag{3}$$

The pseudo-second order equation can be written as:

$$\frac{dq_t}{dt} = k_2(q_e - q_t)^2 \tag{4}$$

where q_e and q_t are the sorption capacity (mol/g) of dye at equilibrium and at a time t, respectively, and k_2 is the rate constant for pseudo-second order sorption (g/(mol·h). For the boundary conditions $t = 0$ to $t = t$ and $q_t = 0$ to $q_t = q_t$, the integrated form of equation (4) becomes:

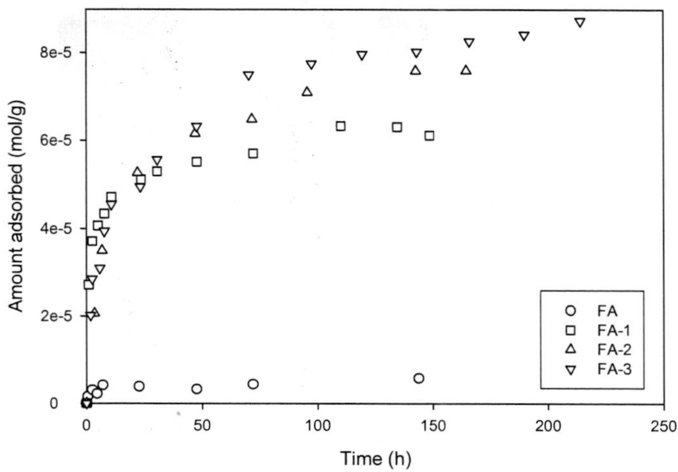

Figure 3. Dynamic adsorption of methylene on fly ash and zeolites.

Table 2. Kinetic parameters of MB adsorption on various treated fly ash samples.

sample	First-order kinetics			Second-order kinetics		
	k1 (h^{-1})	q_e(mol/g)	R^2	k_2(g/mol·h)	q_e (mol/g)	R^2
FA-1	0.365	5.55x 10-5	0.875	9.64x103	5.88x10-5	0.955
FA-2	0.0828	6.98 x10-5	0.962	1.32x103	7.78x10-5	0.992
FA-3	0.0678	7.80x10-5	0.897	1.13x103	8.54x10-5	0.957

$$q_t = \frac{q_e^2 k_2 t}{(1 + q_e k_2 t)} \tag{5}$$

which is the nonlinear form of rate law for a pseudo-second order reaction.

Using the above models, the experimental data were fitted and the results are presented in Table 2. It is seen that the first-order rate expression could not fit quite well to the experiment data evidenced from the correlation coefficients, which are lower than 0.97. In contrast, the second order kinetics will show much better correlation coefficients and most values are larger than 0.95. In addition, the equilibrium adsorption derived from the second-order kinetics is much close to the experimental value, suggesting its applicable to the adsorption kinetics.

Some researchers have conducted kinetic studies on dye adsorption on fly ash. Khare et al. (1987) investigated victoria blue adsorption on fly ash and found that the process followed the first order adsorption rate expression and the uptake of victoria blue by fly ash is diffusion controlled. Mohan et al. (2002) also reported that the adsorptions of crystal violet and basic fuschin on fly ash followed the first-order rate kinetics. While the kinetic studies reported by Acemioglu et al showed that the adsorption process obeyed the pseudo-second-order kinetic model (Acemioglu, 2004), which is similar to this investigation. Our previous investigations also showed that the adsorption of dyes on fly ash and natural zeolite followed the second-order kinetics (Wang et al., 2005c; Wang and Zhu, 2006).

Figure 4 shows the adsorption isotherms of MB on fly ash and its derived zeolites. As seen that the fly ash exhibits much low adsorption capacity. FA-1, FA-2 and FA-3 present the order of adsorption capacity as FA-3 > FA-2 > FA-1. This order is similar to the order of surface area and pore volume.

In order to know the adsorption isotherm, two models were employed. The Langmuir isotherm is based on an assumption that the adsorption occurs at specific homogeneous sites within the adsorbent. The equation is

$$Q_e = \frac{Q^0 K_L C_e}{(1 + K_L C_e)} \tag{6}$$

where Q_e is the adsorbed amount of the dye, C_e is the equilibrium concentration of the dye in solution, Q^0 is the monolayer adsorption capacity and K_L is the constant related to the free energy of adsorption.

The Freundlich isotherm is an empirical equation employed to describe heterogeneous systems. The Freundlich equation is expressed

$$Q_e = K C_e^{1/n} \tag{7}$$

where K and n are Freundlich adsorption isotherm constants, being indicative of the extent of the adsorption and the degree of non-linearity between solution concentration and adsorption, respectively.

Figure 4 shows a comparison of adsorption isotherms for curve fitting of the experimental results. The model parameters from all isotherms obtained from nonlinear regression to the above equations are presented in Table 3. As seen that the Freundlich model is better than the Langmuir model in simulation of the adsorption isotherm as evident from correlation coefficients. This suggests that some heterogeneity in the surface or pores of the fly ash and zeolite will play a role in dye adsorption.

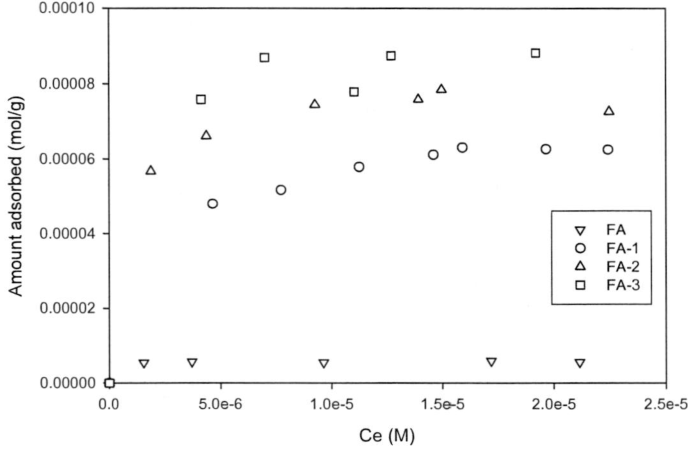

Figure 4. Adsorption isotherm of MB on various adsorbents

Table 3. Parameters of adsorption isotherms derived from curve fitting.

Sample	Langmuir model			Freundlich model		
	K_L (L/mol)	Q^0(mol/g)	R^2	K	1/n	R^2
FA-1	2.15×10^5	8.07×10^{-5}	0.973	4.67×10^{-4}	0.185	0.995
FA-2	5.28×10^5	8.84×10^{-5}	0.943	2.56×10^{-4}	0.111	0.985
FA-3	4.63×10^5	1.42×10^{-4}	0.946	2.71×10^{-3}	0.277	0.963

3.2 Heterogeneous Fenton Oxidation for Decolorization

3.2.1. Effect of pH

The pH value was reported as an important parameter that affects Fenton reaction. Figure 5 presents the decolorization efficiency by Fe_3O_4 over a pH range of 2-7. The decolorization of the dye was pH dependent exhibiting higher decolorization in the range of pH 3-5 at about 100%. The decolorization performance of the magnetite decreased quickly at pH > 5. At pH 7, decolorization efficiency is only 10% after 1 hour. The effect of dye adsorption on magnetite was also investigated. It was found that dye adsorption on the catalyst is about 2-5% and there was a relationship with the pH value. At pH 7, Fe_3O_4 dosage of 2 g/l, and MB concentration is 5 mg/l, the adsorption of dye could be 8%, but the dye adsorption would decrease when pH was about 3. Thus it is deduced that the dye decolorization at low pH is attributed to Fenton oxidation while at higher pH of 7, decolorization is due to adsorption.

The Fenton process is highly efficient at pH 2–4. Zhu et al. (1996) found that the COD removal was the highest under acid condition (pH = 2 ~ 4) and at high pH value, the COD removal remarkably declined. When pH>8, Fe^{2+} ion begins to form floc and precipitates. On the other hand, hydrogen peroxide is also unstable in basic solution and may decompose to give oxygen and water, and finally loses its oxidation ability. Thus hydrogen peroxide and ferrous ions are difficult to establish an effective redox system, and COD removal is less effective.

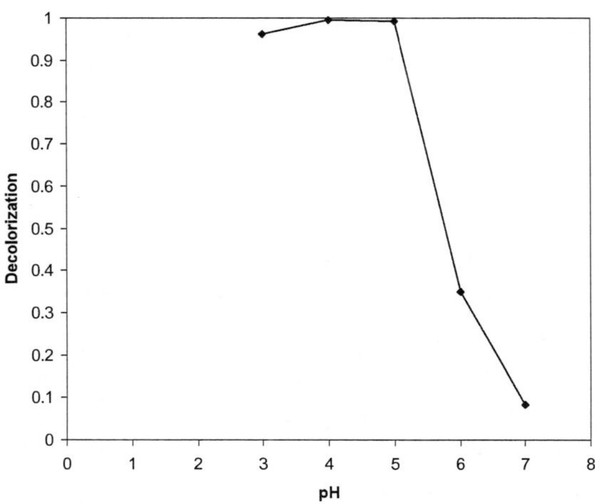

Figure 5. Effect of pH value on decolorization (Fe_3O_4 2 g/l, H_2O_2 18 mM, MB 5 mg/l).

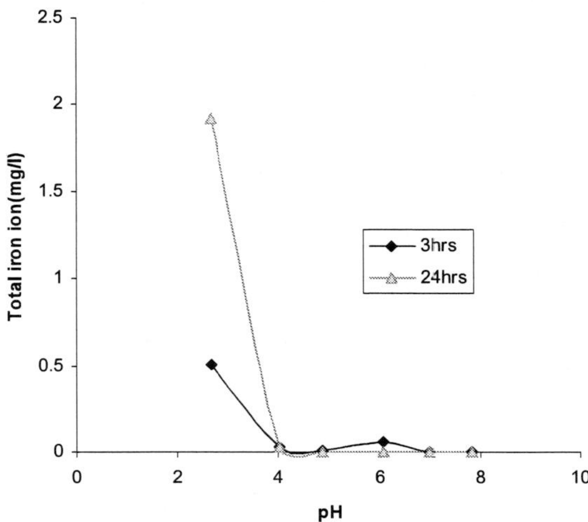

Figure 6. Fe concentration in solution after reaction.

It has been known that homogeneous Fenton oxidation will also occur during the heterogeneous Fenton reaction due to the leaching of Fe ions from solid catalysts. Figure 6 illustrates the leaching of Fe ions from Fe_3O_4 at varying pH values. It is seen that leaching would only occur below pH 4. At high pH (4 - 8), no significant Fe leaching was observed. At pH around 2.7, Fe concentrations at 3 h and 24 hours would be 0.5 and 2 mg/L, respectively. The results suggest that the homogeneous Fenton oxidation in this dye decolorization will not play a significant role.

3.2.2. Effect of Fe_3O_4 dosage

Figure 7 shows the effect of Fe_3O_4 content in solution on dye decolorization. The extent of degradation of the dye increased with increasing Fe_3O_4 dosage. Without Fe_3O_4, dye decolorization was only about 5% after 2 h. When Fe_3O_4 was added to the dye solution the decolorization could arrive at high value around 95%, but needed more time. At 0.1 g/l, the decolorization could reach 90% at 90 minutes. Higher Fe_3O_4 dosage would result in complete decolorization in much less time. At 2.0 g/l Fe_3O_4 dosage, decolorization of 100% would only require 20 minutes.

3.2.3. Effect of H_2O_2 concentration

The effect of hydrogen peroxide was analyzed by varying its initial concentration between 100 and 48000 uM. Figure 8 shows the relationship between decolorization of the dye and initial H_2O_2 concentrations at high (2 g/l) and low (0.2 g/l) Fe_3O_4 dosage. One can see that initial H_2O_2 concentration greatly influenced the decolorization of MB. Without H_2O_2, dye decolorization was only about 2-8% because of adsorption. The results also indicate that the degradation of the dye firstly increased and then dropped down with increasing H_2O_2 concentration whenever the Fe_3O_4 dosage was high or low. When Fe_3O_4 dosage was 2 g/l, the decolorization got the maximum value at H_2O_2 concentration of 800-900 μM, more than 97% decolorization could be achieved after 10 minutes. If H_2O_2

concentrations was increased from 4,400 to 44,000, the decolorization decreased from 60% to 29% (data not shown in Figure 8). When Fe_3O_4 dosage was 0.2 g/l, the decolorization got the maximum value at the H_2O_2 concentration of 300 μM, 90% decolorization could be achieved after 60 minutes.

Several investigations have reported an optimal peroxide concentration in the Fenton oxidation of dyes (Chiou et al., 2006; Bandara et al., 2007; Ramirez et al., 2007a; Ramirez et al., 2007b). Unreacted H_2O_2 will act as a scavenger of OH and produces a less potent perhydroxyl radical, resulting in less dye degradation.

3.2.4. Effect of dye concentration

Figure 9 shows the decolorization efficiency with time at varying dye concentration at low Fe_3O_4 loading (0.2 g/l). The degree of decolorization decreased with increasing dye concentration. After 120 min, the decolorization was 60% at 50 mg/l MB while it could reach 100% at 1 mg/l MB after 90 mins reaction

For the decolorization rate, a first-order reaction was applied,

$$\frac{dC}{dt} = -kC \tag{8}$$

the above equation after integration becomes

$$C = C_0 \exp(-kt) \tag{9}$$

in which C_0 is the initial dye concentration and k is the rate constant.

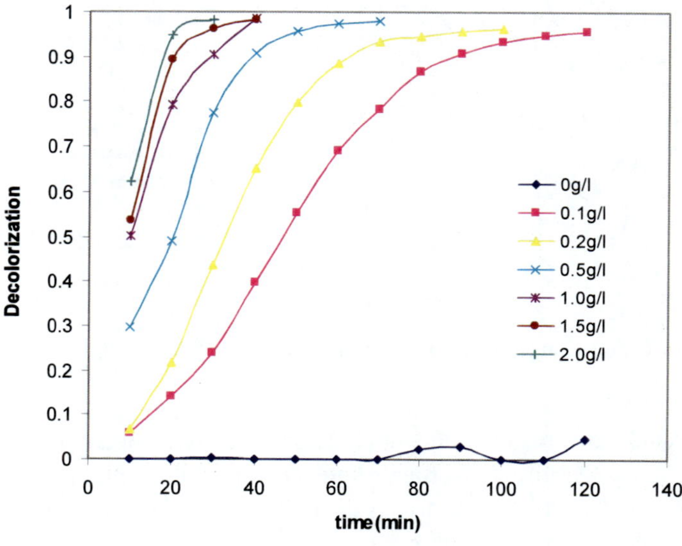

Figure 7. Effect of Fe_3O_4 dosage on dye decolorization (H_2O_2 300μM, pH 2.67, MB 5mg/l).

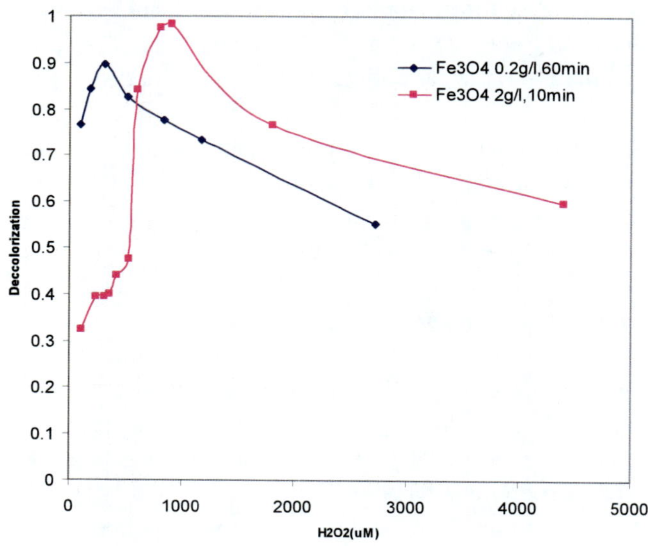

Figure 8. Effect of initial H_2O_2 concentration on decolorization (MB 5 mg/l, pH = 2.68).

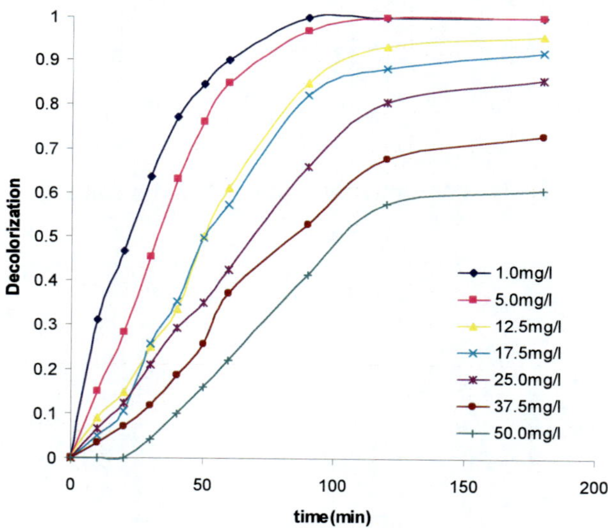

Figure 9. Effect of dye concentration on decolorization (Fe_3O_4 0.2g/l, H_2O_2 300 μM, and pH = 2.68).

Table 4 presents the results of curve fitting to dye decolorization at varying concentrations. It is seen that the first-order kinetics can well describe the dye decolorization rate. The regression coefficients are all higher than 0.94. From the table, it is also shown that the calculated dye concentration is quite close to experimental concentration and the error is within 10%.

Table 4 Parameters of dye decoloration in the first order kinetics.

Dye concentration (mg/L)	Calculated dye concentration (mg/L)	Decoloration rate constant (k, min^{-1})	R^2
1	1.01	0.0359	0.995
5	5.36	0.0266	0.975
12.5	13.6	0.0161	0.959
17.5	19.2	0.0153	0.968
25	26.5	0.0112	0.977
37.5	39.9	$8.25x10^{-3}$	0.970
50	54.3	$6.10 x10^{-3}$	0.942

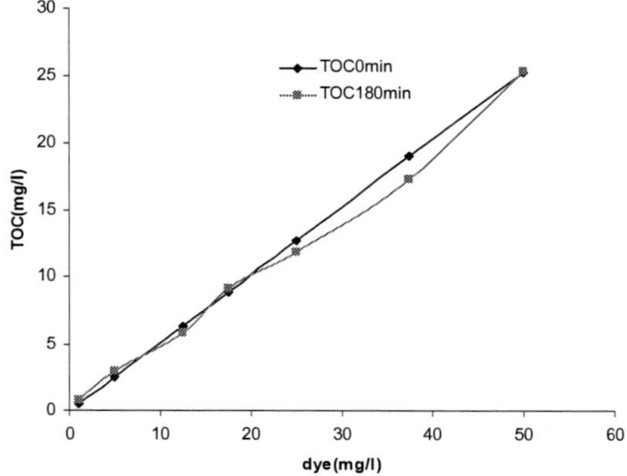

Figure 10. TOC in different dye concentration solution at 0min and 3 hrs (Fe_3O_4 0.2 g/l, H_2O_2 300 μM, and pH = 2.68).

Figure 10 presents TOC measurement of various dye solutions before and after reaction. As shown, in some cases, TOC will not change after reaction, but in other cases, TOC will be slightly reduced after Fenton reaction. The low reduction in TOC will probably be due to the low H_2O_2 concentration used in each run, which is not enough for further organic degradation.

3.2.5. Leaching of iron ion

Figure 11 shows the leaching of Fe from Fe_3O_4 at varying catalyst loadings after oxidation. One can see that Fe leaching rate is higher at higher catalyst loading. After 3 h, Fe concentration in solution was less than 1.5 mg/L. After 24 h, the maximum Fe concentration in solution was 2 mg/L.

Figure 12 also shows the leaching rate of Fe from Fe_3O_4 after dye decolorization at varying dye initial concentrations. It is seen that Fe leaching rate would also increase with dye concentration. After 3 h oxidation, Fe concentration was below 0.9 mg/L. Compared with magnetite loading in solution, the Fe leaching rate was only 0.4% at MB concentration of 50 mg/L.

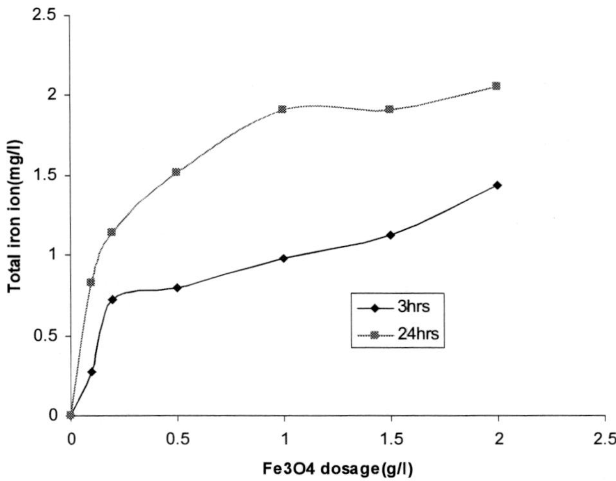

Figure 11 Fe concentration after dye decolorization at different times. ([MB] = 5 mg/L, pH = 2.67, $[H_2O_2]$ = 300 μM.

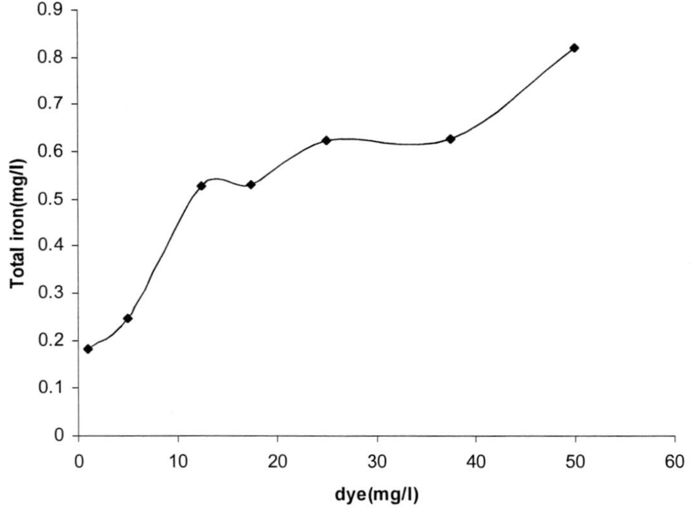

Figure 12. Total iron ion concentration in the dye solution in different dye concentration after 3 hrs (Fe_3O_4 0.2 g/l, H_2O_2 300μM, and pH = 2.68).

3.2.6. Life time of Fe_3O_4 nanoparticles

In practice, it is crucial to evaluate the stability of a heterogeneous catalyst in Fenton oxidation. Thus consecutive experiments were performed with the same sample recovered by filtration after each cycle. Figure 13 shows the performance of Fe_3O_4 in dye decolorization in 6 cycles. As seen, Fe_3O_4 exhibited remarkable stability of performance and dye decolorization efficiency did not reduce after 6 rounds of test. In the first consecutive 5 tests, dye decolorization efficiencies are similar and reached 95% after 50 min. In the 6th test, dye decolorization efficiency is enhanced and reached 95% after 30 min.

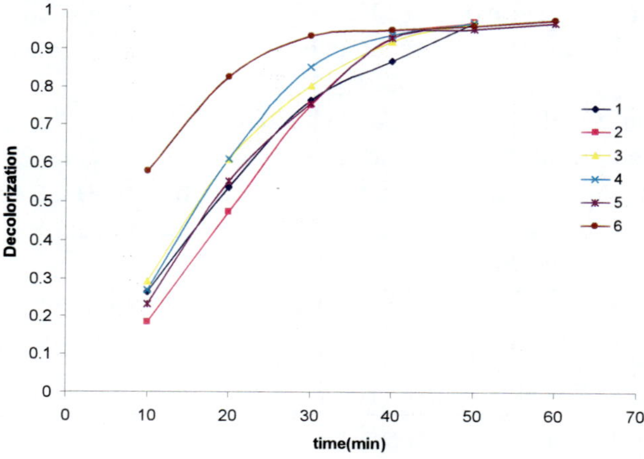

Figure 13 Decolorization efficiency of Fe_3O_4 in 6 consecutive tests. (MB 5mg/l, Fe_3O_4 2g/l, H_2O_2 300 μM, pH = 2.68)

4. Conclusions

Fly ash has been transformed into zeolite P by one-step hydrothermal and two-step fusion and hydrothermal methods. Two-step fusion and hydrothermal treatment can significantly improve the conversion. Aging of fused fly ash after fusion process can also increase the conversion rate, resulting in a fully conversion of fly ash. Adsorption tests show that the converted zeolite P exhibits much higher adsorption capacity. The kinetic adsorption follows the pseudo second-order kinetics and the adsorption isotherm can be well fitted by Freundlich isotherm.

Nanosized magnetite is highly effective in heterogeneous Fenton oxidation for dye decolorization. The dye decolorization rate followed the first-order kinetics and efficiency depends on magnetite loading, H_2O_2 concentration, dye concentration and solution pH. In general, 100% decolorization could be reached at lower loading and H_2O_2 concentration and low pH (2 -5) while the performance would be reduced at pH > 6. At low pH, Fe leaching from magnetite is small and the dye decolorization is mainly due to heterogeneous Fenton oxidation. Moreover, nanosized magnetite exhibited high stability in decolorization performance. After 6 consecutive tests, no reduction in decolorization was observed.

References

Acemioglu, B. (2004). Adsorption of Congo red from aqueous solution onto calcium-rich fly ash. *Journal of Colloid and Interface Science*, *274*, 371-379.

Baldrian, P., Merhautova, V., Gabriel, J., Nerud, F., Stopka, P., Hruby, M. & Benes, M. J. (2006). Decolorization of synthetic dyes by hydrogen peroxide with heterogeneous catalysis by mixed iron oxides. *Applied Catalysis B-Environmental*, *66*, 258-264.

Bandara, J., Klehm, U. & Kiwi, J. (2007). Raschig rings-Fe_2O_3 composite photocatalyst activate in the degradation of 4-chlorophenol and Orange II under daylight irradiation. *Applied Catalysis B-Environmental*, *76*, 73-81.

Chang, H. L., Chun, C. M., Aksay, I. A. & Shih, W.H. (1999). Conversion of fly ash into mesoporous aluminosilicate. *Industrial & Engineering Chemistry Research*, *38*, 973-977.

Chiou, C. S., Chang, C. F., Chang, C. T., Shie, J. L. & Chen, Y. H. (2006). Mineralization of Reactive Black 5 in aqueous solution by basic oxygen furnace slag in the presence of hydrogen peroxide. *Chemosphere*, *62*, 788-795.

Chou, S. S., Huang, C. P. & Huang, Y. H. (2001). Heterogeneous and homogeneous catalytic oxidation by supported gamma-FeOOH in a fluidized bed reactor: Kinetic approach. *Environmental Science & Technology*, *35*, 1247-1251.

Costa, R. C. C., Lelis, M. F. F., Oliveira, L. C. A., Fabris, J. D., Ardisson, J. D., Rios, R. R.V. A., Silva, C. N. & Lago, R. M. (2006). Novel active heterogeneous Fenton system based on $Fe_{3-x}M_xO_4$ (Fe, Co, Mn, Ni): The role of M^{2+} species on the reactivity towards H_2O_2 reactions. *Journal of Hazardous Materials*, *129*, 171-178.

Feng, J. Y., Hu, X. J. & Yue, P. L. (2004). Discoloration and mineralization of orange II using different heterogeneous catalysts containing Fe: A comparative study. *Environmental Science & Technology*, *38*, 5773-5778.

Feng, J. Y., Hu, X. J., Yue, P. L., Zhu, H. Y. & Lu, G. Q. (2003). A novel laponite clay-based Fe nanocomposite and its photo-catalytic activity in photo-assisted degradation of Orange II. *Chemical Engineering Science*, *58*, 679-685.

Ho, Y. S. (2004). Citation review of Lagergren kinetic rate equation on adsorption reactions. *Scientometrics*, *59*, 171-177.

Ho, Y. S. & McKay, G. (1999). Pseudo-second order model for sorption processes. *Process Biochemistry*, *34*, 451-465.

Holler, H. & Wirsching, U. (1985). Zeolite formation from fly-ash. Forschritte der Mineralogies 63, 21-43.

Hsu, T. C. (2008). *Adsorption of an acid dye onto coal fly ash. Fuel*, *87*, 3040-3045.

Hsueh, C. L., Huang, Y. H., Chen, C. Y. (2006). Novel activated alumina-supported iron oxide-composite as a heterogeneous catalyst for photooxidative degradation of reactive black 5. *Journal of Hazardous Materials*, *129*, 228-233.

Khare, S. K., Panday, K. K., Srivastava, R. M. & Singh, V. N. (1987). Removal of victoria blue from aqueous solution by fly ash. *Journal of Chemical Technology and Biotechnology*, *38*, 99-104.

Kumar, P., Mal, N., Oumi, Y., Yamana, K. & Sano, T. (2001). Mesoporous materials prepared using coal fly ash as the silicon and aluminium source. *Journal of Materials Chemistry*, *11*, 3285-3290.

Kuznetsova, E. V., Savinov, E. N., Vostrikova, L. A., Parmon, V. N. (2004). Heterogeneous catalysis in the Fenton-type system FeZSM-5/H_2O_2. Applied Catalysis B-Environmental 51, 165-170.

Makhotkina, O. A., Kuznetsova, E. V., Preis, S. V. (2006). Catalytic detoxification of 1,1-dimethylhydrazine aqueous solutions in heterogeneous Fenton system. *Applied Catalysis B-Environmental*, 68, 85-91.

Mohan, D., Singh, K. P., Singh, G. & Kumar, K. (2002). Removal of dyes from wastewater using flyash, a low-cost adsorbent. Industrial & Engineering Chemistry Research 41, 3688-3695.

Molina, A. & Poole, C. (2004). A comparative study using two methods to produce zeolites from fly ash. *Miner Eng*, *17*, 167-173.

Molina, R., Martinez, F., Melero, J. A., Bremner, D. H. & Chakinala, A. G. (2006). Mineralization of phenol by a heterogeneous ultrasound/Fe-SBA-15/H_2O_2 process: Multivariate study by factorial design of experiments. Applied Catalysis B-Environmental 66, 198-207.

Moura, F. C. C., Araujo, M. H., Dalmazio, I., Alves, T. M. A., Santos, L. S., Eberlin, M. N., Augusti, R. & Lago, R. M. (2006). Investigation of reaction mechanisms by electrospray ionization mass spectrometry: characterization of intermediates in the degradation of phenol by a novel iron/magnetite/hydrogen peroxide heterogeneous oxidation system. *Rapid Communications in Mass Spectrometry*, *20*, 1859-1863.

Neyens, E. & Baeyens, J. 2003. A review of classic Fenton's peroxidation as an advanced oxidation technique. *Journal of Hazardous Materials*, 98, 33-50.

Pignatello, J. J., Oliveros, E. & MacKay, A. 2006. Advanced oxidation processes for organic contaminant destruction based on the Fenton reaction and related chemistry. *Critical Reviews in Environmental Science and Technology*, *36*, 1-84.

Querol, X., Moreno, N., Umana, J. C., Alastuey, A., Hernandez, E., Lopez-Soler, A. & Plana, F. (2002). Synthesis of zeolites from coal fly ash: an overview. *International Journal of Coal Geology*. Elsevier Science B.V., pp. 413-423.

Ramirez, J. H., Costa, C. A., Madeira, L. M., Mata, G., Vicente, M. A., Rojas-Cervantes, M.L., Lopez-Peinado, A. J. & Martin-Aranda, R. M. (2007a). Fenton-like oxidation of Orange II solutions using heterogeneous catalysts based on saponite clay. Applied Catalysis B-Environmental, *71*, 44-56.

Ramirez, J. H., Maldonado-Hodar, F. J., Perez-Cadenas, A. F., Moreno-Castilla, C., Costa, C. A. & Madeira, L. M. (2007b). Azo-dye Orange II degradation by heterogeneous Fenton-like reaction using carbon-Fe catalysts. *Applied Catalysis B-Environmental*, *75*, 312-323.

Shigemoto, N., Hayashi, H. & Miyaura, K. (1993). Selective Formation of Na-X Zeolite from Coal Fly-Ash by Fusion with Sodium-Hydroxide Prior to Hydrothermal Reaction. *Journal of Materials Science*, *28*, 4781-4786.

Wang, S. (2008). *A Comparative study of Fenton and Fenton-like reaction kinetics in decolourisation of wastewater Dyes and Pigments*, *76*, 714-720

Wang, S. B., Boyjoo, Y. & Choueib, A. (2005a). Zeolitisation of fly ash for sorption of dyes in aqueous solutions. Stud. Surf. Sci. Catal, *158*, 161-168.

Wang, S. B., Boyjoo, Y., Choueib, A. & Zhu, Z. H. (2005b). Removal of dyes from aqueous solution using fly ash and red mud. *Water Res*, *39*, 129-138.

Wang, S. B., Li, L., Wu, H. W. & Zhu, Z. H. (2005c). Unburned carbon as a low-cost adsorbent for treatment of methylene blue-containing wastewater. *Journal of Colloid and Interface Science*, *292*, 336-343.

Wang, S. B. & Wu, H. W. (2006). Environmental-benign utilisation of fly ash as low-cost adsorbents. *Journal of Hazardous Materials*, *136*, 482-501.

Wang, S. B. & Zhu, Z. H. (2006). Characterisation and environmental application of an Australian natural zeolite for basic dye removal from aqueous solution. *Journal of Hazardous Materials*, *136*, 946-952.

Wu, J. J., Muruganandham, M., Yang, J. S. & Lin, S. S. (2006). Oxidation of DMSO on goethite catalyst in the presence of H_2O_2 at neutral pH. *Catalysis Communications*, *7*, 901-906.

Yeh, C. K. J., Hsu, C. Y., Chiu, C. H., Huang, K. L., (2008). Reaction efficiencies and rate constants for the goethite-catalyzed Fenton-like reaction of NAPL-form aromatic hydrocarbons and chloroethylenes. *Journal of Hazardous Materials*, *151*, 562-569.

Zhu, W. P., Yang, Z. H. & Wang, L. (1996). Application of ferrous-hydrogen peroxide for the treatment of H-acid manufacturing process wastewater. *Water Research*, *30*, 2949-2954.

In: Buildings and the Environment
Editors: Jonas Nemecek and Patrik Schulz
ISBN: 978-1-60876-128-9

Chapter 5

COMPUTATIONAL PREDICTION OF AIR FLOW AND THERMAL COMFORT IN NATURALLY VENTILATED REAL-SCALE BUILDINGS

D.P. Karadimou, G.M. Stavrakakis and N.C. Markatos*
Computational Fluid Dynamics Unit, School of Chemical Engineering National Technical University of Athens Iroon Polytechniou 9, GR-15780, Greece

ABSTRACT

Natural ventilation is an efficient energy-saving method, used in architecture for the design and construction of low energy-consuming buildings. The present chapter investigates the air flow in and around a naturally ventilated building, which represents a student dormitory, using computational fluid dynamics (CFD) techniques. Turbulence is simulated applying the standard k-ε and the RNG k-ε models, both modified to account for wind and buoyancy forces. The study focuses on single-sided natural ventilation in a typical room located on the ground floor of the student dormitory. Two cases are investigated: rooms with a) two openings, and b) one opening, on the windward side of the room. Internal furnishings are taken into account and the room includes three thermal sources (occupant, TV, PC). Two different mechanisms of natural ventilation are examined: a) displacement ventilation (two openings) and b) mixing ventilation (one opening). The numerical results are compared with those obtained by two well-known empirical models, related to the effective velocity of incoming air and to the height of the neutral level in the case of one opening. It is concluded that the numerical results are in acceptable agreement with those obtained by the empirical models, especially when the standard k-ε model is used. Finally, the mathematical model described is used to evaluate thermal comfort inside the room, and it is found that the case with two openings on the windward side of the room is the best design.

---

* Corresponding author: Corresponding author. Address: School of Chemical Engineering, National Technical University of Athens, Iroon Polytechniou 9, GR-15780 Greece. Tel.: +30210772 3126; fax: +30210772 3228. E-mail address: N.Markatos@ntua.gr

1. INTRODUCTION

The rise of living standards and the deterioration of thermal conditions in the urban environment are mainly responsible for the increase of energy consumption in buildings, relating to the operation of HVAC (Heating, Ventilating and Air-Conditioning) systems. Natural ventilation is an efficient alternative method to reduce energy use in buildings. Not only can the freely available resources of wind and solar energy [1] be used, but it can also create thermal comfort and healthy indoor conditions, provided by an optimum design. Typically, the energy cost of a naturally ventilated building is 40% less than that of an air-conditioned building [2].

The airflow in a naturally ventilated building is the result of pressure differences produced by wind and buoyancy forces. Wind forces are created by the external turbulent flow around the building (atmospheric boundary layer). Buoyancy forces are produced by density differences caused either by the temperature difference between the indoor and the outdoor environment or by thermal sources. Depending on the location of the openings, the buoyancy forces may assist or oppose the wind-driven flow. Furthermore, the opening location strongly affects the air-change rate and the temperature stratification in the interior of the building [3].

Natural ventilation may be classified into (see Figure 1) [4, 5] either : a) cross ventilation (case A, Fig. 1), for which the openings are located at different sides of the building, or b) single-sided ventilation, for which the openings are located at the same side of the building. According to the openings' design, the expected airflow may be classified as displacement ventilation, which takes place when the openings are located at the lower as well as at the upper heights of the building envelope, (case B, Figure 1), or mixing ventilation, which takes place when only one vent exists (see Figure 1, Case C). When the air of the indoor space is warm compared to the environment, then: a) if two vents are open, one near the top and the other near the bottom of the space (case B, Figure 1), cool air enters through the lower opening and warm air flows out through the upper opening, producing strong internal thermal stratification, which characterises displacement ventilation and b) if one vent is open, the incoming cool air enters through the lower part of the opening and the warm air flows out through the upper part of the opening (case C, Figure 1). The incoming cool air descends as a turbulent plume that tends to mix the air within the space, resulting to an almost uniform temperature, which characterises mixing ventilation [1].

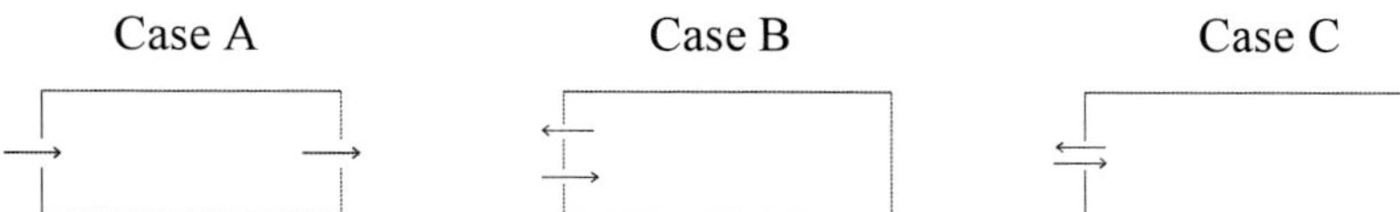

Figure 1. Natural ventilation designs: Cross Ventilation (Case A), and Single-sided ventilation (Cases B and C).

Computational Fluid Dynamics techniques (CFD) are very popular in studying natural ventilation due to the satisfying accuracy they provide and the reliable information on the velocity and temperature distributions they offer. In recent years several CFD studies have been performed for single-sided natural ventilation. Generally, as far as the past CFD studies are concerned, natural ventilation is assumed to be caused by either wind or buoyancy forces or by a combination of them. Additionally, some of the CFD studies are carried out applying a flow domain of the same size as the indoor space, while others are applying an enlarged domain, so as to study both indoor and outdoor air-flow. Schaelin et al [6] simulated the wind and buoyancy driven air-flow through a door-opening using an extended computational domain. Li and Teh [7] performed a similar study of the air-flow produced only by buoyancy forces. Gan [8] calculated the effective depth of incoming air studying a buoyancy-driven flow through a large opening. Another study for buoyancy-driven single-sided ventilation [9] used experimental data for validation purposes. A numerical analysis of the same type of natural ventilation may be found in [10], where an LES model is validated with experimental data for a simple geometry. LES modelling was also used in the case of pilot building models to evaluate both single-sided and cross natural ventilation performance [11]. A validated LES model of wind-driven single-sided natural ventilation in a real-scale building may be found in [12]. Experimental results in that case were obtained by Dascalaki et al [13]. Allocca et al [4] studied the buoyancy effect and the combined effect of wind and buoyancy on single-sided natural ventilation in a realistic geometry using Reynolds-Averaged Navier-Stokes (RANS) turbulence models, together with analytical and empirical models. In cases of complex geometry, such as in [4, 9], RANS turbulence modelling, which was first introduced in ventilation applications in 1983 by Markatos [14], produced reasonably accurate results requiring much less computing time than LES.

The objective of the present chapter is the application of a CFD model to investigate the effect of both wind and buoyancy forces on single-sided natural ventilation in a part of a building of real-scale geometry, such as that found in [4]. The cases considered are: One opening and two (low and high) openings on the windward façade of the building. Turbulent flow is modelled using the standard k-ε and the RNG k-ε model, both modified to account for wind and buoyancy forces. Additionally, two well-known empirical models are applied, which predict the inlet velocity and the neutral plane height (in the case of one opening), found in literature [15, 1], for validation purposes. Finally, the mathematical model described is used for thermal comfort predictions, in terms of temperature variations in the occupied zone, for both cases studied.

2. The Physical Problem and Assumptions

In the present chapter, single-sided natural ventilation is investigated through numerical simulation of the air-flow in and around the lowest apartment of a student dormitory [4], illustrated in Figure 2. Two cases are studied: rooms with a) two openings, one being at the upper and the other at the lower height of the windward side of the building, and b) one upper opening on the windward side of the building. Internal furnishings and three thermal sources (see Figure 3) are also taken into account. Specifically, the room includes internal common furnishings, as well as three heat sources, i.e. a seated occupant in the middle of the room, a personal computer (PC), and a television set (TV), as presented in Figure 3.

The assumptions made for the problem are the following: (a) steady-state incompressible flow of a Newtonian fluid, (b) adiabatic walls (except for heat sources), (c) Constant air properties at 298.5 K, (d) heat transfer at the walls by either conduction or radiation is neglected, (e) neutral atmospheric conditions of the incoming wind, and (f) the effects of neighbouring apartments on the air-flow pattern within the lower room, illustrated in Figure 2, are considered negligible.

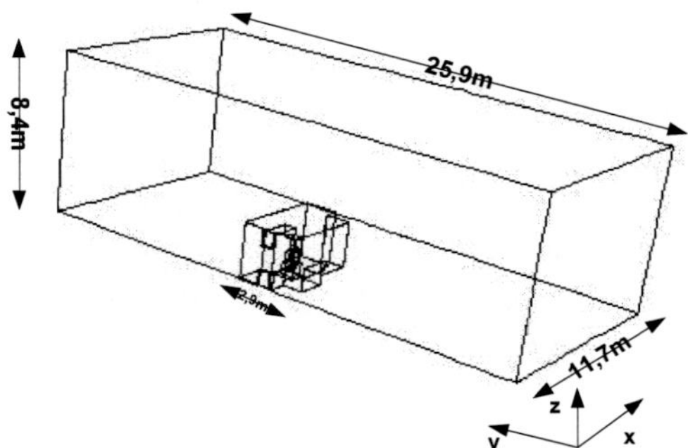

Figure 2. Student dormitory.

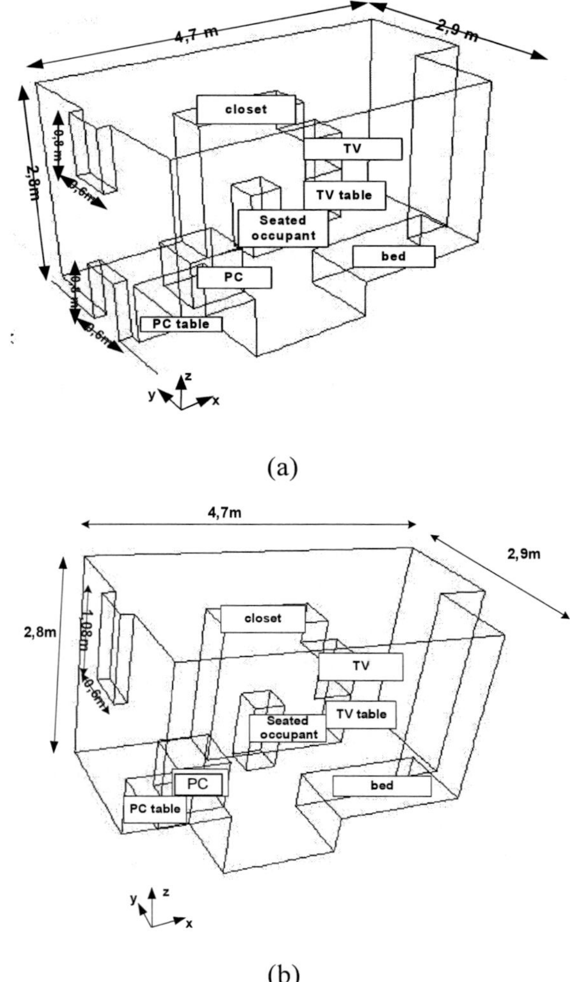

Figure 3. Internal furnishings and thermal sources for the case of: (a) two openings, and (b) one opening, located at the windward façade of the building.

2. Mathematical Modelling

2.1. The Governing Differential Equations and Turbulence Models

The independent variables of the steady-state problem are the three components (*x, y, z*) of a Cartesian coordinate system. The main dependent variables characterizing the flow are the three velocity components (*u, v, w*), pressure *p*, enthalpy h_e, kinetic energy of turbulence *k*, and the turbulence energy dissipation rate ε.

All these dependent variables, with the exception of pressure, appear as the subjects of equations of the general form [16, 17]:

$$\frac{\partial(\rho\varphi)}{\partial t} + div(\rho\vec{u}\varphi - \Gamma_{\varphi} grad\varphi) = S_{\varphi} \tag{1}$$

Where φ: The dependent variable, e.g. velocity components in three directions (*u, v, w*), enthalpy (h_e), *k* and ε, or 1 for the continuity equation.

ρ: Fluid density.

$\vec{u}$: Velocity vector.

Γ_{φ}: The "effective" exchange coefficient of φ

S_{φ}: Source/Sink rate per unit volume

The flow is assumed to take place under steady-state conditions, thus the general conservation equation (1) for all dependent variables is implemented, for steady state simulations:

$$div(\rho\vec{u}\varphi - \Gamma_{\varphi} grad\varphi) = S_{\varphi} \tag{2}$$

The pressure variable is associated with the continuity equation, in anticipation of the so-called pressure-correction equation [17], which is deduced from the finite-domain form of the continuity equation. Further details may be found, for example, in [19-21].

Two turbulence models were used in the present chapter, together with the Boussinesq approximation for buoyancy effects: (a) the standard k-ε model [19, 20], (b) the RNG k-ε model [22]. The modifications applied for buoyancy forces may be found in references [21] and [23]. The models use the logarithmic "wall functions" near solid surfaces ($11.5 < y^+ < 150$), where y^+ is the dimensionless distance of the first grid-node from the wall) [19]. The validity of Boussinesq approximation was tested here by repeating runs using variable density (as a perfect-gas-law function of temperature), and was found adequate.

2.2. Formulation of Equations

The numerical procedure followed is based on the Finite Volume Method (FVM) [16, 17, 21] provided by a general CFD code, i.e. Fluent version 6.3.26. The space (and time when necessary) dimensions are discretized into finite intervals and the variables are correspondingly computed at only a finite number of locations in three-(or four-) dimensional space, i.e. at the so-called "grid points". These variables are connected with each other by algebraic equations, derived from their differential counterparts by integration over the control volumes defined by the above-mentioned intervals, applying the first order upwind scheme for the convective terms [21, 33].

2.3. Computational Domain

The problem follows the general theoretical aspects of the flow around a bluff body. Consequently, specific techniques found in literature [5, 24, 25] are applied concerning the computational domain. Thus, it is extended around the building (see Figure 4), in order to account for external flow phenomena (stagnation, separation, reattachment) correctly, hence to obtain reliable solutions of the distribution of dependent variables (u, v, w, k, ε, p, h_e) through the openings, as well as in the internal space of interest [24, 25]. Consequently, the solution domain is of a downstream length of 12H (H: Building's height), an upstream length of 5H, a lateral length of 5H on both sides of the building and a height of 6H. The computational domain constructed for both cases studied is presented in Figure 4. It is seen that only the middle lowest room has been taken into account, following the assumption (f) in section 2.

2.4. Boundary Conditions

2.4.1. Velocity inlet

The incoming wind velocity profile, for an urban terrain, is given by the equation below [4], and is applied on the windward side of the computational domain (see Figure 3):

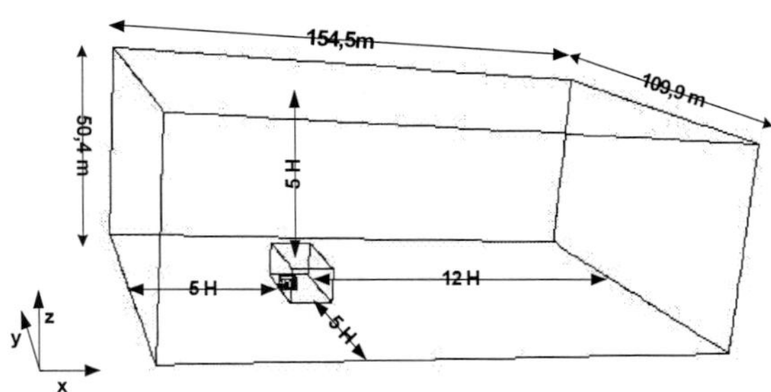

Figure 4. Computational domain

$$V_h = 0{,}35 \cdot Vmet \cdot h^{0{,}25} \tag{3}$$

where V_h is the wind speed at an arbitrary height, h, and $Vmet$ is the wind speed at the height of the weather-station sensor. Numerical simulations were performed for different values of wind speed taken by the weather station [4], *Vmet*, in the range of 2-10 m/sec. Additionally, inlet conditions for turbulence components are considered uniform and defined as follows [26]:

$$k = \frac{3}{2}\left(U_{avg} \cdot T_i\right)^2 \tag{4}$$

$$\varepsilon = C_\mu^{3/4} \frac{k^{3/2}}{l_t} \tag{5}$$

where U_{avg} is the mean inlet velocity, taken as the mean value of the results obtained by equation (3), T_i is the turbulence intensity, considered 5% [27] for a fully developed turbulent flow, C_μ is the viscosity empirical constant, and l_t is the turbulence length scale. Inlet temperature is imposed as a fixed value equal to 298.5 K, since neutral atmospheric conditions are considered. Further details may be found in [25].

2.4.2. Outlet

In case of natural ventilation, the outlet boundary may be handled using Neumman conditions for all φ variables [25], i.e. $\frac{\partial \varphi}{\partial x_j} = 0$ $(\vec{j} \,//\, \vec{n})$ (where $\vec{j}$ is the velocity direction vector and $\vec{n}$ is the normal-to-surface vector). This means that no back flow occurs through the outlet and no profile change happens along the flow direction. Therefore, to ensure this situation the outflow boundary is placed far away downstream of the building, as discussed earlier (section 2.3).

2.4.3. Walls

The flow field in the vicinity of walls is significantly obstructed and is one of the major parameters that influence the propagation of the bulk flow. Thus, correct modelling of the flow near walls is of great importance and can be simulated using methods summarized in [21, 24, 25]. The velocity is set equal to the wall surface's velocity (no-slip condition), while shear stresses are calculated using "wall-functions" [19, 20]. Concerning the thermal boundary conditions of the walls, they include the location and magnitude of heat release in the space from the occupant, and the electrical equipment. This is specified by identifying the cells within which this heat release takes place, together with the rate of heat transfer. In the present chapter, this is imposed applying a Dirichlet condition for heat output at each surface of the heat sources. Given the total heat output, Q_{total}, of the heat sources [4], the surface heat release is calculated using the following equation:

$$\sum_{i=1}^{n_{hs}} Q_{hs} A_{hs,i} = Q_{total} \tag{6}$$

where, Q_{hs} is the heat released by the surface of the heat source, *hs* is the heat source (occupant, PC, or TV), $A_{hs,i}$ is the area of the *i*-th side of the heated obstacle, and n_{hs} is the number of heated sides of the heat source. Consequently, Q_{hs} ($J/(m^2 \cdot s)$) is calculated using equation (6), and imposed as a fixed value at each heated surface.

2.4.4 Symmetry boundaries

This condition implies a zero velocity component normal to the symmetry plane, thus zero flow through these boundaries, whilst normal spatial gradients of the other variables are also zero. In short: $\frac{\partial \varphi}{\partial x_j} = 0,\ u_j = 0\ (\vec{j} // \vec{n})$, where $\vec{j}$ is the velocity direction vector and $\vec{n}$ is the normal-to-surface vector. This boundary condition is applied at the lateral external planes and the upper plane of the computational domain (see Figure 4), and is used for numerical consistency and convergence acceleration purposes [25].

2.5. Computational Details

The domain of interest is divided into a number of control volumes. Over a control volume about the central grid-point, P, the set of differential equations for φ is integrated to give discretized algebraic equations of the form [9]:

$$a_P \varphi_P = \sum_I a_I \varphi_I + S_{\varphi,P} V_P \tag{7}$$

Where: a_I represents coefficients, *I* stands for the neighbouring nodes around the central one, *P*, $S_{\varphi,P}$ is the source term of the transferred variable φ, and V_P is the grid-cell volume. In solving the discretized equations [16], the sum of absolute normalised residuals RES over a control volume about the central node P is defined as:

$$RES = \sum_{all\ grid\ cells} \left| \sum_I a_I \varphi_I + S_{\varphi,P} V_P - a_P \varphi_P \right| \tag{8}$$

Computations were stopped when $RES = 10^{-5}$ for all variables except enthalpy, for which the procedure stops when $RES = 10^{-9}$. Convergence is considered to be obtained when the total residual error in the field is less than 0.1% of the reference values for each variable, which are the mass flow rate, the inlet momentum flux and the inlet energy flux. Numerical solution of the variable φ is obtained by a point implicit (Gauss-Seidel) linear equation solver, combined with an algebraic multigrid method [28]. To ensure convergence, linear relaxation [28] was employed for all variables.

The CPU time required to obtain full convergence when using the optimum grids (792,520 cells) was 12-22h and 14-24h, depending on the wind speed, for the case of one opening and two openings, respectively. Computations were performed on a Windows XP PC (Intel Core 2 Duo, 1.66 GHz CPU and 1GB of RAM).

3. Empirical Models

3.1. Infiltration Velocity

Phaff and de Gids [15] studied the effects of both wind and buoyancy forces and developed the following empirical model, to predict the inlet velocity through a single vent:

$$v_{eff} = \sqrt{C_1 Vmet^2 + C_2 H \cdot \Delta T + C_3} \tag{9}$$

where v_{eff} is the velocity in the middle of the available infiltration area, $Vmet$ is the wind velocity recorded by a weather station, H is the height of the opening, C_1=0.001 is a wind speed constant, C_2=0.0035 is a buoyancy constant and C_3=0.01 is a turbulence constant.

3.2. Neutral Level

The neutral level through one opening is defined as the height where the interior pressure equals the ambient pressure, and thus the value of the average air velocity there becomes zero [1]. Higher internal pressure above this level drives air extraction and lower internal pressure below this level leads to infiltration. This mechanism forms the bi-directional flow presented in Figure 5.

For very low values of air speed, the neutral level can be calculated through the following equation [1]:

$$Z = H - \frac{\Delta p}{\rho g} \tag{10}$$

where Z is the height of neutral level, H is the opening height, Δp is the pressure difference between indoor and outdoor space at the height H, ρ is the air's density, and g is the gravitational acceleration.

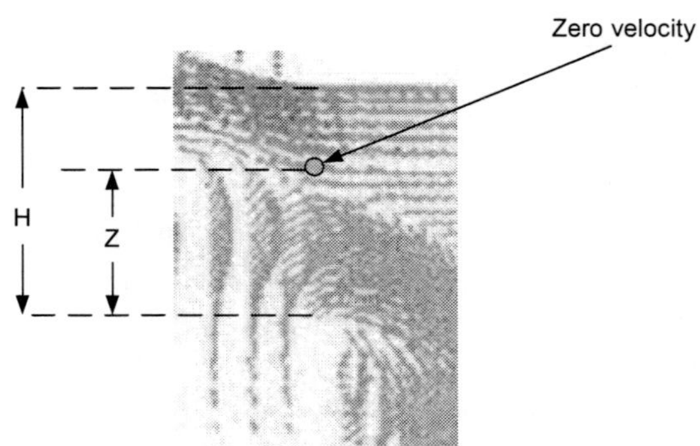

Figure 5. Bi-directional flow through a single vent

4. Results and discussion

4.1. Spatial Discretization

For the numerical solution, a multi-block non-uniform structured grid [25, 29] for each of the two cases is used. Grid independency is tested, by repeating the simulation for a gradually increased grid-cell density. For this reason, the embedded grid method [25, 29] is applied, according to which the grid is locally refined around critical areas, such as the thermal sources and the opening area. For example, the grid after the refinement around the human model, the TV, and the PC becomes as in Figure 6. The major advantage of this technique is the utilization of the results from coarser grids obtained in previous runs as initial values, thus accelerating convergence. This is achieved by interpolation of the previous results into the areas of additional cells. Three grid sizes are tested and, as revealed in Figure 7, for the x-velocity component, the optimum spatial discretization is that of 792,520 cells. The optimum grid for the problem considered is presented in Figure 8.

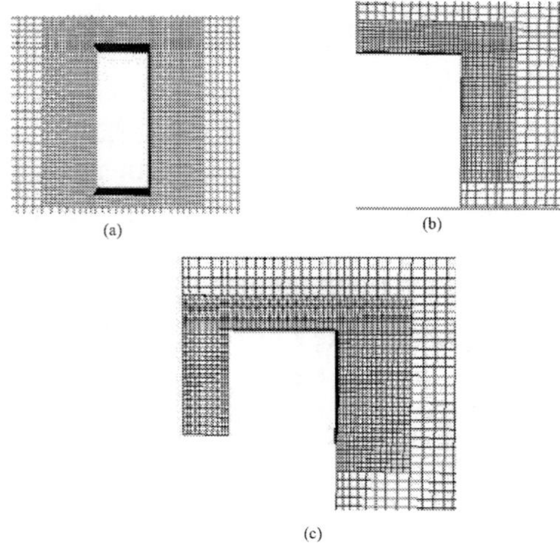

Figure 6. Embedded grid method. Grid refinement around the: (a) human model, (b) the TV, and (c) the PC.

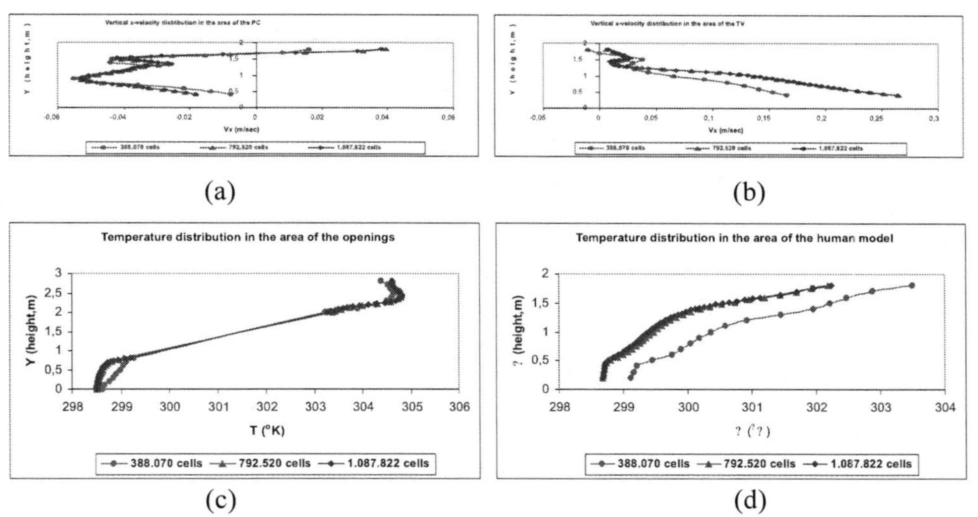

Figure 7. Vertical x-velocity distribution in the area of: (a) the PC, and (b) the TV. Vertical temperature distribution in the area of: (c) the openings, and (d) the human model.

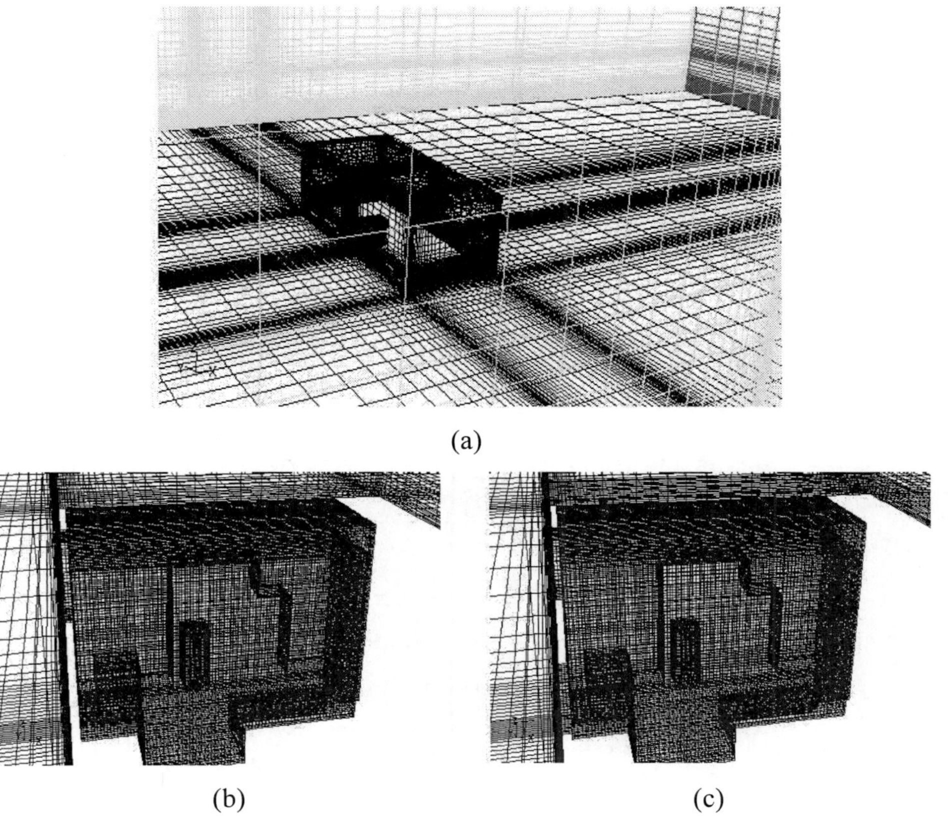

Figure 8. Optimum spatial discretization: (a) External domain; Internal domain with (b) two openings, and (c) one opening.

4.2. Validation of the CFD Model

In Figure 9, the numerical results obtained by both turbulence models are presented, together with the results from the empirical model of Phaff and deGids (eqn. 9). The empirical equation (9) was developed to describe the case of a single opening [15], located at the lower part of the building. However, in the present chapter, results obtained for both cases, i.e. of one and two openings, are presented in the same origin system in order to reveal and describe the differences that occur between the two geometries. Hence, the CFD model is validated, by comparing the results obtained only for the case of one opening. It is shown (Figure 9) that the results obtained by the numerical simulation deviate by approximately 18.6% and 41.4%, in terms of mean absolute value of relative error, using the standard k-ε and the RNG k-ε model, respectively, for a wind speed range $Vmet \in [2,10]$. It is observed that the infiltration velocity calculated, for the geometries studied, is similar at low wind speeds, i.e. 2-6 m/s, while it is higher in the case of two openings as the wind speed increases (6-10 m/s). This means that the air suction obtained is more intense in comparison with the case of only one opening. It should be mentioned, however, that although the width of the openings is similar for both cases, the surface area is different, thus further investigation is needed for reliable conclusions concerning the inflow velocity. Further details of the performance of the turbulence models are tabulated in Table 1.

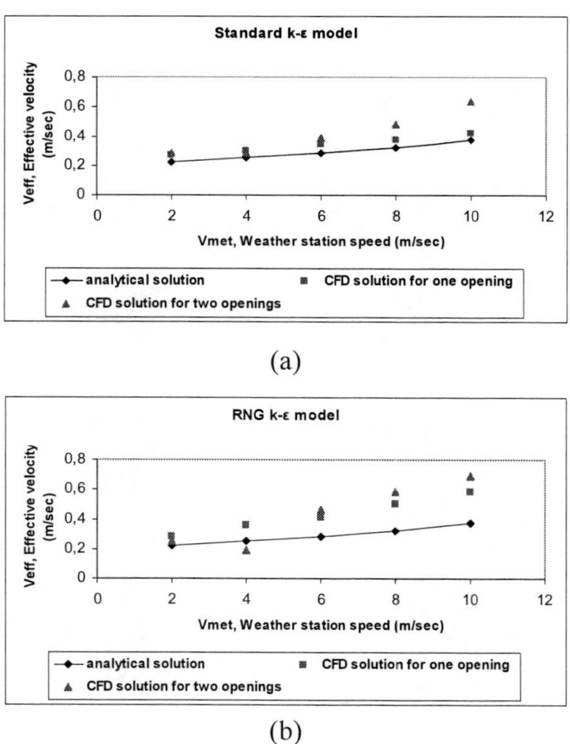

Figure 9. Results obtained by the CFD model for the cases of one and two openings (windows) on the windward façade using the: (a) standard k-ε, and (b) RNG k-ε model, in comparison with the results from eqn. 9, for various *Vmet* values.

Table 1. Numerical results obtained by both turbulence models and by eqn.9 for different wind speeds recorded by the weather station (*Vmet*).

Single-opening room						
	Standard k-ε model			RNG k-ε		
Vmet (m/s)	v_{eff}(m/s) Eqn. 9	v_{eff}(m/s) CFD	Mean absolute relative error*	v_{eff}(m/s) Eqn. 9	v_{eff} (m/s) CFD	Mean absolute value of relative error*
2	0.22	0.27	22.7	0.21	0.28	33.3
4	0.25	0.30	20.0	0.26	0.36	38.5
6	0.28	0.34	21.4	0.30	0.41	36.7
8	0.32	0.37	15.6	0.34	0.50	47.1
10	0.37	0.42	13.5	0.39	0.59	51.3
		Mean value	18.6%		Mean value	41.4%

$$* \ Mean\ absolute\ relative\ error\ for\ each\ Vmet = \left| \frac{v_{eff\,(eqn.9),Vmet} - v_{eff\,(CFD),Vmet}}{v_{eff\,(eqn.9),Vmet}} \right| \cdot 100\%$$

As far as the neutral level height is concerned, it is estimated following a regression technique utilizing the numerical results, through the window centreline, to yield a polynomial function of height; the height of minimum absolute velocity value is then calculated following an iterative procedure (Newton-Raphson). Neutral level heights obtained by both turbulence models are presented in Figure 10. A decrease of the predicted neutral level height is observed, as the wind speed increases, while the mean discrepancy of the predicted value, compared to that obtained by the empirical model (eqn. 10) is approximately 15.2% and 14.2% using the RNG and the standard k-ε model, respectively. The aforementioned differences are presented in Table 2.

It is concluded that the discrepancy obtained using the RNG k-ε model is now lower, compared with that by the standard k-ε model. However, both mean discrepancies are small, and both models produce a satisfying prediction. It must be mentioned that the discrepancy decreases as the wind speed increases, being minimum for the maximum wind speed. In summary, both models are considered satisfying, and they lead to acceptable agreement with the empirical models. Especially, when the standard k-ε model is applied, minimum difference occurs concerning the infiltration velocity (Figure 9), while the observed difference referring to neutral level height may be considered acceptable. It should also be kept in mind that eqn. 9 is, of course, not error-free either.

The predicted flow field, in terms of velocity and temperature distributions, is depicted in Figures 11 and 14, respectively, for both cases considered, at the longitudinal plane of the domain, using the standard k-ε model, for a wind speed *Vmet*=2m/s. In the case of two openings, the air is infiltrated through the bottom vent of the windward wall and is extracted through the upper vent, leading to the displacement of warm air masses from the occupied zone (see Figure 11(a)), and thus strong internal temperature stratification is created (see Figure 14(a)). In the case of a single opening, the air is infiltrated through the lower part (below neutral level height) of the vent, mixing with the internal air mass and finally extracted through the upper part (above neutral level height) of the vent (see Figure 11(b)). As expected, the predicted flow field is characterized by more uniform, and higher, temperature, compared with the previous case, in the occupied zone (approximately 33 ^{0}C) (see Figure 14 (b)). At higher wind speeds, (see Figurs. 12(a), 12(b)) the effective depth increases, as the incoming air descends deeper in the indoor space of the room, and thus it mixes more mass of internal warm air. In the case of two openings the direction of the incoming air changes as the wind speed increases, while in the case of one opening the direction of the incoming air is the same at all wind speeds studied (see for example Figurs. 13(a), 13(b)). Three dimensional temperature distribution, in the case of displacement ventilation, is illustrated in Figure 15, where the heated areas due to the presence of heat sources are revealed. The expected mean flow around the building is illustrated in Figure 16, where separation and reattachment zones on the windward and leeward sides, respectively, are shown. The produced flow field clearly reveals the so-called "horse-shoe" vortex, which usually occurs in flows around bluff bodies [24].

4.3. Thermal Comfort Study

Evaluation of internal thermal conditions is based on two criteria recommended by the American Society of Heating, Refrigerating and Air-Conditioning Engineers (ASHRAE) Standard [30] for thermal comfort:

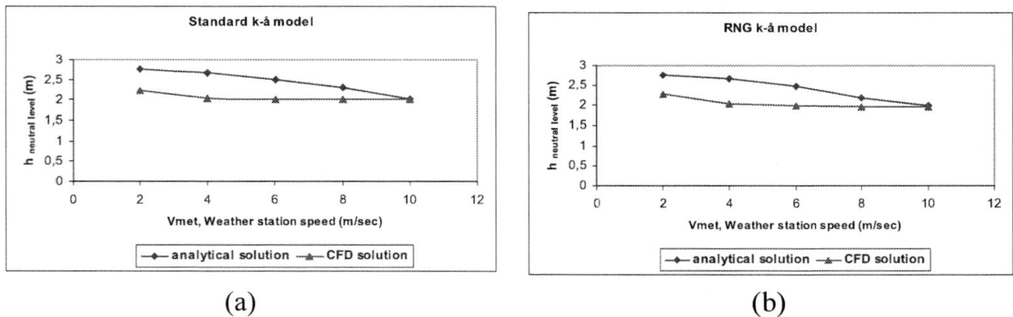

(a) (b)

Figure 10. Neutral level height calculated by: (a) the standard k-ε model, and (b) the RNG k-ε model.

Table 2. Neutral level height calculated by both turbulence models.

Room with a single opening					
Vmet (m/sec)	Neutral level height (m) Eqn. 10	Neutral level height (m) standard k-ε	Mean absolute relative error* (standard k-ε)	Neutral level height (m) RNG k-ε	Mean absolute relative error (RNG k-ε)
2	2.75	2.23	18.93	2.28	17.17
4	2.66	2.03	23.75	2.03	23.46
6	2.49	2.00	19.92	1.98	19.45
8	2.28	2.00	12.60	1.96	10.01
10	1.98	2.01	0.78	1.97	0.64
		Mean value	15.2%		14.2%

$$^{*}\,Mean\ absolute\ relative\ error = \left|\frac{N.L.H._{(eqn.9),Vmet} - N.L.H._{(CFD),Vmet}}{N.L.H._{(eqn.9),Vmet}}\right| \cdot 100\%$$

where $N.L.H.$: Neutral Level Height

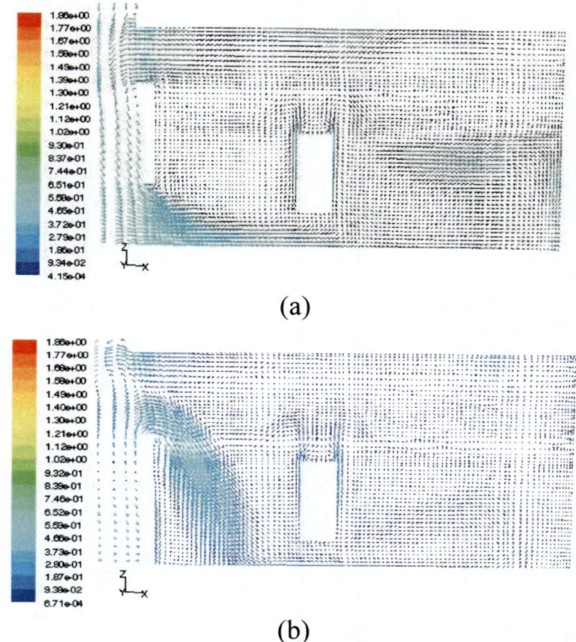

(a)

(b)

Figure 11. Velocity vectors at the longitudinal plane of the domain for a room with: (a) one opening, and (b) two openings, at weather station speed 2m/sec.

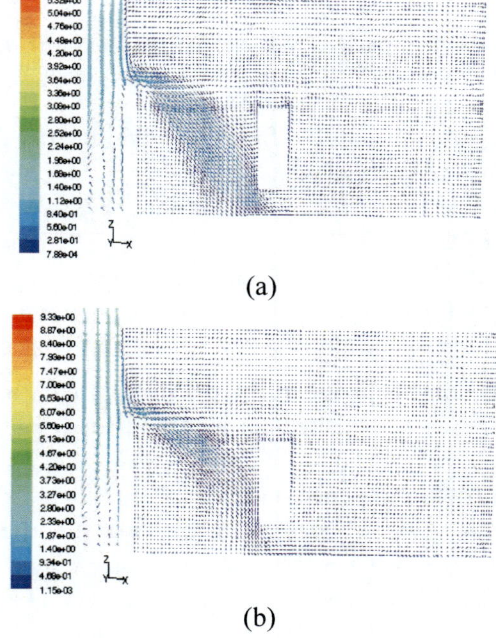

(a)

(b)

Figure 12. Velocity vectors at the longitudinal plane of the domain for a room with one opening at weather station speed: (a) 6m/sec, and (b) 10m/sec.

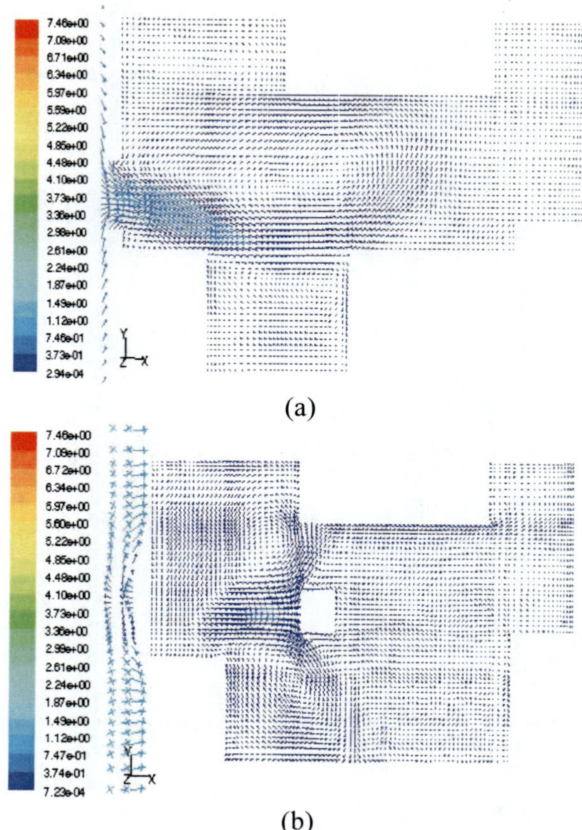

Figure 13. Velocity vectors at the floor-plane of the domain for a room with: (a) one opening, and (b) two openings, at weather station speed 8m/sec.

1st Criterion: Vertical air temperature difference within the occupied zone, measured at the 0.1m and 1.7m levels shall not exceed 3°C.

2st Criterion: The surface temperature of the floor for people wearing typical indoor footwear shall be between 18°C and 29°C.

In Figure 15, temperature difference in the occupied zone is depicted (0.1-0.7m), and in the case of the room with two openings the first recommendation is satisfied at most of the examined wind speeds, with the exception of lower values of wind speed ($Vmet \in [2,4]$). In these cases, warm air mass remains at low level and thermal comfort conditions cannot be established. The second criterion is satisfied for all values of wind speed studied (see, for example, Figure 14(a)).

In the case of the room with one opening the first criterion is satisfied only for the highest value of external wind speed $Vmet = 10m/\sec$. In this case the effective depth [8] of the infiltrated air is greater, and thus better mixing in the occupied zone is created. As a result, thermal comfort conditions are established in the room. The second criterion is not satisfied, as mean internal temperature exceeds 29 ^{0}C (see, for example, Figure 14(b)). It may be concluded that in the case of two windows, the displacement ventilation taking place is able to create thermal comfort conditions inside the room in most cases studied, as has been discussed also in previous investigations [31, 32]. Finally, it is observed that both mechanisms of single-sided ventilation (displacement, mixing ventilation) are more efficient for relatively high external wind speeds.

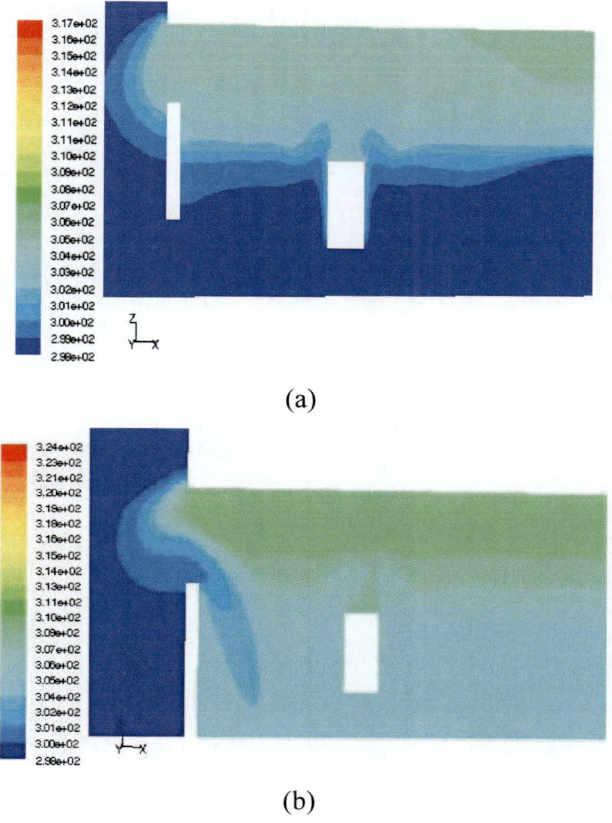

Figure 14. Temperature contours at the longitudinal plane of the domain for a room with: (a) one opening, and (b) two openings.

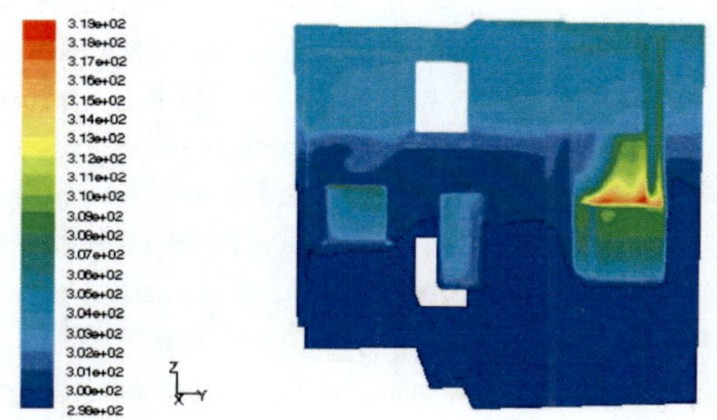

Figure 15. Three dimensional temperature distribution in the case of displacement ventilation.

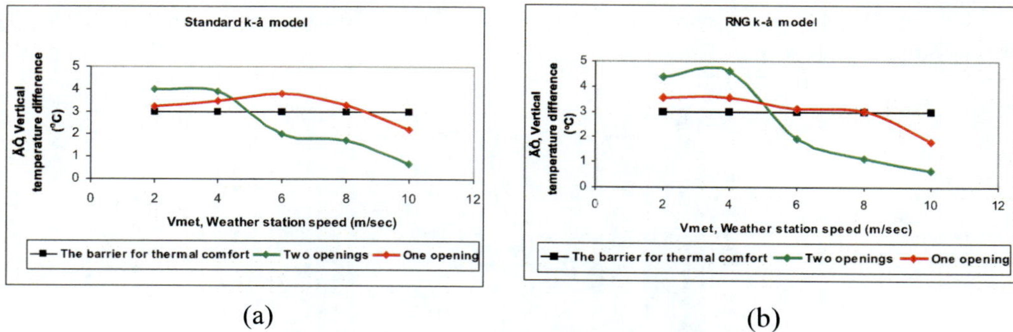

(a) (b)

Figure 15. Vertical temperature difference in the occupied zone in both cases studied, using: (a) the standard k-ε model, and (b) the RNG k-ε model. The barrier for thermal comfort ΔT=3 °C is also presented.

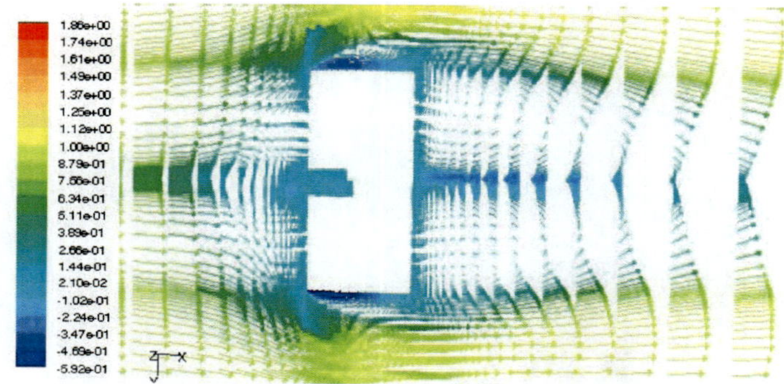

Figure 16. Formation of the "horse-shoe" vortex around the building.

5. Conclusion

The proposed CFD model is considered satisfying, especially when the standard k-ε turbulence model is used, as it provides acceptable agreement with the results obtained by well-known empirical models, i.e. discrepancies of only about 18.6% and 15.2%, referring to mean infiltration velocity and the height of neutral level, respectively.

In the case of the room with two (upper and lower) openings the mechanism of displacement natural ventilation takes place. The average temperature inside the room is lower than that of the room with one opening, while the direction of incoming air changes as the wind speed increases. On the other hand, in the case of the room with one upper opening the mechanism of mixing natural ventilation takes place. As the speed of incoming air increases a decreased height of neutral level is calculated and a greater effective depth is observed. The direction of the infiltrated air is the same for all the wind speeds studied, whilst the average temperature inside the room is higher in comparison with the case of the room with two openings.

As far as the openings' design is concerned, it is observed that single-sided natural ventilation does not contribute to the creation of thermal comfort conditions at low wind speeds ($Vmet \in [2,4] m/\sec$) for both cases studied, while at higher wind speeds, displacement ventilation (two openings) performs better than mixing ventilation (one opening), due to the creation of thermal comfort conditions in most cases of wind speed studied. Finally, it may be concluded that a residential building involving more than one openings located at the lower and the upper heights of the windward facade may prove to be a valuable design idea, as it meets thermal comfort recommendations, especially in areas of relatively strong winds. However, other thermal comfort indices [25] should also be tested, in a future work, in order to ensure optimum architectural design, with thermal sensation due to both skin-bulk flow heat transfer and draught, as for example in [5], taken into account.

ACKNOWLEDGEMENT

This work was sponsored by the State Scholarships Foundation of Greece.

REFERENCES

[33] Linden, PF. *Ann Rev Fluid Mech*, 1999, 31, 201-238.
[34] Energy Consumption Guide 19, *Energy Efficiency Office/HMSO*, London, 1993
[35] Hunt, GR; Linden, PF. *J. Fluid Mech*, 2004, 527, 27-55.
[36] Allocca, C; Chen, Q; Glicksman, LR. *Energy Buildings*, 2003, 35, 785-795.
[37] Stavrakakis, GM; Koukou, MK; Vrachopoulos, MGr; Markatos, NC. *Energy Buildings*, 2008, 40, 1666-1681.
[38] Schaelin, AJ; van der Maas, AJ; Moser, A. *ASHRAE Transactions*, 1992, 98, 319-328.
[39] Li, K; Teh, SA. *Indoor Air*, 1996, 2, 1027-1032.
[40] Gan, G. *Energy Buildings*, 2000, 31, 65-73.
[41] Papakonstantinou, KA; Kiranoudis, CT; Markatos, NC. *Energy Buildings*, 2000, 33, 41-48.
[42] Jiang, Y; Chen, Q. Int *J Heat Mass Transfer*, 2003, 6, 973-988.
[43] Jiang, Y; Alexander, D; Jenkins, H; Arthur, R; Chen, Q. *J Wind Engng Ind Aerodyn*, 2003, 91, 331-353.
[44] Jiang, Y; Chen, Q. *J Wind Engng Ind Aerodyn*, 2001, 89, 1155-1178.
[45] Dascalaki, E; Santamouris, M; Argiriou, A; Helmis, C; Asimakopoulos, DN. *Energy Buildings*, 1996, 24, 155-165.
[46] Markatos, NC. Computer analysis of building ventilation and heating problems; In Proc. Int. *Conf. on Passive and Low Energy Architecture*; Pergamon Press: 1983, pp. 667-675.
[47] Phaff, JC; De Gids, WF; Ton, JA; Van der Ree, DV; Schijndel, LLM. The ventilation of buildings: Investigation of the consequences of opening one window on the internal climate of a room; In Report C 448, TNO Institute for Environmental Hygiene and Health Technology (IMG-TNO); Delft, Netherlands, March 1980.
[48] Versteeg, HK; Malalasekera, W. An introduction to computational fluid dynamics – The finite volume method; Longman Group Ltd: Essex, England, 1995.

[49] Pozrikidis, C. Kluwer academic publishers, Norwell, Massachusetts, USA, 2001, 50-110.
[50] Patankar, SV; Spalding, DB. Int J Heat Mass Transfer, 1972, 15, 1787-1806.
[51] Markatos, NC. Applied Mathematical Modelling, 1986, 10, 190-220.
[52] Launder, BE; Spalding, DB. Lectures in Mathematical Models of Turbulence; Academic Press: London, England, 1972.
[53] Markatos, NC; Malin, MR; Cox, G. Int. *J Heat Mass Transfer*, 1982, 25, 63-75.
[54] Yakhot, A; Orszag, S. *J Scientific Computing*, 1986, 1, 1-51.
[55] Gan, G. Numerical Heat Transfer, Part A: *Applications*, 1998, 33, 169-189.
[56] Ferziger, JH; Peric, M. Computational methods for fluid dynamics-third edition-; *Springer-Verlag*, NY, 2002.
[57] Stavrakakis, GM; Stamou, AI; Markatos, NC. Evaluation of thermal comfort in indoor environments using Computational Fluid Dynamics (CFD), In: Harris RG, Moore DP (editors), Indoor work and living environments: health, safety and performance. Nova Science Publishers Inc., in press, ISBN: 978-1-60741-375-2.
[58] Evola, G; Popov, V. *Energy Buildings*, 2006, 38, 491-501.
[59] Bartzanas, T; Kittas, C; Sapounas, AA; Nikita-Martzopoulou, Ch. *Biosystems Engineering*, 2007, 97, 229-239.
[60] Fluent Inc. Fluent User's Guide, Version 6.2. Lebanon, NH, USA: Fluent Inc., 2005.
[61] Jones, PJ; Whittle, GE. *Building Environment*, 1992, 27, 321-338.
[62] Thermal environmental conditions for human occupancy, ANSI/ASHRAE 55-1992.
[63] Lin, Z; Chow, TT; Fong, KF; Wang, Q; Li, Y. Int *J Refrigeration*, 2005, 28, 276-287.
[64] Q. Chen, *Energy and buildings*, 2004, 36, 1197-1209.
[33] N.C. Markatos, *Revue de l'Institut Francais du Petrole*, 1993, 48, 631-661.

In: Buildings and the Environment
Editors: Jonas Nemecek and Patrik Schulz
ISBN: 978-1-60876-128-9

Chapter 6

ESTABLISHMENT OF NEGATIVELY-CHARGED INDOOR AIR CONDITIONS AND THEIR BIOLOGICAL EFFECTS

Takemi Otsuki[1*], Kazuaki Takahashi[2], Akinori Mase[2], Takashi Kawado[2], Muneo Kotani[2], Yasumitsu Nishimura[1], Megumi Maeda[1], Shuko Murakami[1], Naoko Kumagai[1], Hiroaki Hayashi[1], Ying Chen[1], Yoshie Miura[3], Takashi Shirahama[4], Michiharu Yoshimatsu[4,6] and Kanehisa Morimoto[5]

[1] Department of Hygiene, Kawasaki Medical School, 577 Matsushima, Kurashiki 7010192, Japan,

[2] Comprehensive Housing R&D Institute, Sekisui House, Ltd., 6-6-4 Kabutodai, Kizugawa 6190233, Japan,

[3] Division of Molecular and Clinical Genetics, Department of Molecular Genetics, Medical Institute of Bioregulation, Kyushu University, 3-1-1 Maidashi, Higashi-ku, Fukuoka 8128582, Japan,

[4] Artech Kohboh Co. Ltd., 176 Nippacho, Yokohama 2230057, Japan.

[5] Department of Social and Environmental Medicine, Osaka University Graduate School of Medicine, 2-2 Ymadaoka, Suita 5650871, Japan, and

[6] This author died on September 22, 2007. All authors mourn the loss of this associate.

ABSTRACT

Indoor air condition may be involved to the health status□Among various factors in indoor air conditions, few reports were published concerning negatively charged-particles. In this review, we detaile the biological effects of negatively-charged indoor air conditions, particularly on the psycho-neuro-endocrino-immune network. The short-

* Corresponding author: Department of Hygiene, Kawasaki Medical School 577 Matsushima, Kurashiki 7010192, Japan Tel +81 86 462 1111, Fax +81 86 464 1125 E-mail takemi@med.kawasaki-m.ac.jp

term (2.5 hours) and mmedium-term (2 weeks) abidance of negatively-charged air particle dominant room resulted activation of natural killer cell activity induced by recurrent transient activation of interleukin 2. These results indicate that negatively-charged indoor air condition is better for immune status and may improve daily lives.

INTRODUCTION

People living in developed countries are exposed to many physical and mental stresses caused by the urban environment, including air, water and ground pollution[1-6], as well as stresses resulting from the workplace and human relationships [7,8]. In particular, sick-building syndrome (SBS) [9-13] and multiple chemical sensitivity syndrome (MCS) [14-18] are typical health disorders associated with indoor environments. These diseases are mainly caused by indoor air aldehydes and volatile organic compounds (VOCs) that affect human psycho-neuro-endocrino-immune (PNEI) networks [19-23]. The concept of a PNEI network has evolved over the past few decades and is based on the close relationship between immunoregulation and brain functions [24-26]. Particular emphasis is given to circuits involving immune cell products, the hypothalamus-pituitary-adrenal axis, and the sympathetic nervous system. There is increasing evidence that brain-born cytokines play an important role in brain physiology and the integration of the PNEI network. It is thought that the various complaints and symptoms associated with SBS and MCS result from an impairment of the PNEI network and that psychological stabilization may improve symptoms in patients with SBS and MCS [19-26], although avoidance of chemicals is the most important preventive method for these disorders. Measures for reducing VOCs and aldehydes have been developed, and concentrations of these chemicals have been reduced through legislation in Japan and other well-developed nations.

Various trials concerning the domiciliary environment have been developed to support and promote human health in indoor conditions, such as those involving the use of fewer chemicals in a building [27]. The study of air electrical charge represents another trial. A few studies have demonstrated that indoor air represents a negatively charged electric condition and that this condition may be beneficial for human health [28-30]. For example, charcoal coating on walls and ceilings with positive charging produces indoor air that is negatively charged. Negatively-charged indoor air conditions are recognized in the commercial sector and many companies that are manufacturing and selling negative-ion producers and air purifiers are emphasizing that negatively-charged indoor air conditions are better for human health [31,32]. In addition, it is known that there are relatively high negatively-charged air particles in forests and nearby water-falls. The beneficial immunological effects of a forest tour (forest bathing) over a period of a few days have been reported, which may be the result of these negatively-charged air conditions [33-35]. However, there is no conclusive evidence concerning the biological effects of negatively-charged indoor air conditions [36].

Thus, in this chapter we detail the establishment of negatively-charged indoor air conditions and investigate their biological effects.

PRINCIPLE OF MAKING NEGATIVELY- CHARGED INDOOR-AIR CONDITIONS

As shown schematically in Figure 1, the generation of negatively-charged indoor air conditions involved painting a charcoal coating made by fine charcoal powder onto the walls and ceilings of a room. A charcoal coating designated as Health Coat® was produced by Artech Kohboh Co. Ltd. In addition, forced negatively-charged air conditions were created by applying an electric voltage (72 volts) between the backside of the walls of a room and the ground, and using a circuit for generating negatively-charged air conditions. This condition was made by the selective sorption of floating positively-charged air particles on the surfaces of walls and ceilings since these surfaces were positively charged. Thus, these conditions were recognized as a relative reduction of positively-charged air particles, rather than a production of negatively-charged indoor air particles.

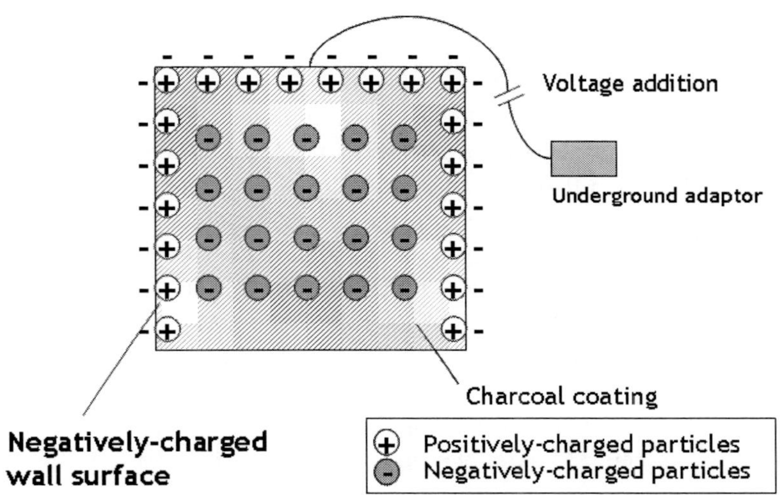

Figure 1. Schematic representation of the creation of negatively-charged indoor air conditions. Rooms were painted with a charcoal coating on the walls and ceiling, and an underground adaptor and additional voltage were then used to create a negative electrical charge on the surface of the walls. Since the positively-charged air particles were then adsorbed to the surface of the walls, the relative indoor air conditions were dominated by negatively-charged particles.

As described below, procedures to analyze the biological effects of this negatively-charged indoor-air condition were performed by two sets of experiments.

The first set was designed to investigate the short-term effects. Sixty healthy volunteers were admitted to a control room (CR; there was no difference between negatively- and positively-charged air particles) or experimental room (ER; negatively-charged air being the dominant condition) and changes of various biological markers that related to the PNEI network were analyzed between CR and ER. The CR and ER were built in a wide underground laboratory of the Comprehensive Housing R&D Institute, SEKISUI HOUSE, Ltd., Kizugawa- City, Kyoto Prefecture, Japan, as shown in Figure 2. The area and volume of the laboratory was approximately 539 m^2 and 1564 m^3, respectively, and that of the experimental rooms was 9.1 m^2 and 22.8 m^3, respectively. All of the 120 healthy volunteers

(HVs) were Japanese, each of whom provided written informed consent. Sixty HVs (age; mean ± S.D. = 44.08 ± 6.50, Male : Female = 30 : 30) were admitted to a CR and the remaining 60 HVs (Age; mean ± S.D. = 44.70 ± 6.36, Male : Female = 30 : 30) were admitted to an ER, each for a period of 2.5 hours. The experiments were performed during 18–30 November, 2005 at the Comprehensive Housing R&D Institute, SEKISUI HOUSE, Ltd., Kizugawa City, Kyoto Prefecture, Japan.

The second set of experiments examined the relative medium-term effects of negatively-charged indoor air conditions. The CR and ER were built in a dormitory as shown in Figure 3. All subjects were company members of Sekisui House, Ltd. Ten subjects came to this dormitory for their company's training for three months from business offices situated in seven different prefectures in Japan, and another five were members of the Comprehensive Housing R&D Institute, Sekisui House, Ltd., situated 1 km from the dormitory. All subjects were male, and the mean age and standard deviation were 27.7 ± 2.0 years. The experiments were performed from October 6, 2007 to March 13, 2007. Subjects received training or engaged in daily business practices during the daytime, and then returned to the dormitory for dinner, leisure time, bathing, and sleeping. Subjects joined the experiments only after their informed consent was obtained. There were no advantages or disadvantages associated with whether company members agreed or declined to join the experiments. These studies were approved by the Ethics Committee of Kawasaki Medical School. For these second experiments, subjects were initially admitted to the pre-living room, which was the same as the CR, for two weeks. They then moved to the CR, and although air conditions were the same, subjects were not notified that they were admitted to the CR or ER. After a stay of two weeks, they were again moved to the next room. This final room was the ER, and subjects were not told that they were in the experimental room. The reason why all the subjects were sequentially admitted to the CR and ER is that there was a limited number of healthy subjects. Thus, a comparative study was not carried out. We checked the biological responses in each subject before and after admission to the ER. The CR was the first step in the experimental protocol because it was better to avoid any long-term (more than two weeks) effects of the ER.

INDOOR AIR CONDITIONS IN THE CR AND ER

Indoor air conditions such as temperature, humidity, air pollutants and electrical charge were measured for all experimental rooms in the two sets of experiments. Temperature and humidity were measured using a temperature/humidity data logger (TR-72STM, T&D Corporation, Matsumoto, Japan). These two parameters were continuously monitored during the entire experimental periods. For the entire duration of the two sets of experiments, there was no difference between CR and ER regarding temperature and humidity.

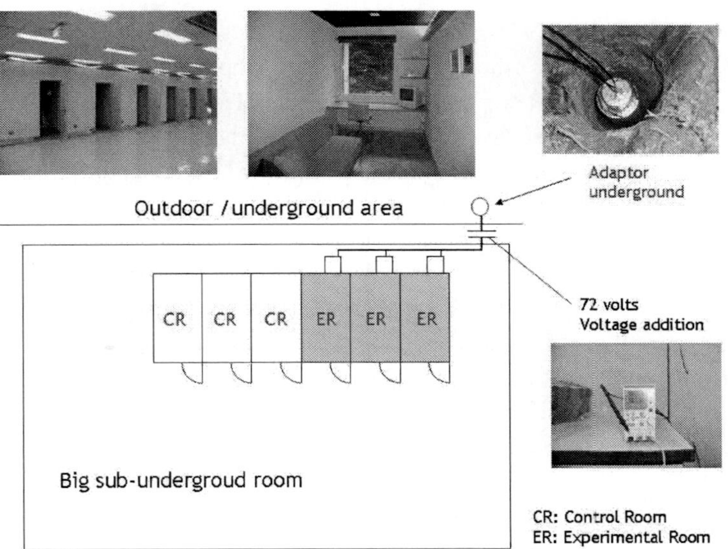

Figure 2. Schematic representation and external/internal view of experimental (ER) (and also control (CR)) rooms used in the first, short-term exposure experiments. The ER and CR were constructed in a huge underground laboratory room as shown. The outer and inner appearances of both rooms were similar. However, the ER was fitted with devices comprising an underground adaptor and additional voltage to create negatively-charged indoor air conditions.

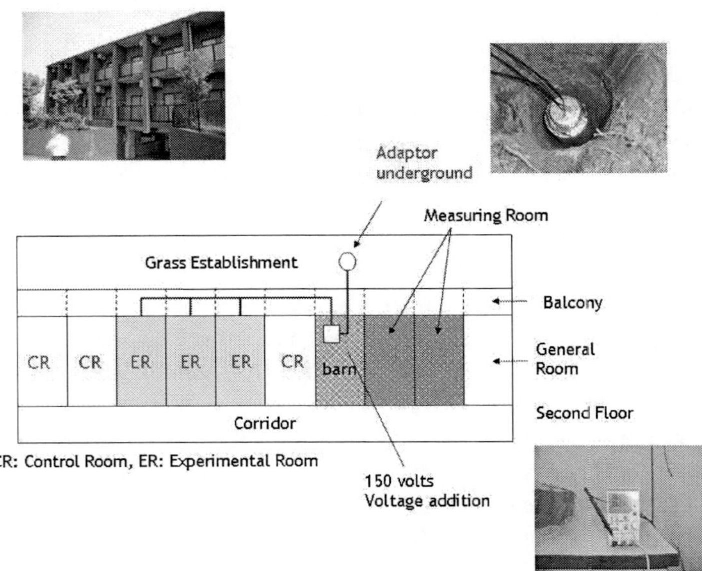

Figure 3. Schematic representation and external view of the dormitory used for the second, medium-term exposure experiments. Devices to create negatively-charged indoor air conditions were the same as those used for the first experiments and are shown in Figures 1 and 2.

Air pollutants designated by the guideline concerning indoor room concentrations for indoor air pollutants, released by the Japanese Ministry of Health, Labor and Welfare and covering levels of formaldehyde, acetaldehyde, toluene, xylene, styrene, ethyl benzene, paradichlorobenzene and total VOC, were measured using an active sampling method. All concentrations of air pollutants showed levels below the legal range and there were no differences between CR and ER for both experiments. Electrical charge in these rooms was measured using an Ion Counter (EB-1000TM, Eco Holistic Inc., Suita, Japan). Electrical charge in two representative rooms (Room C for the ER and Room D for the CR) was monitored every morning, and all rooms were monitored twice a week during the entire experimental period in the first experiment. In the second set of experiments, the electrical charges in the two groups of rooms (all six rooms) were monitored once a week for the entire experimental period. The results showed that there was an approximate 500-particle (/1,000 mm^3) higher concentration of negatively charged than positively charged air particles during the entire experimental periods of both experiments. There were no differences in negatively charged air particles between CR and ER, but the numbers of positively charged air particles in ER were reduced in both experiments. This difference resulted in negatively-charged indoor air conditions.

BIOLOGICAL MONITORING

The following parameters were monitored before and after admission to the ER or CR.

1. General conditions: Blood chemistry including liver and kidney functions, blood sugar and lactic acid, and peripheral blood count were measured using peripheral blood samples. In addition, blood pressure and pulse rate were measured at pre- and post-admission to the CR and ER.
2. Stress markers: Blood cortisol and salivary cortisol, chromogranin A, amylase, and secretory immunoglobulin (Ig) A were measured as stress markers. Saliva was collected by SalivetteTM (Sarstedt Ag & Co., Nümbrecht, Germany) according to the manufacturer's instructions. Briefly, each subject kept the cotton part of the Salivette in his/her mouth for 3 min. The cotton was then frozen at -20°C until centrifugal collection of the saliva. The values of each item were corrected by the total protein of the individual sample.
3. Parameters related to the autonomic nervous system: The autonomic nervous system was examined through a Flicker test, stabilometer, and heart rate monitoring for 3 min. The Flicker test was performed using a Handy Flicker HFTM (Neitz Instruments Co. Ltd., Tokyo, Japan), and flicking frequencies of red, green and yellow colors were monitored. A Gravicoder GS-7TM (Anima Inc., Tokyo, Japan) was used as a stabilometer and the Romberg ratio was used as a measure of body sway. The ratio was calculated using the whole trajectory of body sway during 30 sec while standing with eyes closed divided by that recorded with eyes open. Heart rate was monitored using a Heart Rate Monitor S810iTM (Polar Electro, Kempele, Finland) for 3 min. During monitoring, subjects were lying on a bed and remained at rest. The frequency-domain parameters of heart rate variability such as low-frequency power [LF], high-frequency power [HF], and low-/high-frequency ratio [LF/HF] were analyzed using MemCalc/Tarawa software (G.M.S. Co. Ltd., Tokyo, Japan).
4. Immunological parameters: Serum levels of Ig E and Ig A, cytokines related to the Th1/Th2 balance (Interferon (IFN)-γ, tumor necrosis factor (TNF)-α, Interleukin (IL)-2, IL-4, IL-6 and IL-10) were evaluated. The individual samples for cytokine measurement were applied to the Cytometric Bead Array of Human Th1/Th2 Cytokine Kit II (CBA, BD Bioscience, San Jose, CA, USA) and measurements were made using FACSCalibur

flow-cytometry (BD Bioscience) according to the manufacturer's instructions 37,38). Samples that showed less than the lower limitation of analytical methods for cytokines and Igs, 1/10 the value of minimum values among whole measurable samples, were substituted instead of using "0" or the designation "unmeasurable".

5. Blood viscosity: Blood viscosity was measured using a Micro-Channel Flow Analyzer MC-FAN (MC Laboratory Inc., Tokyo, Japan) according to the manufacturer's instructions [39,40]. Briefly, a peripheral heparinized blood sample (100 μl) was put into the instrument and flowed through the micro-channel chips, which are a model for capillary vessels, and the flowing time was recorded. The flowing blood sample could also be visualized using the CCD camera attached to the microscope's objective.

In addition to the above-mentioned parameters measured after 2.5 hours of abidance, natural killer (NK) cell killing activity, urine 8-hydroxydeoxyguanosine (8-OHdG) concentration, urine 17-hydroxycorticosteroids (17OHCS) and specific Ig E classes for 26 kinds of typical allergens (MAST26 test, SRL Inc., Tachikawa, Tokyo, Japan) were also added to the 2-week study because it was thought that these newer parameters might be altered following a week-based experimental period, in contrast to one that is hour-based. The 26 allergens in MAST26 were as followings: *Dermatophagoides farinae*, house dust, epithelium of cats and dogs, *Phleum pretense*, *Anthoxanthum odorathum*, *Ambrosia artemisiifolia*, mugwort, *Cryptomeria japonica*, *Penicillium viridicatum*, *Cladosporium*, *Candida*, *Alternaria*, *Aspergillus*, wheat, soybean, rice, *Thunnus*, salmon, shrimp, crab, cheddar cheese, milk, beef, chicken meat, and egg albumin.

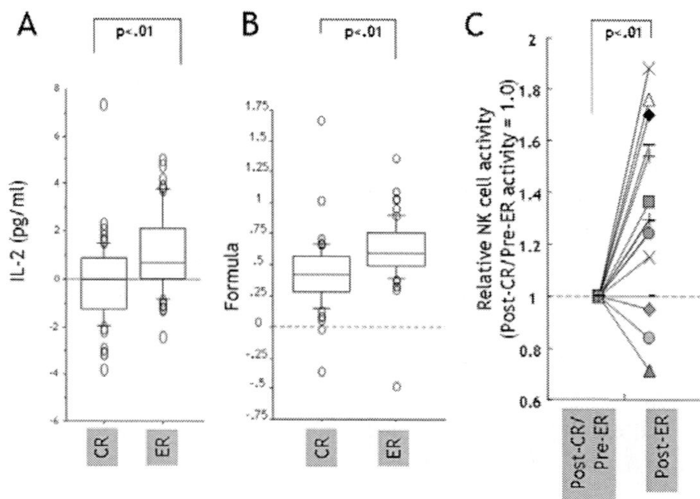

Figure 4. [A] Statistically significant differences for changes (pre-abidance to a CR or ER room compared with post-abidance) in the serum level of IL-2. Changes in the level of IL-2 in subjects that entered the ER were higher than those of the CR. [B] The formula constructed from the multiple/stepwise regression test using data obtained from the first set of experiments showed that the subjects who entered ER and CR were clearly divided. On the basis of variables used in this formula, serum levels of IL-2 and blood viscosity were considered the most important parameters that were affected by negatively-charged indoor air conditions. [C] The relative NK activities of individual subjects are shown such that the post-CR/pre-ER period is given the value of 1.0. The Wilcoxon Signed Ranks test indicated there is a significant increase of NK activity following ER abidance.

Modification of Biological Parameters

In the first set of experiments involving short-term abidance, pre-abidance and post-abidance values of the various parameters were compared between CR and ER using the Mann-Whitney *U* test of StatFlex version 5.0 software for Windows (Artech Co. Ltd., Osaka, Japan), which yields results that are compatible with SPSS software. Changes of IL-2, IL-4, blood viscosity and the RR-interval of heart rate showed statistical significance. Among these parameters, the change of IL-2 levels was the most significant. Subjects that entered ER showed increased IL-2 levels compared with those that entered CR, as shown in panel A of Figure 4.

In addition, a multiple/stepwise regression test was used to derive the following formula to evaluate subjects that entered the ER:

Biological response values = 0.498 + 0.0005[salivary cortisol] + 0.072 [IL-2] + 0.003 [standard deviation of heart rate (fluctuation of heart rates)] – 0.013 [blood viscosity] – 0.0009 [blood sugar] + 0.017 [pulse rate]

Using the above formula, the values of variables in individual subjects that entered the CR and ER were re-calculated. Thereafter, as shown in panel B of Figure 3, values from subjects that entered the ER were significantly higher than those of CR subjects. In the above formula, as the constant numbers for IL-2 and blood viscosity were higher than other parameters, these two parameters were considered to be related significantly to ER conditions. In other words, these two parameters were particularly affected by negatively-charged indoor air conditions.

In the second set of experiments, statistical significance was analyzed after taking into consideration the limited number of subjects. This statistical analysis involved the use of the Wilcoxon Signed Ranks test to compare values of pre- and post-admission to the ER for individual parameters.

The results revealed that only NK activity showed statistically significant differences, as shown in panel C of Figure 3. NK activity varied from subject to subject and changed during the pre-CR time, post-CR/pre-ER time, and post-ER time. However, when we considered the post-CR/pre-ER time as 1.0 in each subject and relative changes of post-ER were calculated, there was a significant increase of NK activity, although four subjects showed no decreased activities. Unfortunately, NK activity was the only parameter that showed statistical significance among all the parameters examined. However, NK activity is very important for tumor immunity and in combating other pathogens and viruses. Negatively-charged air conditions may be helpful in preventing cancers and infections when people are continuously living in a house with negatively-charged air conditions.

Analysis for Mental Status and Mood

All subjects were examined before and after exposure to the ER for various biological parameters related to the PNEI network. An anxiety score using the POMS (Profile Of Mood Status) questionnaire was also recorded [41,42]. In addition, subjects answered a questionnaire before admission for the assessment of daily lifestyle, as developed by Morimoto [43-45].

Our results revealed that there was no change of mood, mental status or anxiety scores, particularly in the first experiment. In addition, there was no correlation between status of lifestyle and biological modifications induced by ER. These results were similar to those of the second set of experiments. In particular, all subjects in the second set received hard job-training and any changes of mood or anxiety scores would have been dependent on the personal characteristics of subjects and their response to this type of training.

Discussion and Future Analyses

The investigations outlined in these studies produced important results. First of all, it should be noted that the experimental rooms utilized in these studies were used to estimate only the effects of negatively-charged air conditions on parameters indicative of human health. Temperature, humidity and air pollutants were controlled, and there was no difference between the CR and ER groups other than air charge. In addition, these experimental rooms contained no air pollutants. Thus, it is important to appreciate that investigations using these rooms do present the biological effects of negatively-charged indoor air conditions. Negatively-charged indoor air conditions are recognized in the commercial sector, and many companies that manufacture and sell negative-ion producers and air purifiers emphasize that negatively-charged indoor air conditions are better for human health [31,32]. However, there is no scientifically conclusive evidence concerning the biological effects of negatively-charged indoor air conditions. As a first step in obtaining data concerning the biological effects of negatively-charged indoor air conditions, the NCEI network was analyzed in the first set of experiments by placing subjects in negatively-charged indoor air conditions for a short-time and monitoring biological markers of human health. The investigation could then move to the relatively medium-term experiments used as the second set of experiments of this study. Devices to make experimental negatively-charged indoor air conditions were applied to the dormitory and subjects spent the night-time in these conditions for two weeks.

The results showed that the serum IL-2 level was the most sensitive and variable parameter that reflected the effects of negatively-charged indoor air conditions for short-term abidance. The slight increase of IL-2 can be translated as a slight activation of general immune status. However, this change will return to the pre-abidance level because changes of cytokine at this level are not expected to persist for periods that last from several hours to days 46-48). However, a very important finding was detected by the second set of experiments. In these experiments, subjects that had stayed two weeks in negatively-charged indoor air conditions during the night-time showed significant enhancement of NK cell activity. It is well known that NK cells can be activated by IL-2. At this point, the two sets of experiments were connected. As shown in Figure 5, short-term exposure to negatively-charged indoor air conditions led to IL-2 activation and elevation during the short exposure period. However, IL-2 is returned to normal or pre-exposure levels if subjects are exposed repeatedly as with individuals of experiment 2 (during the night-time), and NK cells are additionally activated by repeated exposure to slightly elevated IL-2 levels before ceasing their activation [49-53]. NK cells then show a constitutive activated status after (at least) a period of one week. This hypothesis is supported by the combined results of the two sets of experiments presented in this study.

What directions should be taken by future investigations? One direction would be to examine the biological effects of negatively-charged indoor air conditions for longer time periods such as months or years. However, such an experimental protocol may not be easy to analyze statistically because of the limited number of subjects that would be involved in such experiments. In addition, designating control subjects seems to be difficult because there may be many factors that can have an impact on results, such as lifestyle, occupation, stresses, level of exercise, eating habits, mental situations and diseases, including allergies and hypersensitivity. Thus, this direction may involve monitoring people actually living in their own house with a fitted device to make negatively-charged indoor air conditions. This can be achieved because there were no significant biological effects from the two sets of experiments performed over a period of hours or days.

Another direction for future studies would be to investigate cellular and molecular settings. Some effects on immunocompetent cells by exposure to negatively-charged air conditions were detected by the two sets of experiments, i.e., enhancement of IL-2, which is produced by various immunocompetent cells such as T cells and antigen presenting cells, and NK cell activation. Thus, *in vitro* experiments could be performed using freshly isolated immunocompetent cells from healthy or pathological subjects, including patients with cancers or allergies, or immunological cell lines can be utilized, once an *in vitro* exposure device is developed to suite these experiments. In addition, experiments using animals may be helpful to analyze the biological effects of negatively-charged indoor air conditions, especially when considering that it would be easy to build experimental rooms for animals such as rats or mice given that such rooms have been built for humans.

These forthcoming experiments may result in a better understanding of negatively-charged indoor air conditions and might help spread the use of this device in practical situations involving social environments.

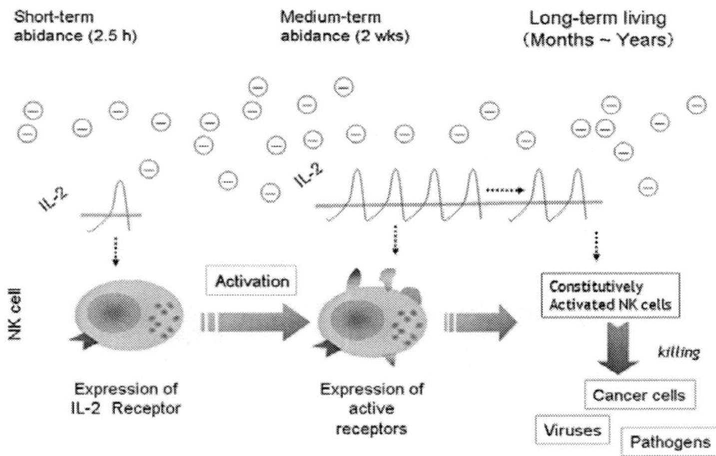

Figure 5. Schematic representation of the speculative effects of negatively-charged air conditions on the human immune system, including the short-term activation of IL-2 and repeated and chronic stimulation of IL-2 to NK cells. NK cells are stimulated recurrently and maintain a higher activity that prevents cancerous and infectious diseases.

ACKNOWLEDGEMENTS

The authors thank Ms. Tamayo Hatayama, Megumi Miyata, Hiromi Takami, Ritsuko Nishiguchi, Keiko Murata, Minako Katoh, Naomi Miyahara, Shoko Yamamoto, Eri Yamada, Mayu Sakamoto, Mr. Ryoichi Matsuse (Ikagaku Co. Ltd., Kyoto, Japan, for the measurement of most parameters), Masahiro Okumura (GMS Co. Ltd., Tokyo, Japan, for LF/HF analysis) and Kazuhiko Yamada (for management of the dormitory) for their technical assistance, and Dr. Keiko Nakatsuji for her medical assistance. This study was supported by KAKENHI grants (19659153) and Kawasaki Medical School Project Grants (19-603T and 20-603).

REFERENCES

[1] Fullerton, DG; Bruce, N; Gordon, SB. Indoor air pollution from biomass fuel smoke is a major health concern in the developing world. Trans R Soc Trop Med Hyg. 2008, 102, 843-51.

[2] Shea, KM, Truckner RT; Weber RW; Peden DB. Climate change and allergic disease. J Allergy Clin Immunol. 2008, 122, 443-53.

[3] Carpenter, DO. Environmental contaminants as risk factors for developing diabetes. *Rev Environ Health*, 2008, 23, 59-74.

[4] Marcogliese, DJ. The impact of climate change on the parasites and infectious diseases of aquatic animals. *Rev Sci Tech*, 2008, 27, 467-84.

[5] Huq, A; Whitehouse, CA; Grim, CJ; Alam, M; Colwell, RR. Biofilms in water, its role and impact in human disease transmission. *Curr Opin Biotechnol*, 2008 Jun, 19(3), 244-7.

[6] Chalupka, S. Environmental health: an opportunity for health promotion and disease prevention. *AAOHN J*, 2005, 53, 13-28.

[7] Bonde, JP. Psychosocial factors at work and risk of depression: a systematic review of the epidemiological evidence. *Occup Environ Med*, 2008, 65, 438-45.

[8] Bender, A; Farvolden, P. Depression and the workplace: a progress report. *Curr Psychiatry Rep*, 2008, 10, 73-9.

[9] Laumbach, RJ; Kipen, HM. Bioaerosols and sick building syndrome: particles, inflammation, and allergy. *Curr Opin Allergy Clin Immunol*, 2005, 5, 135-9.

[10] Burge, PS. Sick building syndrome. *Occup Environ Med*, 2004, 61, 185-90.

[11] Tsai, YJ; Gershwin, ME. The sick building syndrome: what is it when it is? *Compr Ther*, 2002, 28, 140-4.

[12] Hodgson, M. Sick building syndrome. *Occup Med*. 2000, 15, 571-85.

[13] Horvath, EP. Building-related illness and sick building syndrome: from the specific to the vague. *Cleve Clin J Med*, 1997, 64, 303-9.

[14] Spencer, TR; Schur, PM. The challenge of multiple chemical sensitivity. *J Environ Health*, 2008, 70, 24-7.

[15] Koch, L; Rumrill, P; Hennessey, M; Vierstra, C; Roessler, RT. An ecological approach to facilitate successful employment outcomes among people with multiple chemical sensitivity. *Work*, 2007, 29, 341-9.

[16] Rossi, G; Nucera, E; Patriarca, G; Manicone, PF; Raffaelli, L; Pescolla, A; Berardi, D; Perfetti, G. Multiple chemical sensitivity: current concepts. Int *J Immunopathol Pharmacol*, 2007, 20, s5-7.

[17] Cooper, C. Multiple chemical sensitivity in the clinical setting. *Am J Nurs*, 2007, 107, 40-7;

[18] Lacour, M; Zunder, T; Schmidtke, K; Vaith, P; Scheidt, C. Multiple chemical sensitivity syndrome (MCS)--suggestions for an extension of the U.S. MCS-case definition. Int *J Hyg Environ Health*, 2005, 208, 141-51.

[19] Jorgensen, C; Sany, J. Modulation of the immune response by the neuro-endocrine axis in rheumatoid arthritis. *Clin Exp Rheumatol*, 1994, 12, 435-41.

[20] Ottaviani, E; Valensin, S; Franceschi, C. The neuro-immunological interface in an evolutionary perspective: the dynamic relationship between effector and recognition systems. *Front Biosci*, 1998, 3, d431-5.

[21] O'Sullivan, RL; Lipper, G; Lerner, EA. The neuro-immuno-cutaneous-endocrine network: relationship of mind and skin. *Arch Dermatol*, 1998, 134, 1431-5.

[22] Besedovsky, HO; Rey, AD. Physiology of psychoneuroimmunology: a personal view. *Brain Behav Immun*, 2007, 21, 34-44.

[23] Citterio, A; Sinforiani, E; Verri, A; Cristina, S; Gerosa, E; Nappi, G. Neurological symptoms of the sick building syndrome: analysis of a questionnaire. *Funct Neurol*, 1998, 13, 225-30.

[24] Bocchietto, E; Paolucci, C; Breda, D; Sabbioni, E; Burastero, SE. Human monocytoid THP-1 cell line versus monocyte-derived human immature dendritic cells as in vitro models for predicting the sensitising potential of chemicals. Int *J Immunopathol Pharmacol*, 2007, 20, 259-65.

[25] Mitchell, CS; Donnay, A; Hoover, DR; Margolick, JB. Immunologic parameters of multiple chemical sensitivity. *Occup Med*. 2000, 15, 647-65.

[26] Dietert, RR; Hedge, A. Chemical sensitivity and the immune system: a paradigm to approach potential immune involvement. *Neurotoxicology*. 1998 Apr, 19(2), 253-7.

[27] Center for Environment, Health and Field Science, Chiba University, Chiba, Japan. Available:http://www.h.chiba-u.jp/center/english/top.htm (the last access 2009/1/5)

[28] Krueger, AP; Reed, EJ. Biological impact of small air ions. *Science*, 1976;193, 1209-13

[29] Misiaszek, J; Gray, F; Yates, A. The calming effects of negative air ions on manic patients: a pilot study. *Biol Psychiatry*, 1987, 22, 107-10.

[30] Yamada, R; Yanoma, S; Akaike, M; Tsuburaya, A; Sugimasa, Y; Takemiya, S; Motohashi, H; Rino, Y; Takanashi, Y; Imada, T. Water-generated negative air ions activate NK cell and inhibit carcinogenesis in mice. *Cancer Lett*, 2006, 239, 190-197.

[31] Peak Pure Air.com. Available: http://www.peakpureair.com/ (last accessed 2009/1/5)

[32] Negative Ion Generators. Available: http://www.negativeiongenerators.com/ (last accessed 2009/1/5)

[33] Li, Q; Morimoto, K; Kobayashi, M; Inagaki, H; Katsumata, M; Hirata, Y; Hirata, K; Shimizu, T; Li, YJ; Wakayama, Y; Kawada, T; Ohira, T; Takayama, N; Kagawa, T; Miyazaki, Y. A forest bathing trip increases human natural killer activity and expression of anti-cancer proteins in female subjects. *J Biol Regul Homeost Agents*, 2008, 22, 45-55.

[34] Li, Q; Morimoto, K; Kobayashi, M; Inagaki, H; Katsumata, M; Hirata, Y; Hirata, K; Suzuki, H; Li, YJ; Wakayama, Y; Kawada, T; Park, BJ; Ohira, T; Matsui N; Kagawa,

T; Miyazaki, Y; Krensky, AM. Visiting a forest, but not a city, increases human natural killer activity and expression of anti-cancer proteins. Int *J Immunopathol Pharmacol.* 2008, 21, 117-27.

[35] Li, Q; Morimoto, K; Nakadai, A; Inagaki, H; Katsumata, M; Shimizu, T; Hirata, Y; Hirata, K; Suzuki, H; Miyazaki, Y; Kagawa, T; Koyama, Y; Ohira, T; Takayama, N; Krensky, AM; Kawada, T. Forest bathing enhances human natural killer activity and expression of anti-cancer proteins. Int *J Immunopathol Pharmacol*, 2007, 20, s3-8.

[36] Takahashi, K; Otsuki, T; Mase, A; Kawado, T; Kotani, M; Ami, K; Matsushima, H; Nishimura, Y; Miura, Y; Murakami, S; Maeda, M; Hayashi, H; Kumagai, N; Shirahama, T; Yoshimatsu, M; Morimoto, K. Negatively-charged air conditions and responses of the human psycho-neuro-endocrino-immune network. *Environ Int*. 2008, 34, 765-72.

[37] Cook, EB; Stahl, JL; Lowe, L; Chen, R; Morgan, E; Wilson, J; Varro, R; Chan, A; Graziano, FM; Barney, NP. Simultaneous measurement of six cytokines in a single sample of human tears using microparticle-based flow cytometry: allergics vs. non-allergics. *J Immunol Methods*, 2001, 254, 109-18.

[38] Aoe, K; Hiraki, A; Murakami, T; Murakami, K; Makihata, K; Takao, K; Eda, R; Maeda, T; Sugi, K; Darzynkiewicz, Z; Takeyama, H. Relative abundance and patterns of correlation among six cytokines in pleural fluid measured by cytometric bead array. Int *J Mol Med*, 2003, 12, 193-8.

[39] Katayama, Y; Horigome, H; Murakami, T; Takahashi-Igari, M; Miyata, D; Tanaka, K. Evaluation of blood rheology in patients with cyanotic congenital heart disease using a microchannel array flow analyzer. *Clin Hemorheol Microcirc*, 2006, 35, 499-508.

[40] Lee, CY; Kim, KC; Park, HW; Song, JH; Lee, CH. Rheological properties of erythrocytes from male hypercholesterolemia. *Microvasc Res*. 2004, 67, 133-8.

[41] Fukuda, H; Ichinose, T; Kusama, T; Yoshidome, A; Anndow, K; Akiyoshi, N; Shibamoto, T. The relationship between job stress and urinary cytokines in healthy nurses: a cross-sectional study. *Biol Res Nurs*. 2008, 10, 183-91.

[42] Satoh, N; Sakai, S; Kogure, T; Tahara, E; Origasa, H; Shimada, Y; Kohoda, K; Okubo, T; Terasawa, K. A randomized double blind placebo-controlled clinical trial of Hochuekkito, a traditional herbal medicine, in the treatment of elderly patients with weakness N of one and responder restricted design. *Phytomedicine*, 2005, 12, 549-54.

[43] Nakayama, K; Morimoto, K. Relationship between, lifestyle, mold and sick building syndromes in newly built dwellings in Japan. Int *J Immunopathol Pharmacol*, 2007, 20, s35-43.

[44] Li, Q; Morimoto, K; Nakadai, A; Qu, T; Matsushima, H; Katsumata, M; Shimizu, T; Inagaki, H; Hirata, Y; Hirata, K; Kawada, T; Lu, Y; Nakayama, K; Krensky, AM. Healthy lifestyles are associated with higher levels of perforin, granulysin and granzymes A/B-expressing cells in peripheral blood lymphocytes. *Prev Med*, 2007, 44, 117-23.

[45] Tsukinoki, R; Morimoto, K; Nakayama, K. Association between lifestyle factors and plasma adiponectin levels in Japanese men. *Lipids Health Dis*, 2005, 4, 27.

[46] Bindon, C; Czerniecki, M; Ruell, P; Edwards, A; McCarthy, WH; Harris, R; Hersey, P. Clearance rates and systemic effects of intravenously administered interleukin 2 (IL-2) containing preparations in human subjects. *Br J Cancer*. 1983, 47, 123-33.

[47] Sano, T; Saijo, N; Sasaki, Y; Shinkai, T; Eguchi, K; Tamura, T; Sakurai, M; Takahashi H; Nakano, H; Nakagawa, K; Hong, WS. Three schedules of recombinant human interleukin-2 in the treatment of malignancy: side effects and immunologic effects in relation to serum level. *Jpn J Cancer Res*, 1988, 79, 131-43.

[48] Konrad, MW; Hemstreet, G; Hersh, EM; Mansell, PW; Mertelsmann, R; Kolitz, JE; Bradley, EC. Pharmacokinetics of recombinant interleukin 2 in humans. *Cancer Res*, 1990, 50, 2009-17.

[49] Shau, HY; Golub, SH. Signals for activation of natural killer and natural killer-like activity. *Nat Immun Cell Growth Regul*, 1985, 4, 113-9.

[50] Rosenberg, SA; Lotze, MT. Cancer immunotherapy using interleukin-2 and interleukin-2-activated lymphocytes. *Annu Rev Immunol*, 1986, 4, 681-709.

[51] Herberman, RB. Natural killer cells. *Prog Clin Biol Res*, 1989, 288, 161-7.

[52] Grimm, EA; Owen-Schaub, L. The IL-2 mediated amplification of cellular cytotoxicity. *J Cell Biochem*, 1991, 45, 335-9.

[53] Warren, HS. NK cell proliferation and inflammation. *Immunol Cell Biol*, 1996, 74, 473-80.

In: Buildings and the Environment
Editors: Jonas Nemecek and Patrik Schulz
ISBN: 978-1-60876-128-9

Chapter 7

ENERGY SAVINGS IN HVAC SYSTEMS OF SPANISH SCHOOL BUILDINGS

José Antonio Orosa*
Department of Energy and Marine Propulsion.
University of A Coruña, Paseo de Ronda 51,
15011, A Coruña, España.

ABSTRACT

A Coruña, located in northwestern Spain, has a mild climate with a high relative humidity as a consequence of winds. Public buildings, like Spanish school buildings, present the highest energy consumption in air conditioning during the winter season, but during the spring the heating system is employed only if indoor conditions are below certain temperature and relative humidity values. Some energy saving methods could be employed to reduce or, in some cases, replace the heating systems and, as a consequence, reduce energy consumption. Some of these methods are centred in adequate design and application of an HVAC system while others are focused on the study of heat and mass transfer through building envelopes. A clear example of these methods was shown when permeable coverings [1, 2] were employed to reduce relative humidity peaks in office buildings, but there are other parameters, such as thermal inertia, that could help us to compare indoor ambience with respect to energy savings. In this sense, actual software, such as HAM tools, could be employed to simulate indoor conditions and phenomena of material and energy transfer through building envelopes and their effect on indoor conditions. This chapter describes a study of methods to reduce the energy consumption in different kinds of schools buildings by means real sampled data and HAM tool simulations. Results showed that parameters such as thermal inertia could be influenced principally by solar heat gains, changing the buildings' time constant. Other parameters like air changes per hour or permeable coverings present a clear enhancement of indoor ambiences.

* Corresponding author: Tel.: +34981167000 ext. 4320; Fax:+349811167107; E-mail address: jaorosa@udc.es

1. INTRODUCTION

A Coruña, located in northwestern Spain, has a mild climate with a high relative humidity as a consequence of winds. Public buildings, such as Spanish school buildings, present the highest energy consumption in air conditioning during the winter season, but during the spring the heating system is employed only if indoor conditions fall below certain temperature and relative humidity values. Some energy saving methods could be employed to reduce or, in some cases, replace the heating systems and, as a consequence, reduce energy consumption. Some of these methods are centred in adequate design and application of an HVAC system, while others focus on the study of heat and mass transfer through building envelopes. A clear example of these methods was shown when permeable coverings [1, 2] were employed to reduce relative humidity peaks in office buildings, but there are other parameters, like thermal inertia, that could help us to compare indoor ambience with respect to energy savings.

The system with the highest thermal inertia has the lowest energy requirement because, in times of abundance (due to solar irradiation, e.g.), energy is stored in internal and external building constructions. This energy is transferred back into the zone when the indoor temperature decreases, thus limiting the heat needed from the heating system without significantly affecting thermal comfort [3]. The fact that a high time constant for the thermal processes in a building decreases the energy requirement for heating makes it possible to choose a higher winter temperature or change the working conditions of the HVAC system, such as a change from intermittently controlled heat pumps to continuous capacity controlled heat pumps [4]. Other investigations are centred on characterization of thermal inertia as a part of the installation of a system of summer cooling by means of nighttime cooling ventilation [5], and others focus on the study of passive solar buildings [6, 7, 8] .

To understand and predict these effects, software tools were employed, but these underestimated the energy consumption of buildings by up to 60% because the energy models ignore moisture. In this sense, whole building performance can only be realistically evaluated by accounting for the HAM interactions. Through its Energy Conservation in Buildings & Community Systems Program, the International Energy Agency launched Annexes 17 and 41 [9], a working group to address issues surrounding whole building HAM response. The group involved more than 50 researchers from 28 institutes and more than 20 countries as this Energy Department [10]. This Annex 41 tested building simulation software like H-tools and HAM Tools from the Chalmer Institute of Technology to simulate heat and mass transfer through building envelopes. These tools are employed to simulate indoor conditions as phenomena of material and energy transfer through building envelopes and their effects on indoor conditions considering heat gains like occupation, solar heat, illumination and air renovation and infiltrations between others. Once this software is tested, new conclusions could be obtained. In the present chapter, a study of the methods to reduce energy consumption in different kinds of school buildings by means real sampled data and HAM tool simulations is done.

2. Materials

2.1. Tiny Tag Data Loggers

Temperature and relative humidity were measured using an Innova 1221 data logger equipped with a temperature transducer MM0034 based on thermistor technology, and a humidity transducer MM0037 incorporating a light emitting diode (LED), along with a light sensitive transistor, a mirror, a cooling element and a thermistor.

2.2. Air Changes

One of the components of the measuring apparatus was a multi-gas monitor. The ventilation rate was performed using the concentration decay method, measuring SF_6 as tracer gas with a Brüel&Kjaer multi-sampler.

3. Methods

In this sense, two school buildings were sampled during different seasons to relate indoor conditions with weather, heat and moisture balances and an HVAC system.

ASHRAE Standard 1992 [11] indications and Burch [12, 13] simulations showed that the massiveness of an exterior wall reduces the heating and cooling requirements of buildings, provided the room air temperature floats above the thermostat set point in heating and below the thermostat in the cooling season. The floating temperature occurs more frequently in mild climates and during the spring. Special interest presents the spring season because then the HVAC system could be fully substitute by some passive methods. In consequence, this paper will study thermal inertia effect of two buildings with high and low wall density during the spring season of the mild weather of A Coruña (Spain).

The week period selected for this study was a weekend and a holiday to identify the appropriate indexes of thermal stability for building envelopes. The focus must consider the solar heat gain and heat storage of building walls under conditions of natural ventilation [14]. During this unoccupied period, occupation and excessive air renovations will not interfere with the sampled data and an easy environmental simulation could be done.

After tests, the simulation of indoor conditions modifications will be proposed for energy savings. In this sense, location of the mass in relation to the insulation has a large effect on the deviation between measured energy use and steady state analysis. In high-density buildings where mass was outside the insulation, the measured energy use closely matched that predicted by steady-state analysis but not when the insulation was outside the mass [11]. In consequence, parameters like thermal inertia, air renovation and internal coverings were simulated showing a clear different behaviour in each condition.

Conclusions let us understand the passive methods that could let us reach better indoor condition during the first hour of occupation of the morning and evening class periods.

3.1. Schools

Two schools are sampled and simulated. One of the areas of the older school was built in 1890, and the other part was built in 1960. The new school was built in 1999. As a consequence, the old school has 0.43 m of stone and 0.5 cm of concrete in the indoor side of the wall. The wall of the new building consists of layers of insulation, brick, concrete and plaster arranged symmetrically with respect to the middle of the wall and reaching 0.30 m of total thickness. The classroom sampled in the old building is located on the second floor and has a volume of 210 m^3, while the classroom in the new building is located on the first floor with a volume of 150 m^3.

Both of these buildings present a working period from February to June and an unoccupied period during the weekends and holidays. In those periods, classrooms are under natural ventilation and a central heating system was not employed. The active period ends in June and, as a consequence, energy savings is not considered during the summer period. Furthermore, during extreme conditions of winter, these schools are not open and, as a consequence, the heating system works only when the indoor conditions exceed the thermal comfort during winter and spring.

3.2. Indoor and Outdoor Sampling Conditions

The indoor and outdoor humidity and temperature have been monitored simultaneously in the most representative classroom of each school during part of the winter and spring seasons. Both schools have purely adventitious ventilation. Transducers were hung in the middle of the classrooms. Data has been gathered in Tiny tags data loggers that can store 7,600 readings.

Air infiltration was measured by tracer gas method employing SF_6 as tracer gas.

3.3. Thermal Comfort and Indoor Air Quality Indexes

Local thermal comfort has been evaluated in terms of the parameter PD, Percentage of Dissatisfied Persons, through equation 1 obtained by Toftum et al. (1988) and Simonson et al. [15-18].

$$PD = \frac{100}{1 + \exp(-3.58 + 0.18 \cdot (30 - t) + 0.14 \cdot (42.5 - 0.01 \cdot P_v)} \tag{1}$$

where t is the temperature (°C) and Pv the partial water vapour pressure (Pa). An acceptable environment would be that in which less than 15% of the occupants are dissatisfied.

The Indoor Air quality has been evaluated through the so-called Acceptable Indoor Air Quality parameter, Acc. Equation 2 was proposed by Fang et al. (1988).

$$Acc = 0.033 \cdot H + 1.622 \tag{2}$$

where H is the air enthalpy (kJ/kg).

3.4. Ham Tools Simulation

The mathematical model employed in this simulations is the result of whole building Heat, Air and Moisture (HAM) [19, 20, 21] balance and depends on moisture generated from occupant activities, moisture input or removed by ventilation, and moisture transported and exchanged between indoor air and the envelope [10].

The mathematical model is based in the numerical resolution of the energy and moisture balance through the building. In accordance with the next equations [20], the heat flow presents a conductive and a convective part as we can see in equation 3 and described in equations 4 and 5.

$$q = q_{conductive} + q_{convective} \tag{3}$$

$$q_{conductive} = -\lambda \frac{\partial T}{\partial x} \tag{4}$$

$$q_{convective} = m_a \cdot c_{pa} \cdot T + h_{evap} \tag{5}$$

Where:

λ is the thermal conductivity (W/mK)
T is the temperature (°C)
m_a is the density of moisture flow rate of dry air (kg/m2s)
c_{pa} is the specific heat capacity of the dry air (J/kg K)
h_{evap} is the latent heat of evaporation (J/kg)

The moisture flow transfer was separated in liquid and vapour phases as we can see in equations 6 and 7.

$$m_l = K \cdot \frac{\partial P_{suc}}{\partial x} \tag{6}$$

m_l is the density of moisture flow rate of vapour phase (kg/m^{2}s)
K is the hydraulic conductivity
P_{suc} is the suction pressure (Pa)

The vapour phase was divided in diffusion and convection as we can see in equation 7.

$$m_v = -\delta_p \cdot \frac{\partial p}{\partial x} + m_a \cdot x_a \tag{7}$$

Where

δ_p is the moisture permeability (s)

x_a is the water vapour content (kg/kg)

The mass airflow through the structure driven by air pressure differences across the structure is showed in equation 8.

$$m_a = r_a \cdot \rho_a \tag{8}$$

Where

r_a is the density of the air flow rate (m^3/m^2s)

ρ_a is the density of the material (kg/m^3)

The finally energy and moisture balance are showed in equations 9 and 10.

$$-\frac{\partial}{\partial x} q = c \cdot \rho_o \cdot \frac{\partial T}{\partial t} \tag{9}$$

$$-\frac{\delta}{\partial x} m = \frac{\partial w}{\partial t} \tag{10}$$

Where

ρ_o is the density of the dry material (kg/m^3)
c is the specific heat capacity of the material (J/kg K)
w is the moisture content mass by volume (kg/m^3)
t time (s)
x space coordinates (m)

Finally, the numerical model, based on a control volume method present lumped the thermal capacity C in the middle of the total thickness d/2 and, in consequence, the thermal resistances for one half is showed in equations 11, 12 and 13.

$$R = \frac{d/2}{\lambda} \tag{11}$$

$$R_p = \frac{d/2}{\delta_p} \tag{12}$$

$$R_{suc} = \frac{d/2}{K_{suc}} \tag{13}$$

The obtained discretized heat and moisture balance equations are showed in equations 14 and 15.

$$\frac{T_i^{n+1}-T_i^n}{\Delta t}=\frac{1}{C^n}\cdot\left\{\left[\frac{(T_{i-1}-T_i)}{R_{i-1}+R_i}+\frac{(T_{i+1}-T_i)}{R_{i+1}+R_i}\right]-h_{evap}\cdot\left[\frac{(p_{i-1}-p_i)}{R_{p,i-1}+R_{p,i}}+\frac{(p_{i+1}-p_i)}{R_{p,i+1}+R_{p,i}}\right]\right\}\cdots$$
$$+\begin{Bmatrix}m_a\cdot c_{pa}\cdot(T_{i-1}-T_i)^n, m_a>0\\ m_a\cdot c_{pa}\cdot(T_1-T_{i+1})^n, m_a<0\end{Bmatrix} \quad (14)$$

$$\frac{w_i^{n+1}-w_i^n}{\Delta t}=\frac{1}{d}\cdot\left\{\left[\frac{(p_{i-1}-p_i)}{R_{p,i-1}+R_{p,i}}+\frac{(p_{i+1}-p_i)}{R_{p,i+1}+R_{p,i}}\right]-\left[\frac{(P_{suc,i-1}-P_{isuc,i})}{R_{suc,i-1}+R_{suc,i}}+\frac{(P_{suc,i+1}-p_{suc,i})}{R_{suc,i+1}+R_{suc,i}}\right]\right\}\cdots \quad (15)$$
$$+\begin{Bmatrix}6.21\cdot 10^{-6}\cdot m_a\cdot(p_{i-1}-p_i)^n, m_a>0\\ 6.21\cdot 10^{-6}\cdot m_a\cdot(p_i-p_{i+1})^n, m_a<0\end{Bmatrix}$$

where *i* is the objective node and *i*+1 and *i*-1 are the preceding and following node and n and n+1 the previous and corresponding time steps.

To solve these balance equations room models were created from the individual Building Physics Toolbox [22, 23]. HAM tools library is a Simulink model upgrade version of H-Tools with a similar structure and specially constructed for thermal system analysis in building physics. The library contains blocks for 1-D calculation of heat, air and moisture transfer through the building envelope components and ventilated spaces. The library is the part of IBPT-International Building Physics Toolbox, and available for free downloading [24].

This library presents two main blocks: a building envelope construction (walls, windows) and thermal zone (ventilated spaces), which are enclosed by the building envelope. Component models provide detailed calculations of the hydrothermal state of each subcomponent in the structure; according to the surrounding conditions to witch it is exposed.

In Figure 1 we can see the principal blocks employed for a building simulation. There we can see a block that represents the different exterior/interior walls, floor, roof and windows components. These constructions are defined respect they physical properties (density of the dry material and open porosity), thermal properties (specific heat capacity of the dry material and thermal conductivity) and moisture properties (sorption isotherm, moisture capacity, water vapour permeability and liquid water conductivity) in accordance with the BESTEST structure. Other parameters are considered in the heat and moisture building balance like, for example, internal gains (convective gains, radioactive gains and moisture gains), air change and heating/ cooling system.

Classroom characteristics are defined in the thermal zone block indicating the surface areas, orientations and tilts of each wall. Room volume, solar gain to air and initial temperature is adjusted so. Thermal model of the classroom is based on the WAVO model described by de Witt (2000) [25] and developed by the assumptions that long wave radiation is equally distributed over the walls; the room air has the uniform temperature, the surface coefficients for convection and radiation are constant and, finally, that all radioactive heat input is distributed is such a way that all the surfaces except windows absorb the same amount of that energy per unit of surface area.

In this paper, wall construction, indoor air renovation and internal gains were adjusted in accordance with measured data and simulated for three days of unoccupied period and determine simulations accuracy. Finally, a weather database was done in accordance with meteorological sampled data to be introduced. This database was loaded before simulation adjusting the time step and study period.

3.5. Time Constant

The time constant is normally found from a slow cooling down period with a constant low outdoor temperature as (heat capacity)/(heat loss factor) [3]. This method is based on a seasonal steady state energy balance on the building as a whole or on a particular building zone. The thermal inertia is introduced in terms of the utilisation factor that shows the part of energy gains (solar irradiation and others) that can be stored in building construction to be transmitted into the zone when needed, as we can see in equation 16.

$$Q_{heat} = Q_{loss} - \eta Q_{gain} \tag{16}$$

Where

Q_{heat} is the heat requirement (W)
Q_{loss} is the heat loss (W)
Q_{gain} is the heat gain (W)
η is the utilisation factor, having a value between 0 and 1.

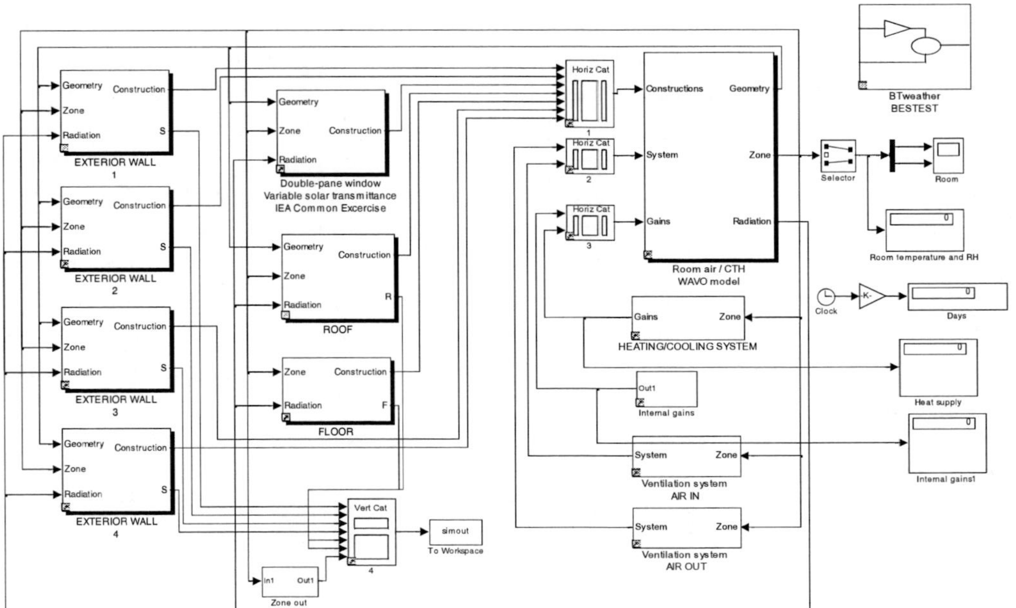

Figure 1. Matlab blocks for buildings simulations.

The utilisation factor η is a function of the building periodic time constant and the ratio Q_{gain}/Q_{loss}. The time constant is defined in the standard by the equation 17.

$$\tau = \frac{\sum C}{\sum H} \tag{17}$$

C is the sum of thermal capacity C of each construction based on a 24 hour periodic response.

H is the sum of heat loss factor of each construction, ventilation and air leakage.

As [3] recommends, when we want to work in a more precise way, the logarithm of the temperature difference in-outdoors is taken and matched to a straight line by the method of least squares. The time constant is the inverse of the coefficient for the independent variable (time) given by this curve fit. In consequence, after test our simulations with real sampled data; both buildings were simulated under constant weather conditions with the aim to determine building time constants.

4. Results

4.1. Outdoor Conditions

Figures 2 and 3 show the outdoor conditions during the unoccupied period.

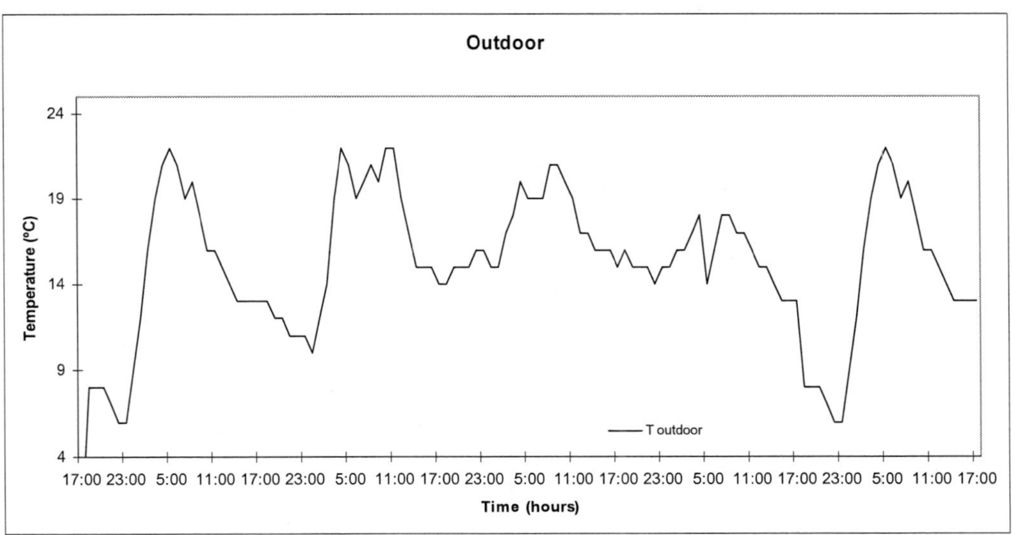

Figure 2. Outdoor sampled temperature.

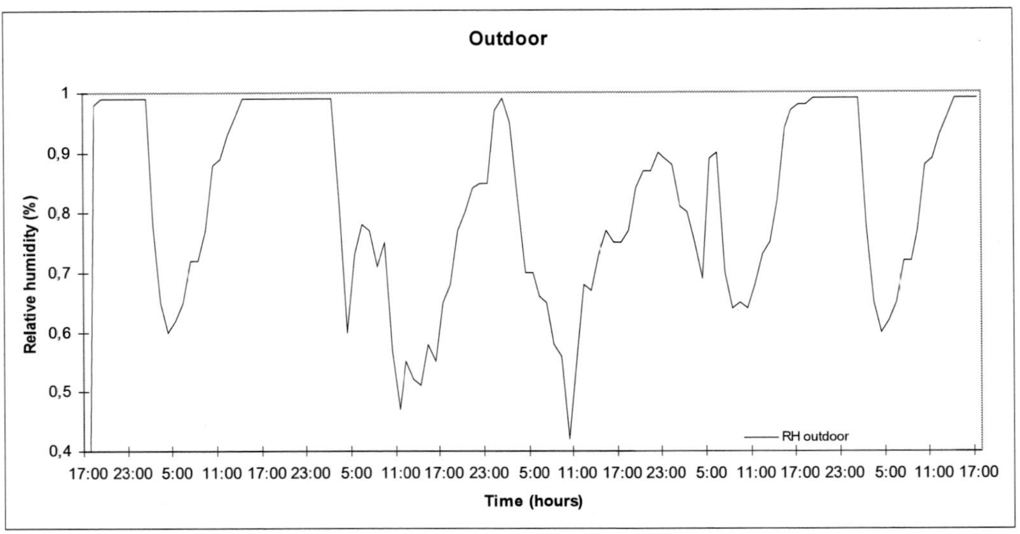

Figure 3. Outdoor sampled relative humidity.

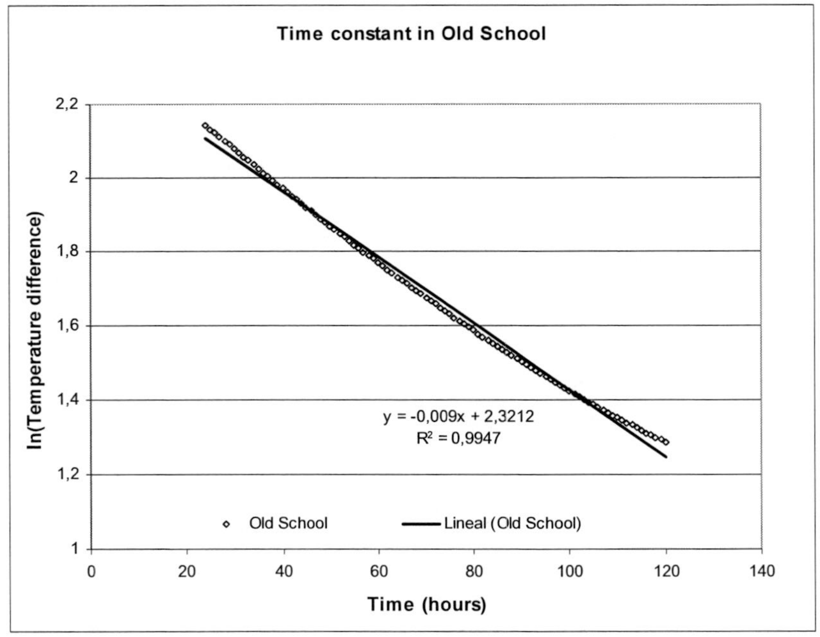

Figure 4. Time constant determination for old school building.

4.2. Thermal Inertia and Solar Gain: Time Constant Determination

Figures 4, 5, 6 and 7 represent the logarithm of the temperature difference between indoor and outdoor temperatures when building is under constant weather conditions. Its linear regression constants will give us the time constant of each building.

Time constant in old schools =111

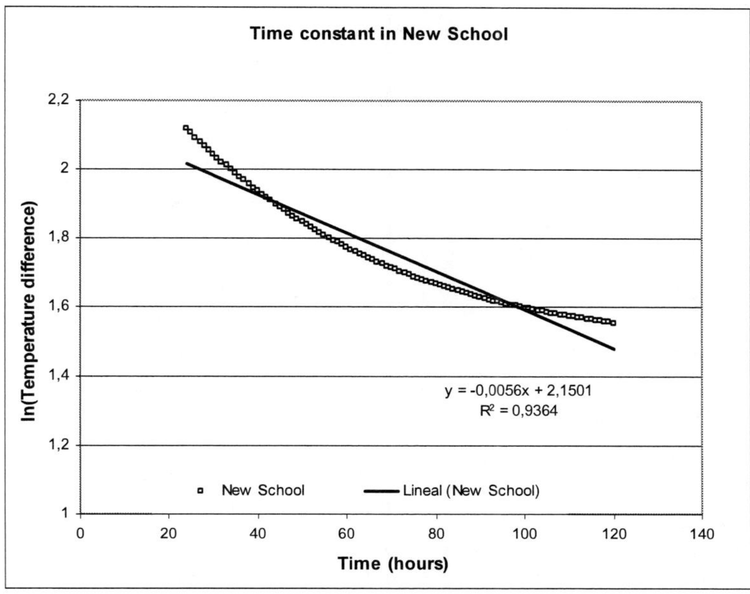

Figure 5. Time constant determination for new school building.

Time constant in new school=178

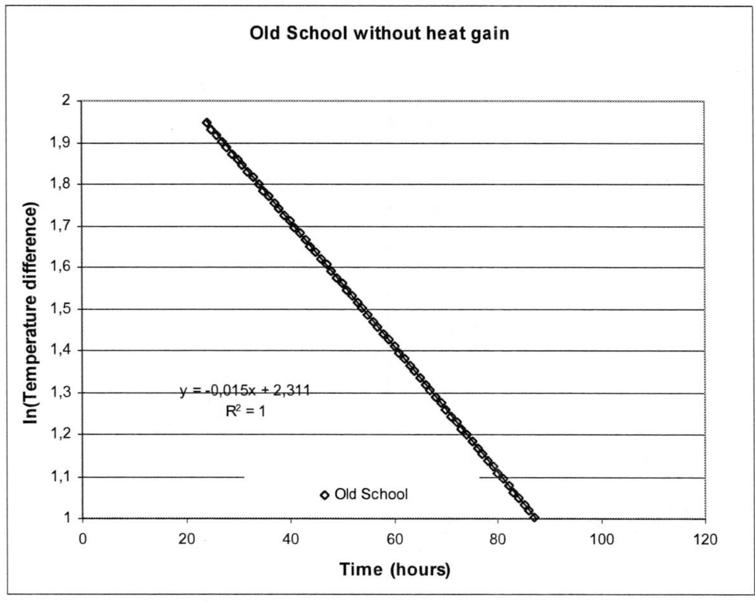

Figure 6. Time constant determination for old school building without heat gains.

Time constant in old school without solar heat gain= 66.6

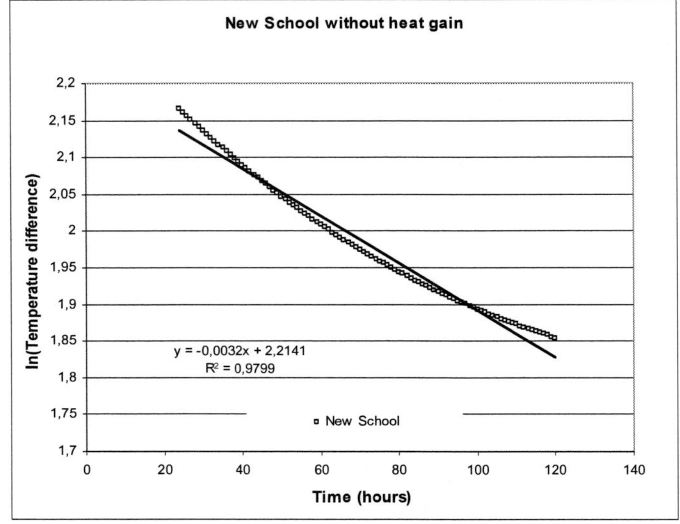

Figure 7. Time constant determination for new school building without heat gains.

Time constant in new school without solar heat gain= 36.9

4.3. Air Changes

Sampled air renovations were changed from 0.7 during the unoccupied period for the old school and 0.6 in the new school to a lower value of 0.4 air renovations with the aim of observant the effect of weather on indoor conditions.

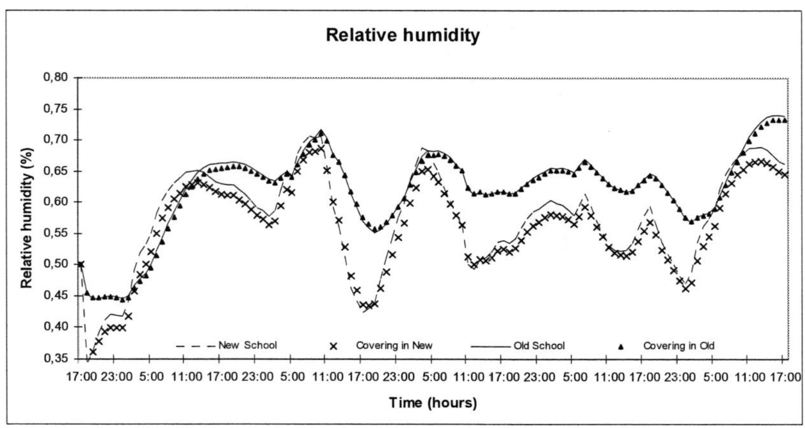

Figure 8. Relative humidity when air renovations were reduced.

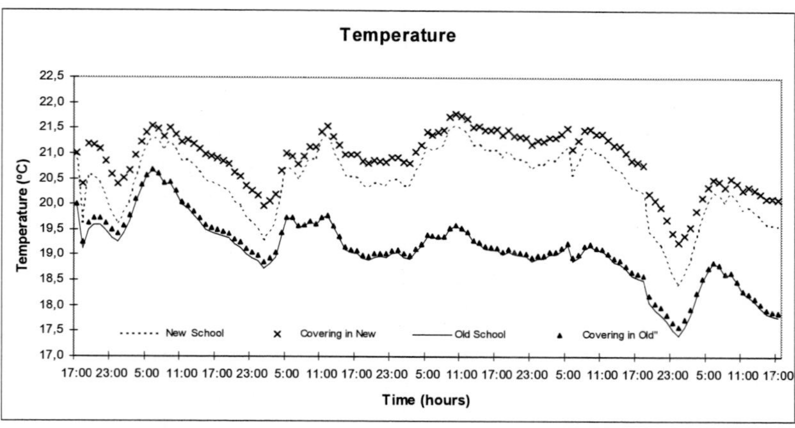

Figure 9. Temperature when air renovations were reduced.

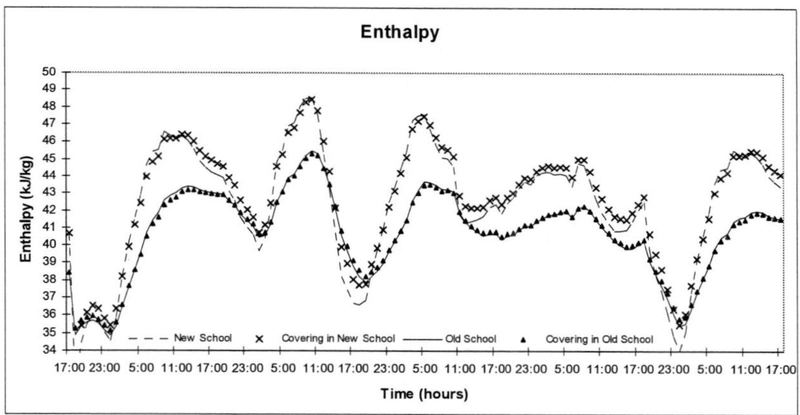

Figure 10. Enthalpy when air renovations were reduced.

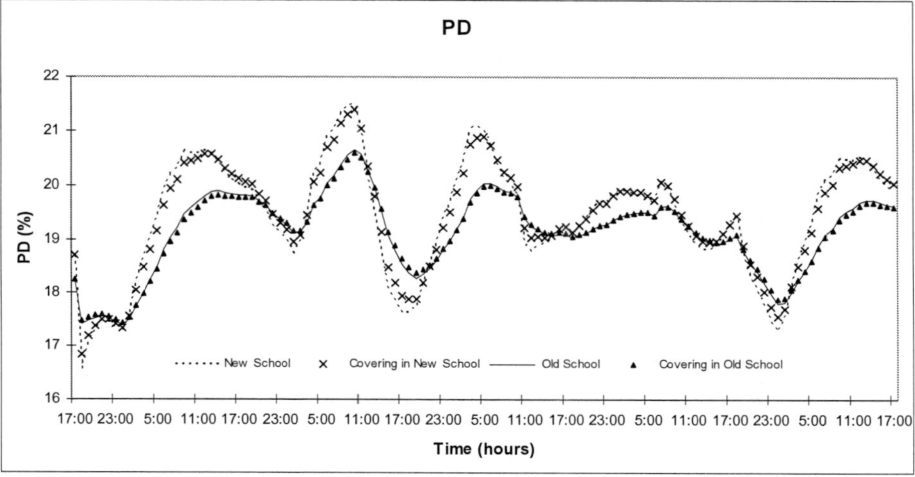

Figure 11. PD when air renovations were reduced.

4.4. Heat and Mass Transfer Through Building Envelope

Now, the internal covering was changed in both buildings with the aim to improve indoor conditions. The permeable covering selected to substitute the concrete internal covering was 1.5 cm of wooden panel that could control the high indoor relative humidity, especially in the old school building.

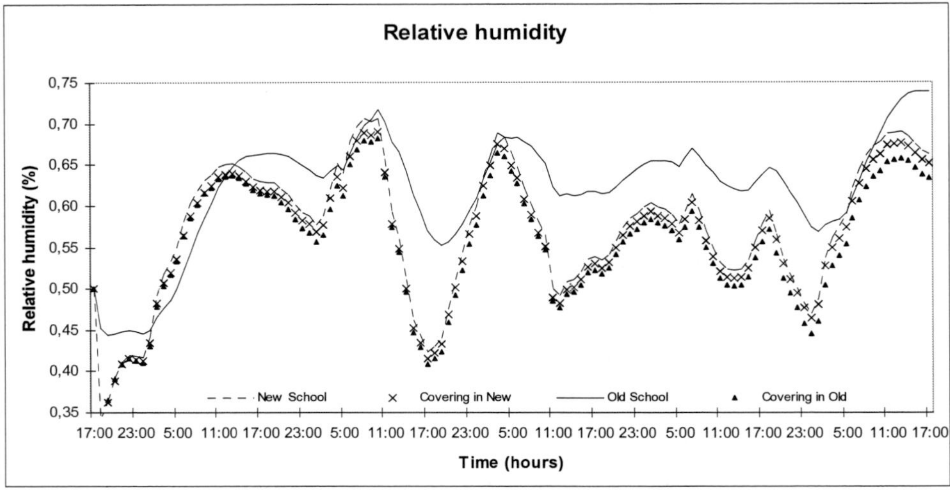

Figure 12. Relative humidity with and without internal coverings.

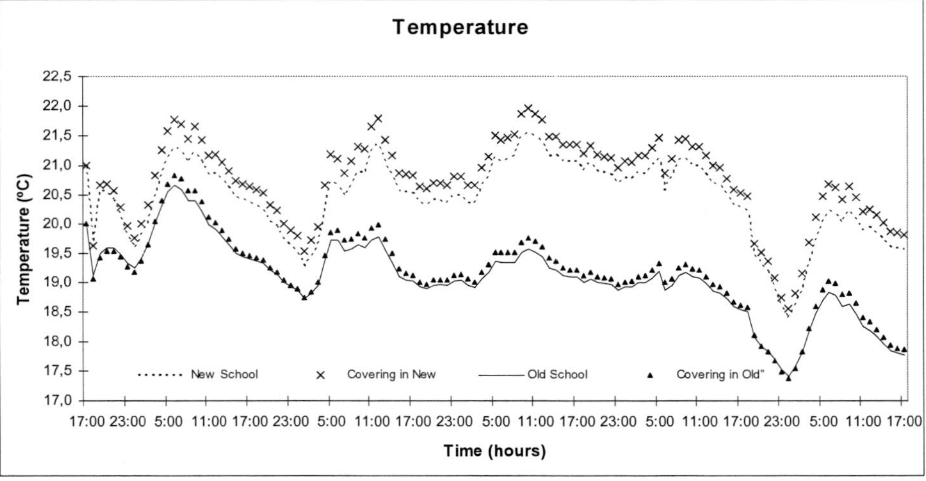

Figure 13. Temperature with and without internal coverings.

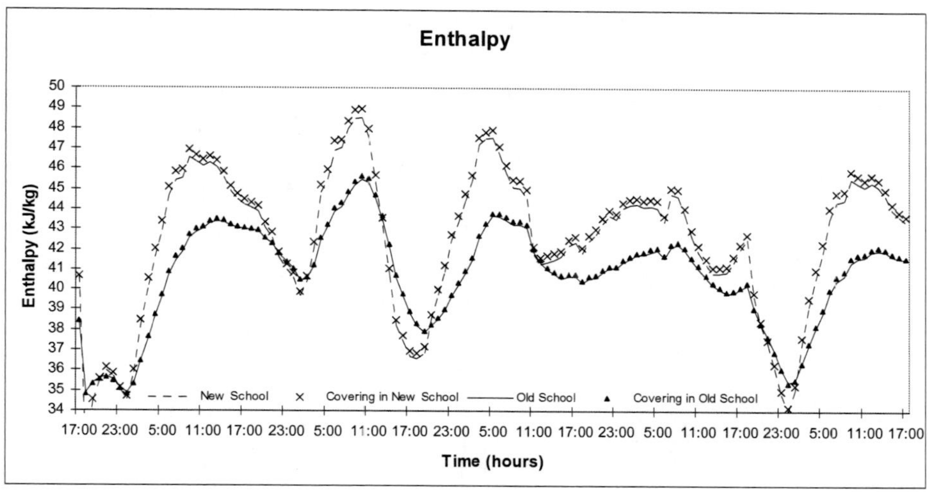

Figure 14. Enthalpy with and without internal coverings.

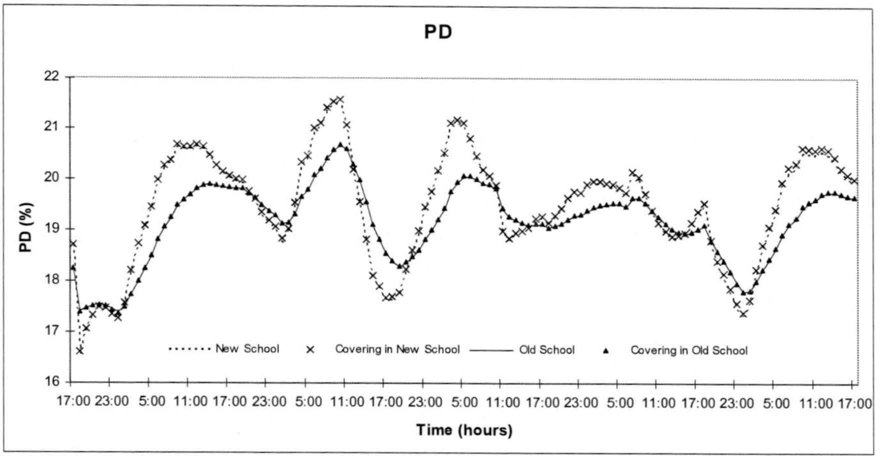

Figure 15. PD with and without internal coverings.

5. DISCUSSION

As we can see in Figures 2 and 3, A Coruña present a mild climate but during all the year exist a certain high relative humidity, nearly 80%, related with some health problems, building maintenance and energy consumption.

As was commented previously, simulation and sampled indoor data was compared obtaining a high approach between real and simulated ambiences. Once tested that simulations, the two schools were simulated under a constant weather conditions of 10 °C and 80% of relative humidity to obtain the time constant, after a linear regression of the logarithm difference temperature respect outdoor ambience.

Results of Figures 4 and 5 showed a time constant of 111 for the old and 178 in the new school with an adequate correlation factor in each case. This value shows a higher thermal inertia of the new school than the old. Other simulations were done under different indoor ambience temperatures obtaining the same value. The explanation of this effect is related with heat solar heat gains that the new building experiments respect the old as a consequence of the classroom way and the presence of another nearer buildings that interfere in that heat gain. To understand this solar effect, if we simulate this same process but without this heat gain another time constants were obtained in Figures 6 and 7. Once again adequate linear regressions were obtained with correlation values of 0.98 and 1 and time constants of 36.9 and 66.6 for new and old schools respectively. Now the old school present a higher thermal inertia respect the new as a consequence of the purely effect of wall thickness and heat transmission properties.

Once obtained, these results and other changes were proposed. For example, old school presents a high air renovation during the unoccupied period as a consequence of the air leakages. These air renovations reach values upwards of the 0.5 usually obtained in closed ambiences. When the air renovations of the two buildings are changed to 0.4, other curves are shown in Figures 8, 9, 10 and 11. Indoor air humidity in the new and old school exhibit a decrease to more adequate values of 60%. This effect is related to the heat and moisture transfer through the stone walls. The indoor air temperature reaches the same maximum values but exhibit a slow decrease in the new building, reaching higher minimum values than under normal conditions of air renovation, as we can see in Figure 9. This thermal effect will be present in the indoor air enthalpy and, as a consequence, on the percentage of dissatisfied which reaches slightly lower PD maximum value during the night (Figure 11).

Finally, other internal coverings were proposed to change the indoor ambience conditions. As a consequence, wooden paneling was simulated as an internal covering in the two cases. Results are shown in Figures 12, 13, 14 and 15.

In this sense, permeable coverings result in adequate covering to control the extreme relative humidity evolution, especially in the old building. Now we can see that indoor relative humidity was greatly modified in the old school, reaching a value nearer the new building, which did not exhibit an apparent change in the indoor conditions of Figure 12. In Figure 13, indoor temperature exhibits a clear enhancement in the new building, reaching values of 1°C higher than in the previous case. This effect is less intense in the old buildings. As a consequence of this change in indoor air temperature and relative humidity, the percentage of dissatisfied reached nearly the same values than that obtained in normal conditions in the old and in new schools. This is a consequence of the slight changes in indoor air enthalpy, which is nearly constant when the proportion 1°C/5% of relative humidity was maintained. For example, if indoor air temperature exhibits an enhancement of 1°C, relative humidity will exhibit a decrease of 5%, as we can see in Figures 12 and 13 and, as a consequence, there is nearly the same indoor air enthalpy and percentage of dissatisfied in Figures 14 and 15, respectively.

In this last simulation, despite the fact that there is a permeable covering, there was not an intense effect on indoor ambience. This is related to the reduced surface of permeable covering, which represents the only surface of the class to interact with the outdoor ambience, and the high volume of the old classroom (210 m^3).

The enthalpy conditions indicate that an HVAC system is not needed because the indoor air enthalpy under natural ventilation reaches the value of 39kJ/kg, estimated for adequate indoor conditions.

In addition, we can conclude that a reduction in indoor air renovations leads to increments in the indoor temperature of the new building, where the solar heat gain is more important and, as a consequence, to an indoor ambience more insensible to outdoor weather change, as we can see in Figure 9 in the indoor temperature slope after a peak in temperature.

On the other hand, when a permeable covering is proposed, the temperature decrease is parallel and slightly higher than initial conditions, enhancing a reduction in indoor air relative humidity. This increment of indoor air temperature is related to the insulation properties of the wooden paneling.

As we have noted, the importance of these time constants is based on the fact that the building with the lower time constant reacts more quickly to weather changes and variation in internal heat gains than the heavier buildings. Even during a short period of cold weather, heat must be supplied in the lightweight building, whereas such periods can pass without heating in the constructions with higher thermal inertia due to heat stored in the structure from previous warmer periods, in accordance with [3]. Furthermore, the amplitude of temperature fluctuation of the inner surfaces of walls in low time constant buildings under intermittent air-conditioning conditions is 1°C higher than that of the walls of buildings with a higher thermal inertia under continuous air-conditioning conditions, in accordance with [14].

These effects can be summarised in their study of time constant of each variant. In our case, when indoor air renovations are reduced, the time constant for new and old schools is 185 and 112, respectively, as we can see in Table 1.

These values are similar to that of 325, 164 and 31 for heavyweight, massive wood and lightweight walls shown by [3]. Despite this, there is an internal gain from convective heat source in the new building that leads to a greater effect than the wall construction thermal inertia.

Table 1. Time constant for each modification.

	Without heat gain	Initial conditions	Air renovation reduction	Permeable coverings
New	37	178	185	188
Old	67	111	112	115

As a consequence, the energy savings and HVAC system design must be done in accordance with the individual characteristics of each building and not only taking into account meteorological data and general procedures.

6. Conclusions

This chapter samples and simulates different indoor conditions in school buildings with the aim to determine the possibility of energy savings and design considerations for buildings and HVAC systems.

The heat transfer through the night thermal inertia elements is analysed by using a 1D time dependent conduction heat transfer equation that is solved numerically by using HAM tools. The model takes into account in a detailed fashion the inertial heat sources and air renovation. These simulations showed that the old building presents a lower thermal inertia than the new as a consequence of solar heat gain. In consequence, the building with the lower time constant reacts more quickly to weather changes and variation in internal heat gains than the heavier building. Parameters like air renovation and permeable coverings interact over the time constant. If the air renovation is reduced the old building exhibits a slow increment, while the new will exhibit a clear internal temperature increment as a consequence of the heat gain. Finally, the presence of permeable internal coverings like wood paneling allows us to increase the time constant as a consequence of the increment of wall insulation, especially in the old school.

The effect of permeable coverings on the indoor ambience was reduced in the presence of permeable coverings on a small surface. More research is needed to define new design and corrections of HVAC systems taking into account these individual parameters of each building location.

7. References

[1] Orosa, JA; Oliveira, AC. Hourly indoor thermal comfort and air quality acceptance with passive climate control methods. *Renewable Energy*, 2009. 34. 12. 2735-2742.

[2] Orosa, JA; Oliveira, AC. Energy saving with passive climate control methods in Spanish office buildings. *Energy and Buildings,* 2009. 41. 8. 823-828.

[3] Norén, A; Akander, J; Isfält, E; Söderström, O; The effect of Thermal Inertia on Energy Requirement in a Swedish Building-Results Obtained with Three Calculation Models. *International Journal of Low Energy and Sustainable Buildings*, 1999.1.

[4] Karlsson, F; Fahlén, P. Impact of design and thermal inertia on the energy saving potential of capacity controlled heat pump heating systems. *International Journal of Refrigeration*, 2008. 31. 1094-1103.

[5] Roucoult, JM; Douzane, O; Langlet, T. Incorporation of thermal inertia in the aim of installing a natural night time ventilation system in buildings. *Energy and Buildings*. 1999.29. 129-133.

[6] Badescu, V; Sicre, B. Renewable energy for passive house heating II Model. *Energy and Buildings*, 2003. 35. 1085-1096.

[7] Badescu, V; Sicre, B. Renewable energy for passive house heating Part I. Building description. *Energy and Buildings* 2003. 35. 1077–1084.

[8] Krüger, E; Givoni, B. Thermal monitoring and indoor temperature predictions in a passive solar building in an arid environment. *Building and Environment*. 2008. 43. 1792-1804.

[9] Hauer, A; Mehling, H; Schossig, P; Yamaha; M. Cabeza, L; Martin, V; Setterwall, F; International Energy Agency Implementing Agreement on Energy Conservation through Energy Storage. Annex 17. "Advanced Thermal Energy Storage through Phase Change Materials and Chemical Reactions–Feasibility Studies and Demonstration projects". Final Report.

[10] http://www.iea.org. (Accessed August 2009).
[11] ASHRAE handbook—fundamentals. Load and Energy Calculations, Energy Estimating Methods. 1993. Chap. 28
[12] Burch, MD; Chi, J. MOIST A PC Program for Predicting Heat and Moisture Transfer in building Envelopes. NIST Special Publication 917. NIST United States Department of Commerce Technology Administration. National Institute of Standards and Technology.1997.
[13] Burch, DM; Remmert, WE, Krintz, DF.and Barnes, CS. A Field Study of the Effect of Wall Mass on the Heating and Cooling Loads of Residential Buildings (aka Log Home Report). National Bureau of Standards Washington, DC. 20234. Proceedings of the Building Thermal Mass Seminar. Knoxville, TN; 6/2-3/82. Oak Ridge National Laboratory
[14] Ya Feng. Thermal design standard for energy efficiency of residential buildings in hot summer/cold winter zones. *Energy and Buildings*, 2004. 36. 1309-1312.
[15] Simonson CJ; Salonvaara M; Ojalen T. The effect of structures on indoor humidity-possibility to improve comfort and perceived air quality. *Indoor Air*, 2002.12. 243-251.
[16] Simonson CJ, Salonvaara, M, Ojalen, T. Improving indoor climate and comfort with wooden structures. Espoo 2001. Technical Research Centre of Finland. VTT Publications. 2001. 431.200p. + app 91.
[17] Toftum J; Jorgensen AS; Fanger PO. Upper limits for indoor air humidity to avoid uncomfortably humid skin. *Energy and Buildings*. 1998. 28. 1-13.
[18] Toftum J; Jorgensen AS; Fanger PO. Upper limits of air humidity for preventing warm respiratory discomfort. *Energy and Buildings* 1998. 28. 15-23.
[19] Kalagasidis, AS. BFTools. Building physiscs toolbox block documentation. Department of Building Physics. Chalmer Institute of Technology. Sweeden. 2002.
[20] Kalagasidis, AS. HAM-Tools. International Building Physics Toolbox. Block documentation.
[21] Weitzmann, P; Kalagasidis, AS; Nielsen TR; Peuhkuri, R; Hagentoft, C. Presentation of the international building physics toolbox for simulink.
[22] Nielsen, TR; Peuhkuri, R; Weitzmann, P; Gudum, C. (2002). *Modelling Building Physics in Simulink*. BYG DTU Sr-02-03. ISSN 1601-8605.
[23] Rode, C; Gudum C; Weitzmann P; Peuhkuri R; Nielsen TR; Sasic Kalagasidis A; Hagentoft CE. International Building Physics Toolbox-General Report. Department of Building Physics. Chalmer Institute of Technology. Sweden. Report R-02: 2002. 4.
[24] www.ibpt.org. (Accessed August 2009).
[25] Wit, M. WAVO A simulation model for the thermal and hygric performance of a building. Faculteit bouwkunde, *Technische Universiteit Eindhoven*. 2000.

In: Buildings and the Environment
Editors: Jonas Nemecek and Patrik Schulz
ISBN: 978-1-60876-128-9

Chapter 8

TRADITIONAL UNDERGROUND WINE CELLARS OF SPAIN: BUILDING AND MORPHOLOGICAL PROPERTIES AND HYGROTHERMAL BEHAVIOUR – TAKING ADVANTAGE OF SOIL PROPERTIES

F.R. Mazarrón and I. Cañas*
Department of Construction and Rural Roads. Escuela Técnica Superior de Ingenieros Agrónomos. Universidad Politécnica de Madrid, Spain

1. INTRODUCTION

Since the dawn of civilisation, human beings have used underground spaces for living and storing food. There are numerous examples of traditional underground constructions throughout the world, such as homes in North Africa, China, Tunisia and the United States. In Spain, caves were used as houses up to the last century and underground cellars were used for preparing, maturing and conserving wine.

Wine is a "living" product that requires specific hygrothermal conditions for its preparation and aging which affect its final quality. For this reason, in places with fierce climates, wine was traditionally matured in underground cellars since the heat capacity and density of the ground dampened the extreme temperatures of the outside, providing suitable conditions for maturing and conserving wine. However, since the last century, wine production has moved to buildings above ground that require a large energy consumption to maintain suitable conditions of temperature and relative humidity.

With the energy crisis, it is necessary to develop bioclimatic strategies that reduce energy consumption. Traditional underground cellars are a good example of techniques adapted to the environment since they provide optimum conditions for wine aging without energy costs.

* Corresponding author. Address: Escuela Técnica Superior de Ingenieros Agrónomos. Universidad Politécnica de Madrid. Avda. Complutense s/n. 28040 Madrid, España. Tel.: +34 913365767; fax: +34 913363688; E-mail address: ignacio.canas@upm.es

The viability of applying this type of architecture today can be found in Spain where various of its best and most prestigious wines continue to be produced in underground cellars.

Traditional underground constructions have certain advantages in hygrothermal control for maturing wine compared to modern commercial (above ground) wineries. Previous work [1] has shown that there are large differences in the thermal behaviour of underground cellars compared to modern wineries. The conditions inside traditional cellars meet the optimal climatic conditions for maturing wine while this can only be achieved in commercial wineries with air conditioning. The traditional constructions are more stable with a small annual temperature variation and a higher relative humidity level even when commercial wineries are only partially above ground. The thermal behaviour inside a totally above ground commercial winery will be even more unfavourable for the wine aging process, assuming that the energy exchange between a below-grade building wall and the surrounding soil is smaller than that between above-grade walls and the surrounding ambient conditions.

For this reason, traditional underground cellars have certain economical and environmental advantages. Traditional underground cellars provide suitable conditions to achieve high-quality wines by natural energy, so the operational cost of these cellars is lower than that of commercial wineries, because of both the cooling energy savings and the reduced wine losses.

1.1. History of Underground Cellars in Spain

The grapevine was introduced into the Iberian Peninsula various centuries before Christ but it was the Romans who extended its cultivation and introduced the first techniques for producing wine, between the 1st century BCE and the 3rd CE. During the Arabian conquest of the Peninsula (in the 8th century CE), wine production lost its importance due to the Koranic prohibition on the consumption of alcoholic drinks. From the ninth century on, with the Christian reconquest, wine acquired great importance due to its role in religious ceremonies. In this period, wine production was in the hands of the monasteries where wine was usually kept in underground rooms. In later centuries, its consumption and preparation extended among the peasants. Production became generalised and a large number of underground cellars and districts of cellars appeared on the outskirts of villages. Since then, the building of this type of cellar was been maintained until a few decades ago in various wine producing areas in Spain. At present, the increasing depopulation of rural districts and the replacement of the ancestral wine-making techniques by new industrial methods have given rise to the loss of the original use and the consequent dereliction of an important part of this architectural heritage [2].

Throughout their history, underground cellars have had great social importance. As well as being the place for preparing wine, the underground cellars and their surroundings have served as a place for meetings and social gatherings where numerous community activities were carried out. Today, many of them continue to support this function where friends and family members meet daily to converse and drink wine or serve as restaurants or museums.

1.2. Features of the Study Area

The study of the underground cellars was carried out in one of the most important and oldest wine producing areas in Spain, the Ribera del Duero (Figure 1). This area is on a plateau at altitudes of 800 to 1100 metres. It has a continental Mediterranean climate featuring temperatures similar to a continental climate with very high temperatures in the summer, exceeding 30 °C, and very low in the winter when it frequently drops below 0 °C. The thermal oscillations are very important with fluctuations of up to 25 °C in the same day. The rainfall is similar to that of a Mediterranean climate but with a very small annual rainfall (between 400 and 600 mm) with a maximum in winter and a minimum in summer. Given this very unfavourable climate, underground constructions have for centuries been the only solution to provide suitable conditions for the conservation and maturing of wine.

The soil in the area consists of Tertiary sediments. The cellars are usually excavated in clayey or silty clay soils since the presence of a high percentage of fine particles provides suitable waterproofing and facilitates excavation while reducing the risk of cave-ins.

2. Conditions for Maturing Wine

The preparation of wine starts with the harvesting of the grapes. After transport to the cellar, the grapes are pressed to obtain their juice which is then submitted to a fermentation process in which the sugars are transformed into alcohol and carbon dioxide (CO_2). After fermentation, the ageing process is a fundamental step for obtaining high-quality wines, improving their sensorial characteristics. The two most important parameters to be controlled in wine maturing rooms are the temperature and relative humidity. However, other, secondary, factors must also be considered, such as light, vibration and unusual aromas that could negatively affect the wine's properties: ultraviolet rays destroy the organic compounds of the wine; excessive vibration prevents the normal sediment in an ageing wine to settle; the wood in the casks as well as the corks in the bottles are porous materials that transmit outdoor odours to the wine [3].

The bibliography on the hygrothermal conditions suitable for maturing wine is extensive and varied. A temperature that is too high may cause rapid development with alteration of the colour, bouquet and palate as well as increased loss of wine through evaporation, which is especially dangerous in the summer. Excessively low values can cause a very slow development of the wine and tartaric precipitations. The wine must also be protected from thermal oscillations, both daily and annual, since these cause the wine to expand and contract in the barrels, causing handling problems and loss of quality. The relative humidity cannot be too low since this increases losses of the wine through evaporation in the air. However, the relative humidity must not be excessively high since this could increase the appearance of moulds that affect the wine's palate and bouquet.

Figure 1. Location of the study area.

After a wide-ranging bibliographic study, Martin and Canas[1], set three optimal climatic conditions for wine aging: (1) maximum temperature lower than 15 °C, (2) annual temperature variation less than 6 °C, and (3) relative humidity higher than 80%.

3. Constructional Properties of Underground Wine Cellars

The design and properties of the wine cellar decisively affect the final quality of the wine since they determine the temperature, relative humidity, lighting and vibration values throughout the maturing period. The main objective when designing a cellar must be to provide the optimum conditions for the preparation, maturing and conservation of the wine.

Traditionally, the desired conditions were obtained by means of construction systems and good practices inside the cellar. The solutions include 1) the use of thick walls to dampen the changes in outside temperature, placing one or more walls in contract contact with the earth or rock, either by removing part of the sloping land or half burying the maturing room, 2) underground cellars excavated in slopes or below the surface, 3) buildings with very high ceilings to stratify temperatures, separating the hot air from the cold, 4) suitable orientation according to the climate, generally with the entrance and openings facing north to minimise the effects of sunlight, and 5) spraying the ground to increase humidity and cool the atmosphere. However, these traditional practices have been progressively replaced by air-conditioned buildings.

Of the above solutions, underground cellars have been one of the best options in areas with aggressive climates and suitable soil properties since they provide suitable conditions for maturing the wine without energy cost and without the need for handling.

3.1. General Description

The constructional features of underground wine cellars vary slightly depending on the properties of the area in which they are found, especially its orography. A general rule, the cellars are excavated in clay or silty clay ground that facilitates the extraction of the material at the start but which hardens, giving cohesion and waterproofing over time in contact with the air.

The thermal stability inside the cellar depends principally mainly on its depth and on the type of soil and, to a lesser extent, on factors such as ventilation. Thus, when there are slopes or hills near the village centre, the cellars are usually excavated in these since even if it is excavated horizontally, it can reach a great depth and, therefore, more stable conditions. The slope also favours surface water run-off. In more or less flat land where there are no significant variations in level, the cellars are always excavated downward with a sloping access tunnel of varying length depending on the depth of the cellar.

Regardless of the type of excavation (in flat or sloping ground) the traditional underground wine cellars have elements in common: the doorway or access construction, the entrance tunnel (horizontal or sloping) and the cave for preparing and maturing. Usually there is also at least one ventilation chimney, called a *zarcera* (Figure 2).

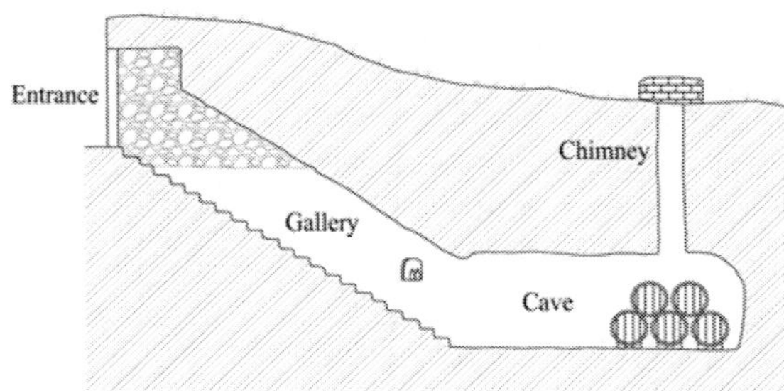

Figure 2. Cross-section of a typical underground wine cellar.

Entrance

There are various types of entrance constructions. The most simple and usual one is the building of a small access hut separating the outside from a small vestibule through which the tunnel is accessed. These elements are covered with material from the excavation and plants to prevent infiltrations and to provide greater thermal stability (Figure 3).

Figure 3. Entrances to traditional wine cellars.

Sometimes the constructions are more complex with a room for meals and meetings or even large constructions called *lagares* (pressing rooms) for pressing the grapes (Figure 4).

Figure 4. Interior of pressing room in a traditional wine cellar.

The façade or access usually faces north to reduce the effect of sunlight and to favour ventilation. It is built of stone although sometimes other materials are used such as wood and adobe. The door is of wood and has various ventilation holes in its upper part, permanently open, with an area of between 200 and 600 cm^2.

Gallery

The access tunnel is usually less than 2 m high and one wide, with variable gradients and lengths, depending on the depth of the cellar. In most cases it is reinforced with stonework at the initial part to prevent collapsing. The steps are directly excavated in the soil (Figure 5). There may be small cavities in the walls, used to store tools or to hold small barrels.

Cave

The tunnel opens into a cave where traditionally the wine is stored for fermenting and maturing (Figure 6). The size, layout and depth varies enormously depending on the properties of the ground and its use. The average depth of this type of wine cellar is usually between 2 and 6 metres below the surface. In cellars excavated in hillsides or at a great depth, it may exceed 15 metres. The layout of the cave may consist of a single room although more usually it has a branched structure with two or more rooms excavated at different times depending on the storage requirements and hardness of the ground. There are cellar rooms less than 5 metres long and others measuring dozens of metres. The walls are not usually lined, leaving the excavated soil visible.

Figure 5. Access tunnel in a typical underground wine cellar.

Figure 6. Detail of the cave interior in a typical underground wine cellar.

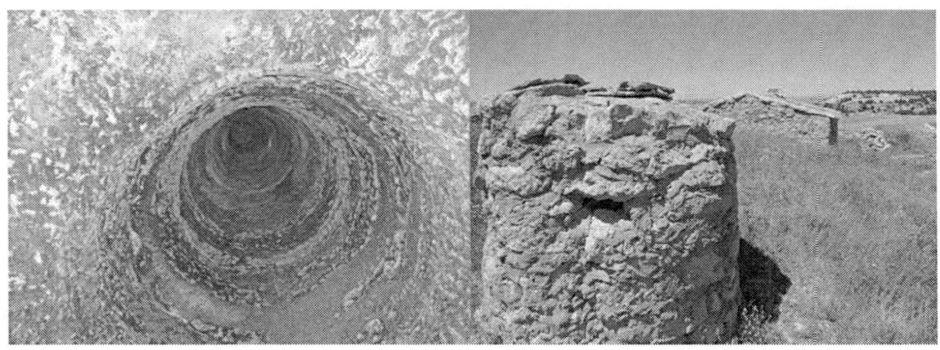

Figure 7. Detail of the interior (left) and exterior structure (right) of a ventilation chimney.

Chimney

In large cellars (>10 metres cave length) there are usually one or more ventilation chimneys, called *zarceras*, connecting the cave to the outside (Figure 7). Their purpose is to ventilate the wine cellar, especially in the autumn when the CO_2 from the grape fermentation process may cause death in a few minutes. The *zarceras* usually have a diameter of between 0.5 and 1 metre and end at the surface in easily identifiable cylindrical or conical structures. In small cellars, these ventilation chimneys are found less frequently.

3.2. Typological Classification

The cellars can be classified by various criteria such as the entrance construction, the layout in plan or the length of the cave.

By type of entrance construction

- With no construction at the entrance. It is a direct access through a door
- With a room. It is a small construction usually built to change the orientation of the entrance. Its purpose is to achieve good ventilation and to prevent sunlight entering the cellar.
- With a picnic area. This area originates in the previous type. The space between the tunnel and the entrance is used as a meeting place, for wine tasting and for meals.
- With a wine press. The space before the tunnel is the place where the grapes are pressed.

By layout

- Wine cellars with a linear cave.
- Wine cellars with a linear cave with small side rooms.
- Branched wine cellars.

By cave length

- Short (<5 m)
- Medium (5-10 m)
- Long (>10 m)

A detailed classification of types of wine cellars can be found in Pardo and Guerrero [2]

3.3. Traditional Building Process

The traditional wine cellars were dug by hand using tools such as picks and shovels. The material was removed in baskets or sacks through the cave entrance or, using pulleys, through the ventilation chimneys.

The building technique varied somewhat depending on whether the cellar was being excavated in sloping or flat ground. In the traditional system for building cellars on sloping ground, a vertical cut was made first that allowed the building of the entrance by excavating horizontally. The tunnel was then dug downward or horizontally. In cellars excavated in flat ground, the entrance was built on the surface and the tunnel always dug downward. In both cases, the entrance was built with rock or other materials, reinforcing the initial part with stone masonry. Throughout the process, the material excavated was deposited on top of the outside of the cave to form a hillock, favouring the run-off of rainwater and preventing its filtration and helping to provide greater thermal stability. Sometimes, grass was planted on top of the cave to facilitate this operation.

The ventilation chimneys were dug simultaneously with the tunnel. Once the tunnel was finished, the cave was excavated according to the requirements and the ground properties. Finally, the finishing touches were given to the exterior of the *zarcera*s and entrance. In most cases, the caves were enlarged in later years, enlarging the existing rooms or building new side rooms.

4. Hygrothermal Behaviour Of Traditional Underground Wine Cellars

The two most important parameters to be controlled during the wine maturing process are temperature and relative humidity. In the maturing cave of an underground wine cellar, the hygrothermal conditions are fundamentally determined by the temperature of the ground in which it is excavated and by the ventilation to the exterior.

4.1. Thermal Behaviour

The factor which most influences the conditions inside an underground wine cellar is the temperature of the surrounding soil. The ventilation in this type of construction is usually reduced so that the temperature in the cellar can be estimated from the ground temperature without distortion.

To explain the thermal behaviour of the ground, we will use one of the oldest and most widely-used methods to estimate the undisturbed soil temperature. This approximation is derived from the time-dependent, one-dimensional temperature solution for heat flow in a semi-infinite solid with a prescribed surface temperature boundary condition and constant thermal diffusivity [4]. An annual sinusoidal ambient temperature profile and an exponentially decaying sinusoidal temperature profile as a function of depth are assumed. Temperature at any given depth in a moment can be estimated on the basis of the following equation [5]:

$$T_{(x,t)} = (T_m) - A_s e^{-x\sqrt{\frac{\pi}{365\alpha}}} \cos\left[\frac{2\pi}{365}\left(t - t_0 - \frac{x}{2}\sqrt{\frac{365}{\pi\alpha}}\right)\right] \quad (1)$$

Where:

$T_{(x,t)}$	Soil Temperature at Depth x and Time t (ºC)
T_m	Average Soil Temperature (ºC)
A_s	Thermal Wave Amplitude (ºC)
x	Depth (m)
t	Day of Year (in days, where t = 0 at midnight on 31 December)
t_0	Phase Constant (days)
α	Apparent thermal diffusivity (m^2/day)

This simplified model has disadvantages such as the difficulty of precisely approximating the annual outside temperature as a sine wave or the unevenness of the soil profile. Nevertheless the application of this equation has provided valid results in the comparisons with experimental results from US [5], Australia [6,7] and other regions [8].

From equation (1) it can be inferred that the wave amplitude decreases exponentially with depth:

$$A_x = A_s e^{-x\sqrt{\frac{\pi}{365\alpha}}} \quad (2)$$

Where:

A_x Thermal Wave Amplitude at Depth x (ºC)

In addition, the phase lag decreases lineally with depth:

$$t - t_0 = \frac{1}{2}\sqrt{\frac{365}{\pi\alpha}}x \quad (3)$$

The above equations can be used to deduce that the ground temperature at any depth depends on the surface temperature of the soil (mean temperature and amplitude) and of the physical properties of the soil profile (conductivity (κ), density (ρ) and specific heat (c)) grouped in the apparent thermal diffusivity (α) considering a homogenous soil:

$$\alpha = \frac{\kappa}{\rho \cdot c} \quad (4)$$

Thus, the soil properties and depth determine the damping and the phase lag of the temperature wave. The greater the depth to which the cave is excavated and the lower the apparent thermal diffusivity of the soil, the greater the stability inside it and the less the external variations will be perceived. Figure 8 is an example in which it can be seen how temperature varies with depth and thermal diffusivity of an ideal surface temperature wave of an average temperature of 12 °C, annual amplitude of 10 °C and phase lag of 20 days.

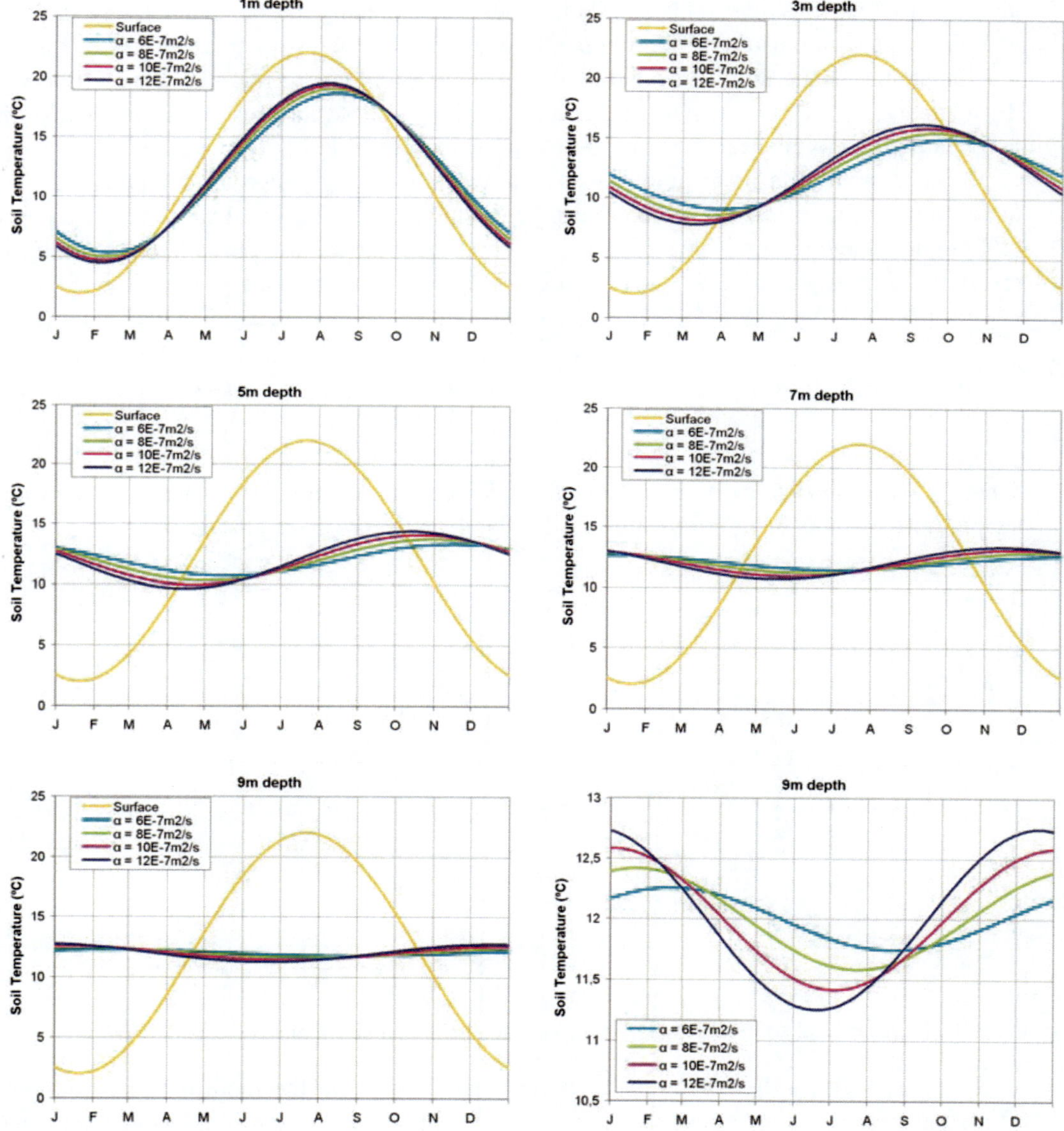

Figure 8. Temperature without distortion of the ground at depths of 1, 3, 5, 7 and 9 m for different apparent thermal diffusivity values (6E-7m^2/s, 8E-7m^2/s, 10E-7m^2/s, 12E-7m^2/s) compared to the surface ground temperature, calculated using equation (1).

As well as damping the amplitude and phase lag of the temperature wave, the oscillations also reduce with depth although this effect is not shown in the above theoretical model. Figure 9 shows experimental data of the ground temperature in a probe taken in the study area, where it can be seen how the daily variations recorded decrease as depth increases. The stability of earth temperatures with respect to daily cycles and the phase lag of the annual wave make the ground a useful heat source in winter and a means of cooling in summer.

All these soil properties affect the temperature in the underground wine cellars. Nevertheless, the exchange of heat with the outside due to the ventilation also affects it, although to a lesser extent. This ventilation increases the daily interior temperature variations and displaces the undisturbed soil temperature graph. Therefore, when equation (1) is applied to estimate the temperature surrounding underground buildings the results obtained show that the temperature of the outside face wall exhibits large departures from the undisturbed temperature due to the heat exchange between the building and its surrounding soil.

According to Mazarron and Canas [9], when studying inner air temperature in the underground wine cellar, equation (1) can be modified to correct the distortion caused by the building itself and its inner air mass, adapting average temperature, phase lag and depth. As a result, the adapted equation for temperature determination inside underground wine cellars is:

$$T_{(x',t)} = (T_m - k) - A_s e^{-x'\sqrt{\frac{\pi}{365\alpha}}} \cos\left[\frac{2\pi}{365}\left(t - t_0' - \frac{x'}{2}\sqrt{\frac{365}{\pi\alpha}}\right)\right] \qquad (5)$$

Where k is a coefficient for correcting the average temperature, t_0' is the day when outer temperature exceeds inner temperature and x' the depth of profile average height.

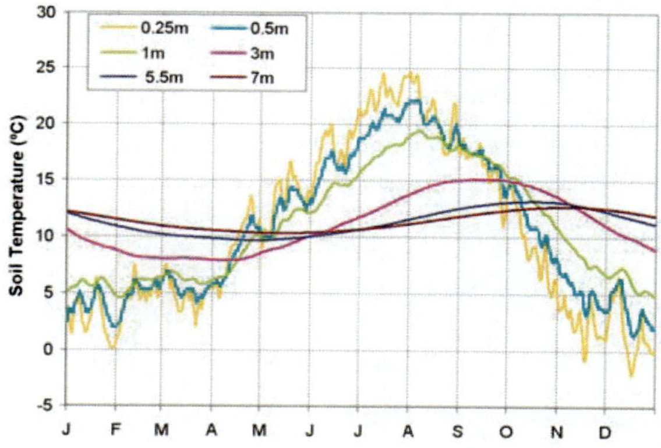

Figure 9. Experimental soil temperature at various depths in Atauta (Spain).

Figure 10 shows the results of applying equation (1) and the modified equation (5) to estimate the interior temperature of a typical underground wine cellar in the Ribera del Duero, used by Mazarron and Canas [9].

The experimental air temperature data from the inside of a typical wine cellar (Figure 11) show the difference in the cellar's response to exterior temperature changes throughout the year, as described by Martin and Canas [10]. The influence of the soil temperature is greater in the summer than in the winter, providing greater thermal stability in the most dangerous part of the year for maturing wine due to the high temperatures outside. This phenomenon is the result of reduced ventilation in the summer.

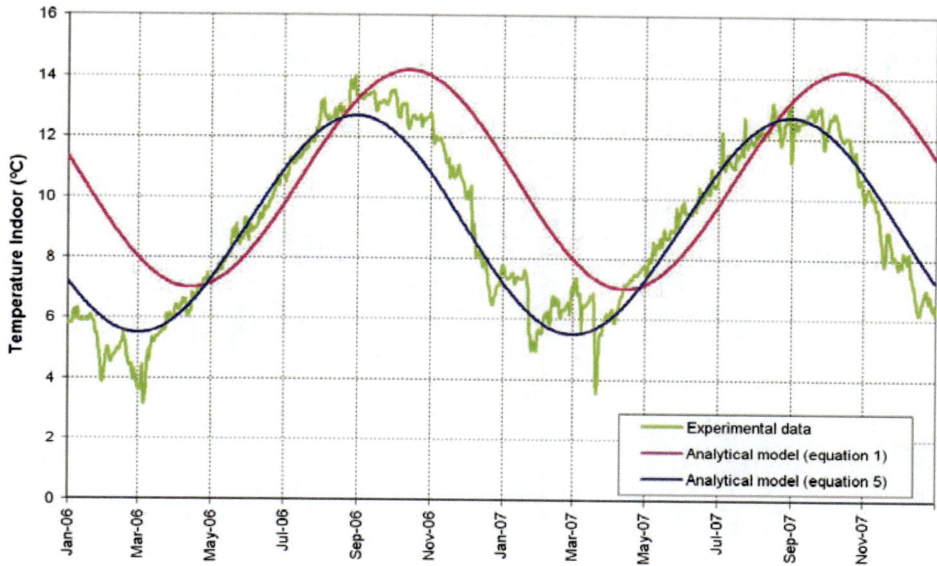

Figure 10. Experimental air temperature inside a typical underground wine cellar compared to that obtained by applying the model proposed by Labs (equation 1) and the model proposed by Mazarron and Canas (equation 5).

It can be concluded that the air temperature inside the cellar is fundamentally conditioned by the soil temperature at the average depth to which the cellar is buried and, to a lesser extent, by the exchange of heat with the outside due to the ventilation.

The average depth to which the cellar is buried has a greater influence on the soil temperature (and, as a result, on the air temperature inside the cellar) then the apparent thermal diffusivity of the soil profile. However, it is the combination of both parameters that determines the damping and phase lag of the exterior temperature curve, providing heat in the winter and cold in the summer, providing a more stable temperature.

After studying 22 traditional wine cellars in various villages in the Ribera del Duero, we found that an average depth of 3 m is usually sufficient to achieve the necessary temperature conditions for maturing wine without the need for air conditioning systems.

4.2. Relative Humidity

Relative humidity fundamentally affects two parameters in the wine maturing process, losses by evaporation and growth of mould.

To know the real effect on wastages, it is necessary to study temperature and relative humidity together. Table 1, extracted from Négre-Francot [11], shows the influence of temperature and relative humidity inside the cellar on wine losses by evaporation. When the relative humidity is low and the temperature is high, the percentage of yearly wine losses can be higher than 7%.

The relative humidity in this type of construction is usually very high, exceeding 80% for most of the year (Figure 11). Further, the period of low relative humidity usually coincides with temperatures far from the maximum so that losses due to evaporation are not very important. However, in many cellars without a ventilation chimney, the relative humidity reaches 100% in summer, producing condensation and increasing the risk of mould growth. The presence of *zarceras* is recommendable to favour ventilation in this type of construction, to prevent the relative humidity from reaching excessive values.

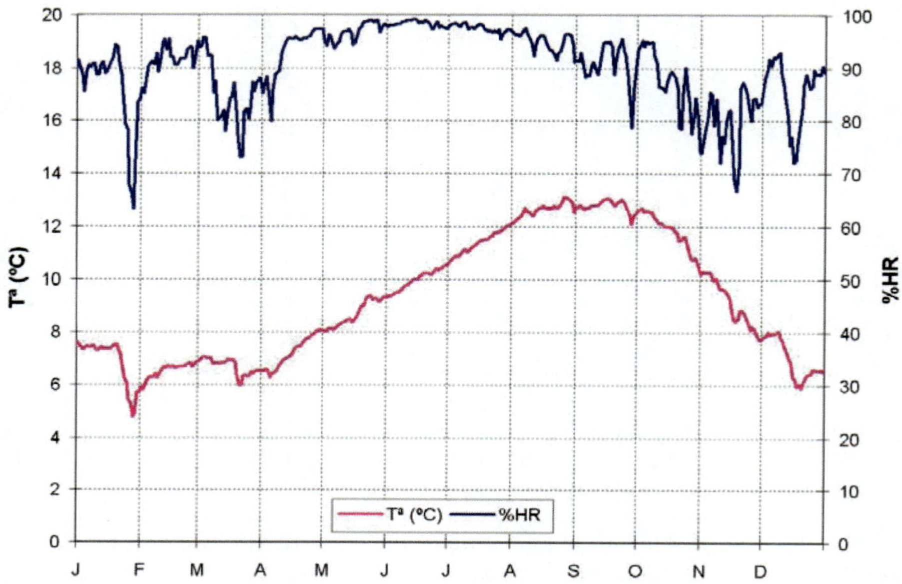

Figure 11. Experimental temperature and relative humidity values for the air inside the cave of an underground wine cellar buried at an average depth of 3 metres.

Table 1. Influence of temperature and relative humidity on yearly wine losses (all values in %) by evaporation [11].

Temperature (°C)	Relative Humidity (%)					
	45	55	65	75	85	95
10.0	4.4	3.9	2.9	2.2	1.4	0.6
12.0	5.0	4.2	3.3	2.5	1.6	0.7
14.0	5.7	4.8	3.8	2.8	1.8	0.8
16.0	6.5	5.4	4.3	3.2	2.0	0.9
18.0	7.4	6.1	4.9	3.6	2.3	1.1

4.3. Ventilation

Ventilation in traditional underground wine cellars is usually poor. However, its importance varies according to the time of year which can affect the interior temperature and relative humidity.

When the outside air temperature is higher than the inside temperature, the colder and heavier inside air accumulates inside the cave, impeding the entry of outside air, regardless of the presence of ventilation chimneys, so that the conditions are very stable (Figure 12).

When the outside temperature is lower than that inside, the colder and heavier outside air enters the cave and displaces the warm air inside, causing greater changes in the interior temperature and relative humidity. The cold outside air enters through the holes in the door, descends through the tunnel and displaces the warm air inside air via the *zarcera* (Figure 13).

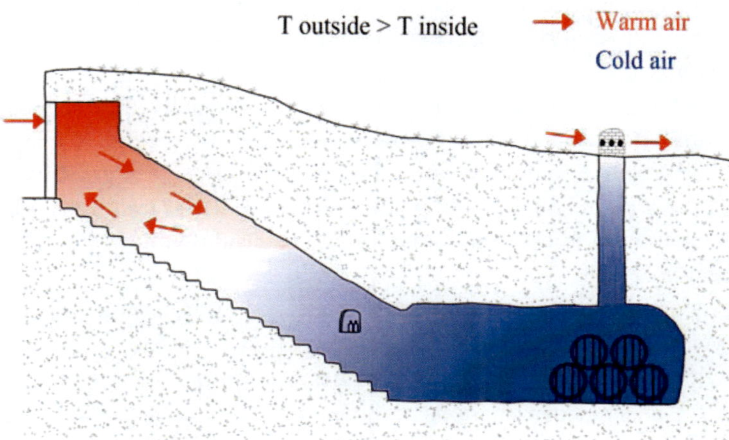

Figure 12. Natural ventilation in the summer in an underground wine cellar excavated below the surface.

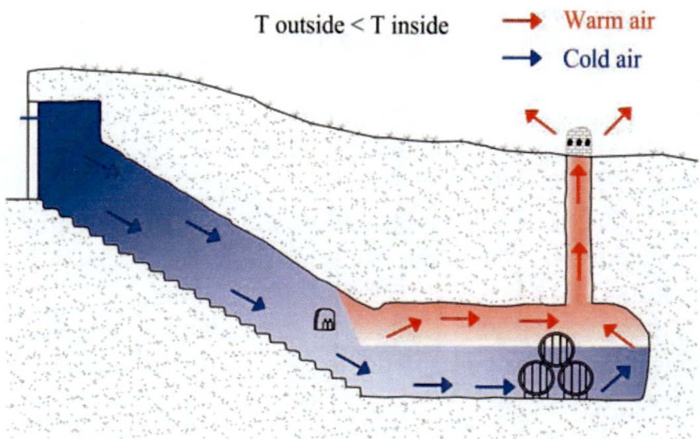

Figure 13. Natural ventilation in the winter in an underground cellar excavated below the surface.

Generally speaking, although temperature variations in the interior are greater when there are ventilation chimneys, they do not exceed excessive values that could negatively affect the quality of the wine. Further, the increased ventilation helps to remove the CO_2 produced during the fermentation of the grapes (when this takes place within the cellar) which, because it is denser than air, accumulates in the interior.

5. Conclusions

The traditional underground wine cellars in the Ribera del Duero are a good example of traditional bioclimatic architecture adapted to the environment.

The temperature inside depends mainly on the soil temperature at the average depth of the cellar so that the most influential parameter on the interior temperature is the depth to which the cellar is excavated; the greater the depth, the greater the interior thermal stability throughout the year. Ventilation from the outside also has an effect, to a lesser extent, especially in the winter.

The relative humidity in underground wine cellars is usually high throughout the year which, together with moderate temperatures, reduces losses of wine due to evaporation.

Excessively high relative humidity values increase the risk of the growth of mould which negatively affects the wine's bouquet and palate. The presence of traditional chimneys called *zarcera*s favours ventilation, reducing relative humidity inside without significantly affecting the interior temperature.

In areas with a continental Mediterranean climate, featuring large temperature variations throughout the year and low rainfall, these constructions provide suitable conditions for maturing wine with no energy cost. This includes an optimal temperature interval, high relative humidity and absence of vibrations, light and odours.

The techniques used in traditional architecture can be applied nowadays to modern constructions, to a make use of the advantages of this system. Its viability is shown by the fact that many of the highest quality wines in Spain are matured in underground wine cellars.

References

[1] Martin, S. & Canas, I. (2006). A comparison between underground wine cellars and above ground storage for the aging of Spanish wines, *Transactions of the Asabe*, 49, 1471-1478.

[2] Pardo, J. M. F. & Guerrero, I. C. Subterranean wine cellars of Central-Spain (Ribera de Duero): An underground built heritage to preserve, Tunnel.Underground Space Technol. 21(2006) 475-484.

[3] Canas, I. & Guerrero, S. (2005). Martin Ocaña, Study of the thermal behaviour of traditional wine cellars: the case of the area of "Tierras Sorianas del Cid" (Spain), *Renewable Energy*, 30, 43-55.

[4] Carslaw, H. S. & Jaeger, J. C. (1959). Conduction of Heat in Solids (second ed.), Oxford University Press, London,.

[5] Labs, K. (1982). Regional-Analysis of Ground and Above-Ground Climate Conclusion, *Underground Space*, 7, 37-65.

[6] Baggs, S. A. (1982). Remote prediction of periodic ground temperatures in Australia using isothermal contour maps, *Underground Space*, *7* 127–132.

[7] Baggs, S. (1983). Remote prediction of ground temperature in Australian soils and mapping distribution, *Solar Energy*, *30* (4) 351–366.

[8] Mihalakakou, G. Santamouris, M. & Asimakopoulos, D. (1992). Modelling the earth temperature using multiyear measurements, *Energy and Buildings*, 19, 1–9.

[9] Mazarron, F. R. & Canas, I. (2008). Exponential sinusoidal model for predicting temperature inside underground wine cellars from a Spanish region, *Energy Build*, *40* 1931-1940.

[10] Martin, S. & Canas, I. (2006). Comparison of analytical and on site temperature results on Spanish traditional wine cellars, *Applied Thermal Engineering*, *26* 700–708.

[11] Négre, E. & Françot, P. (1980). Manual Práctico de Vinificación y Conservación de Vinos. 3rd ed. Barcelona, *José Monteső*.

In: Buildings and the Environment
Editors: Jonas Nemecek and Patrik Schulz
ISBN: 978-1-60876-128-9

Chapter 9

ESTIMATING THE ENVIRONMENTAL EFFECTS ON RESIDENTIAL PROPERTY VALUE WITH GIS

Lubos Matejicek
Institute for Environmental Studies Charles University in Prague,
Faculty of Natural Science Benatska 2, 128 01, Prague, Czech Republic

ABSTRACT

Ongoing research to develop a new generation of decision-making tools for estimating the environmental effects on residential property value has significantly increased the demand for land surface data, information on the state of living environment, and the corresponding need for the advanced use of geographic information systems (GIS). In order to provide the foundation for price estimates, all the existing data focused on building and environment are integrated in the framework of a GIS spatial database. Traditional methods mostly emphasize the relationship between the effects of accessibility to central locations, and ignore location-specific attributes of housing. However, little has been done on high-rise, densely populated residential areas. Thus, the paper aims to investigate the neighboring and environmental characteristics of the selected site. A more advanced approach based on spatial analysis and modeling in the GIS environment is used to manage spatio-temporal data, to process aerial images and satellite images, to import measurements from GPS, to create digital terrain models, to analyze topography together with environmental data, and to visualize the results. The partial results based on the traveling time to the closest services and the origin-destination cost matrixes are derived from the network analysis. The landscape characteristics can be demonstrated on animated sequences showing the flight over a landscape model that is based on the digital terrain model with draped images over it to show the area. The living environment contains a few compartments: air pollution, water pollution, waste management, noise assessment, and monitoring of environmental impacts on population health. Air pollution is estimated by continuous map surfaces, predicting the values of pollutants concentration for the selected site in dependence on the sample points at the air quality monitoring stations. Water pollution covers surface water pollution, drinking water supply and quality, waste water, accidental contaminant spills, and optionally, flood control measures. The waste compartment is focused on the system of municipal waste management with sorting of reusable components of municipal waste, and hazardous chemicals information. Noise assessment covers road traffic noise and air traffic noise. The environmental impacts on population health come out from reports of the national institute of public health. In addition to the main compartments, other data can complement the partial inputs to the system (energy prices, public transport schema, locations of neighbor natural reservation sites and historical sites). In the framework of spatio-temporal analysis, the spatial weighting matrixes are used for prediction of the final rating. The final results are represented by the thematic map layers in the GIS project based on ArcGIS and its extensions. The attached case study shows a scenario in dependence on setting the weight parameters. Unexpected findings are caused by rapid changes in the dense living environment and slow conversion of the reality market in the selected site in Prague, the Czech Republic. For all that, the case study brings better understanding of how the residential site rating depends on various environmental attributes.

Keywords: building; living environment; spatial modeling; GIS.

1. Introduction

Throughout history, the residential environment affects our health and well-being and we, through our activities, reshape the residential environment. This environment is interconnected to natural, social, and economic processes that run on the local, regional, and global scales. But, different neighborhoods or communities have different demographic characteristics, because of the economic and social processes that structure age, and race-ethnicity of the population.

The residence represents an important activity site in the environment where a person spends time. In addition to residence sites, the activity space is larger and depends on other activity sites like workplaces, stores, recreational areas, and the pathways traveled to and from the residential site. The size and shape of an individual person vary depending on its profession, the modes of transportation available and the geographical locations where all the activities take place. The residential activity sites together with others represent the zones where the individuals can be exposed to risk or resources, which affect the residential property value. It can be documented by disease mapping that has made contributions to public health and epidemiology for centuries. A common practice in mapping the residential property value by health events has been to calculate health event rates for political or administrative units like towns or tracts, because population and health outcome data are mostly reported for these areas. An obstacle with this approach is evident in arbitrary partitioning the underlying distribution of population and health events. Also, the risk factors, like population at risk, are not located at individual residential sites. Selected contaminants like air pollution, water pollution, waste, noise and biological agents are present in our living environment in various local and global scales. The temporal and spatial patterns are needed to model the distribution of known or potential risk factors.

2. GIS and Associated Systems

Geographic Information Systems (GISs) are being used in pollution environmental studies to model pollution dispersion and its effects on residential areas. The ability to estimate the environmental effects on residential property values is important for decision-making processes in order to start health public intervention activities [1]. Spatio-temporal modeling, in the framework of GIS, occurs whenever operations attempt to emulate geographic processes in the real world, at selected points in time or over extended periods. In the past, many other modeling applications have been coupled with GIS for high performance in dynamic modeling, remote sensing calculations and database management. But with the increasing power of GIS hardware and software, it is now possible to reconsider this relationship and to develop multifunctional tools [2]. Thus, the spatio-temporal modeling is trying to integrate several different contexts in the GIS world. There are two particularly important meanings.

First, spatial modeling by itself is focused on data models that include a set of expectations about data extended by spatial analysis. The data models are mostly represented by templates into which the data needed for particular applications are fitted together with topological rules and spatial analysis results. For example, the ESRI's geodatabase supports models of topologically integrated feature classes with extensions for complex networks, topologies, relationships among feature classes, and other object-oriented features. The geodatabase is implemented on standard relational databases complemented by spatial database extensions [3]. The way feature classes are often used in GIS, the features correspond to a group or class of real-world features, such as buildings, roads, railways, rivers, bus lines, dominant sources of environmental pollutions, and monitoring networks. In essence, GIS assists to develop the data models and to create map composition and visualization based on data models [4].

Second, temporal modeling deals with representation of processes that modify or transform some aspects of the Earth's surface through time. For example, contemporary environmental pollution forecasts are based on dynamic models of the particle dispersion in air, water and soil. Temporal models of vehicles behavior are used to predict traffic congestions. The strategies of integration spatially distributed environmental models into GIS range from simple linkage through shared data to building temporal models as analytical extensions into fully functional GISs. In case of shared data, two separate systems, the GIS and the simulation system, just exchange data through common files or a shared database [5].

GISs, computer based systems for the management and analysis of geographic data, integrate spatial data that originate from observation and measurement of the earth phenomena referenced to their locations on the earth's surface [6]. Whenever urban planners or environmental authorities use residential registration with address information, consider the location of pollution sources, or look at air quality and water quality reports from monitoring networks, they are working with geographic data. Thus, the GISs are considered as tools applicable to the integration and analysis of many different types of spatio-temporal data by people in different organizational setting asking various questions [7]. Thus, the GIS is related to other types of software that are focused on spatial data and spatial analysis [8].

2.1. GIS Mapping and Standard Digital Data Sources

The process and object views are expressed in the GIS by two main data structures: raster and vector. These data structures have different properties for storage and processing of spatial data. Majority of the recent GISs can manage these data structures, but many various data formats implemented in GISs require advanced data transformation methods [9].

In a raster system, each raster layer matches to a single attribute for a unit of space. For example, elevation or pollutant concentration at a particular place corresponds to a single attribute. Raster data are represented by an array of cells corresponding to particular places. The setting of the cell size affects the size of an object that can be discerned in a digital image, which determine the spatial resolution of the data. Landscape feature such as a single-family detached residence has to be discerned in high resolution raster images like the 10-meter spatial data, because the feature could be smaller than the raster cell size. On the other hand, low resolution is used to be provided for large geographical areas where the limited detail of landscape features is sufficient.

In a vector system, traditional GISs use layers that correspond to a single class of objects with the same dimensionality. Thus, the point layers can represent locations of residential buildings, sources of pollution, a network of air quality monitoring stations or locations of water quality measurements. The line layers represented by a sequence of straight-line segments interconnected with vertexes are used for traffic networks, surface water networks or line sources of pollution. The polygon layers formed by line boundaries can represent residential buildings in a more detailed view, different types of land-cover or land-use. In addition to the spatial location of vector objects, many attached attributes stored in tables contain other information. While the raster topology is based on regularly organized cells in grids, the vector topology has to reflect the neighborhood relationships among various types of vector objects.

In order to effectively store spatial data and their spatial relationship, a few advanced vector formats have been developed during last decades. For example, the ESRI's coverage model has been used for a few years. An extension of the coverage model is the ESRI's geodatabase that support complex networks, relationship among feature classes, and other object-oriented features. Both the models are also implemented on standard relational databases in the framework of the ArcSDE application server [10]. Thus, the individual environmental effects can be studied in dependence on location and time [11]. Mostly, they are represented by main environmental compartments like air pollution, water pollution, landscape, waste management, and noise assessment. The air pollution compartment contains layers focused on emissions, air quality monitoring, and predictions of pollutant concentrations. The water compartment can include layers with monitoring of surface water, drinking water, waste water, accidental contaminant spills, and flood control measures. The landscape compartment usually includes a wide range of thematic map layers focused on land cover, nature conservation, landscape protection, city greenery, and protected areas. The layers dealing with waste management include information about the system of municipal waste management. The noise assessment compartment include mapping of road and air traffic noise.

2.2. Remote Sensing and Field Mapping by GPS

Remote sensing deals with the acquisition of information about the entity without being in physical contact [12]. The data are collected from sensors on satellites or aircrafts. In case of Earth observations, two major groups of sensors can be used for collection of images in dependence on the source of electromagnetic spectrum.

Passive sensors detect natural energy that is emitted by the surface objects. Examples of passive remote sensors include airborne cameras or sensors carried by Landsat 7, Aster, Ikonos, and other existing satellites. Aerial images can assist in the high resolution mapping of residential zones and their surrounding areas. Satellite images offer to collect and transmit data from various parts of the electromagnetic spectrum (red-green-blue visible spectrum, near-middle-far infrared invisible spectrum), which in conjunction with aerial or ground-based sensing provides Earth science research with more information to monitor trends such as evolution of environmental and climatic changes.

Active sensors, on the other hand, emit energy in order to scan objects whereupon passive sensors then detect the radiation that is reflected from the target objects. Examples of these sensors represent 3D surface laser scanners or sensors carried by Radarsat. Radar/Lidar systems offer active remote sensing where the time delay between energy emission and its detection is measured in order to acquire the location and shape of the surface objects. It is often used for digital terrain mapping, which include the construction of a digital terrain model (DTM) [13]. In such a process, points are sampled by remote sensing sensors or by a global positioning system (GPS) from the terrain to be modeled with a certain observation density and accuracy. The terrain surface is then represented by the set of sample points that can be complemented by spatial interpolation if attributes on locations on the digital surface other than sample points is needed. In case of visualization, grid- and triangle-based modeling approaches are widely used than point-based approaches. For grid-base modeling, a grid network needs to be formed through spatial interpolation, if the original data are not in a grid form. For triangle-based modeling, a triangulation network is formed from point-based approaches through a triangulation procedure in order to create the triangular irregular network (TIN). Finally, the digital terrain model together with the draped satellite or the aerial images are included into other GIS layers for landscape analysis and visualization that are linked to the residential property value mapping [14, 15, 16].

3. Mapping of the Environmental Pollution

The success of environmental-related GIS studies depends on having access to accurate, timely, and compatible spatio-temporal data. Map layers focused on living environment mostly complement spatial data sets for a wide range of policy and planning issues. In case of mapping environmental pollution, their value extends well beyond the scope of the original projects for which they were created, and it increases as the data sets are used and updated. In order to use environmental data sets in GIS, the data sets must be captured and linked to the existing map layers in the framework of a foundation spatial database. In many cases, the digital data need to be imported into GIS from exiting information systems. Thus, the complex process of scanning, digitizing, entering field data and address matching can be reduced or even omitted.

The mapping of the environmental pollution together with other spatial data brings together four key elements: the spatio-temporal data stored in spatial database, the representations of that information on map layers, queries about the spatio-temporal data, and the analysis that creates new information and new map layers.

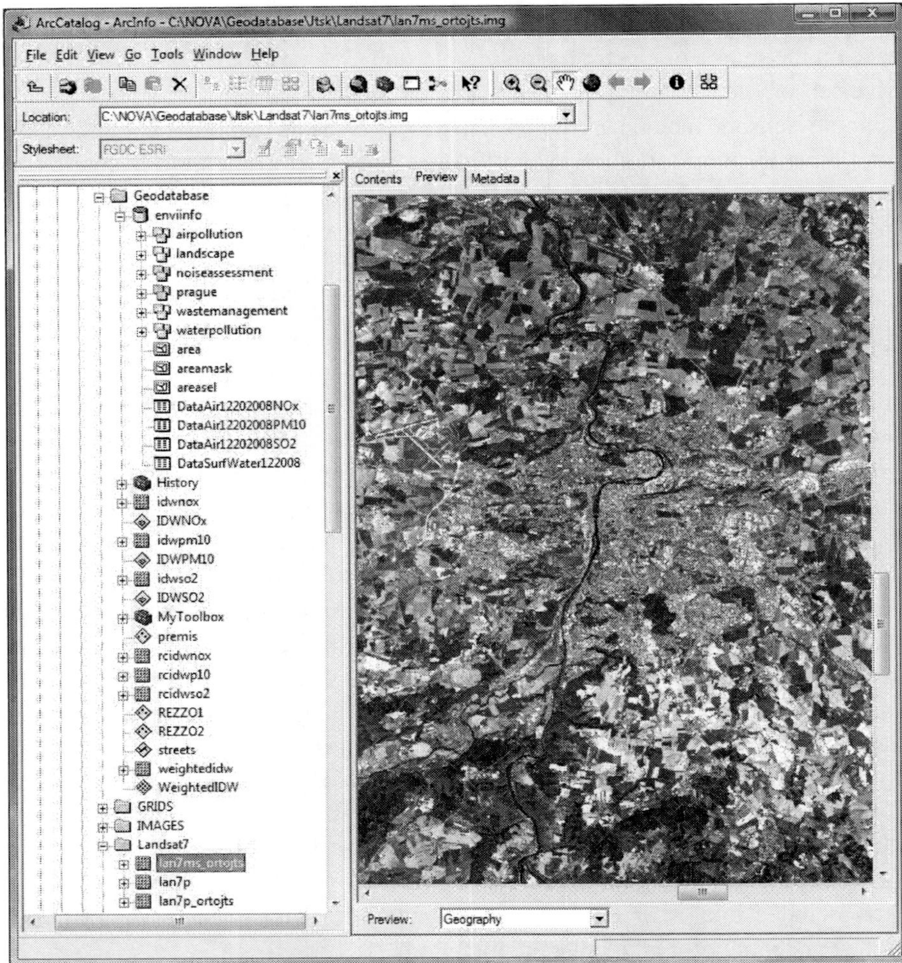

Figure 1. The data structure of the GIS project for estimating of the environmental effects: the list of the map layers is on the left site, the whole city from the Landsat 7 is in the preview mode

4. SPATIAL ANALYSIS AND VISUALIZATION

The environmental effects on residential property value affect agents that produce adverse health outcomes in humans. At residential zones, these agents can be physical (noise), chemical (ozone, nitrogen oxides, sulphur dioxide, particle fraction PM_{10}) or biological (cryptosporidium) in nature. Human population encounters these agents by breathing, eating, and drinking, or coming into physical contact at built environment. The processes by which agents in the living environment produce an adverse health outcome can be modeled as hazard-exposure-outcome processes by risk assessment methods [17]. In the last decade, many models have been carried out in the field of noise impacts [18, 19], air pollution [20, 21, 22, 23], water pollution [24, 25], waste management [26] and land use/land cover changes [27].

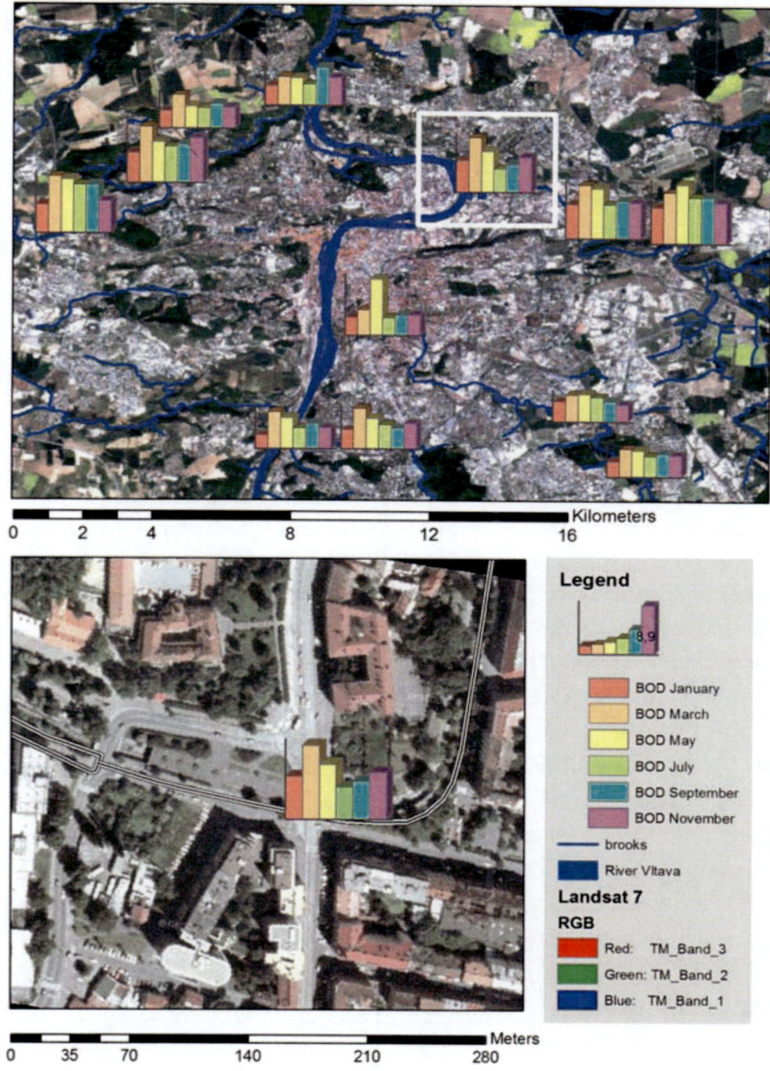

Figure 2. The measurements of water quality at the sampling points linked to the GIS layer

In order to link spatio-temporal data drawn from many sources into the GIS, time and space must be the basis for data integration in a way that models hazard-exposure-outcome processes. Besides the integration of digital map layers focused on the location of residential buildings, hazardous agents need to be identified and their source locations and associated contamination fields must be estimated. In spite of that human experience is an important source of awareness of the harmful effects of many toxicants, there is still the need to carry out long term research in the framework of quantitative risk assessment that deals with the process of characterizing the health effects expected from exposure to an agent and estimating the probability of occurrence of health effects. Many case studies and reports are annually published by U.S. EPA. Subsequent legislation represents the major impetus for conducting risk assessment, but the responsibility for determining acceptable risk is in the range of state authorities and local agencies that implement the legislation. In this case, the methods of hazard identification where the GISs make the strong contribution include spatio-temporal association of adverse environmental residential effects [28, 29, 30].

5. A Case Study: Mapping of Environmental Pollution in Prague with Respect to Environmental Effects on Residential Property Value

The mapping of environmental pollution in the city Prague aimed to evaluate the application of spatio-temporal analysis with estimating the environmental effects on residential property value. The project was created in the framework of ESRI's ArcGIS and the spatio-temporal data were mostly saved in ESRI's geodatabase, Figure 1.

As an example, measurements of air pollution are determined by monitoring of pollutants (SO_2, PM_{10}, NO_x) in the ground-level strata of atmosphere, and measurements of surface water quality are taken at fixed sampling points in the urban surface water network. Information about the ground-level air pollution is based on the public data accessible through the Internet (http://www.premis.cz , December, 2008). The actual configuration of the monitoring network is 15 monitoring stations that measure 12 main pollutants. Data are recorded in one hour period. The data are archived in the database of the Air Quality Information System of the Czech Republic. The surface water quality assessment is performed according to the local standards. The recommended monitoring frequency is 12 samples taken per a year. Surface water is classified into five classes based on quality parameters over a longer period, usually a two-year period in order to have 24 data series (http://envis.prague-city.cz , December, 2008). The selected measurements are stored in the geodatabase (EnviInfo) and linked to the map layers. The network of sampling sites is in Figure 2, which also includes, as an example, measurements of the biological oxygen demand (BOD) in the attached column diagrams.

In case of air pollution, the prediction map layers for selected pollutants (SO_2, PM_{10}, NO_x) were created by examining the relationships between all of the sample points and producing a continuous surface of concentration. In order to produce the predictions maps, there are two main groupings of interpolation techniques: deterministic and geostatistical. In spite of that the geostatistical methods offer not only prediction surfaces but also error or uncertainty surfaces to give an indication of how good the predictions are, the deterministic method Inverse Distance Weighted (IDW) was selected considering to the limited number of the sample points-monitoring stations. This interpolation method forces the resulting surface to pass through the data value at sampling points and assumes that each measured point has a local influence that diminishes with distance. It weights the pollutant concentration at the points closer to the prediction location greater than those farther away. An example of the spatial interpolations based on IDW for PM_{10} concentrations is in Figure 3. Similarly, the spatial interpolation derived from the same sampling locations and based on IDW for NO_x is in Figure 4. In order to explore the results of the PM_{10} prediction map, the map composition in Figure 3 contains the map layers of the monitoring stations, the road network, and the potential sources of air pollution based on the national Air Pollution Sources Register (large and the mid-sized pollution sources: REZZO 1 and REZZO 2). The NO_x prediction map in

Figure 4 is complemented by the map layers of the monitoring stations and the road network that is considered to have the high influence on the NO_x concentration in urban areas together with ozone and other compounds [31].

The next step in Figure 5 demonstrates using of the GIS spatial analysis to produce a suitability map by combining prediction maps of air pollutants (NO_x, PM_{10}, SO_2). The prediction maps are reclassified to a common scale, weighted according to a percentage influence and, finally, combined to produce a map displaying suitable locations for residential building. As an example, the datasets from the air pollution compartment have been selected to document the GIS ability for creating the spatial weighting matrixes for prediction of the final rating. Similarly, the map layers from other environmental compartments dealing with water pollution, waste management, noise assessment, and monitoring of environmental impacts on population health can be rated or complement existing results [32, 33]. Other GIS rating can include the calculation of the traveling time to the closest services and the origin-destination cost matrixes. Multimodal network datasets in the GIS enable to model multiple forms of transport across rail stations, subway stations or bus stops that form the linkages between several different forms of transportation. The temporal changes can be identified by comparison of a few spatial weighting matrixes linked to the selected time events.

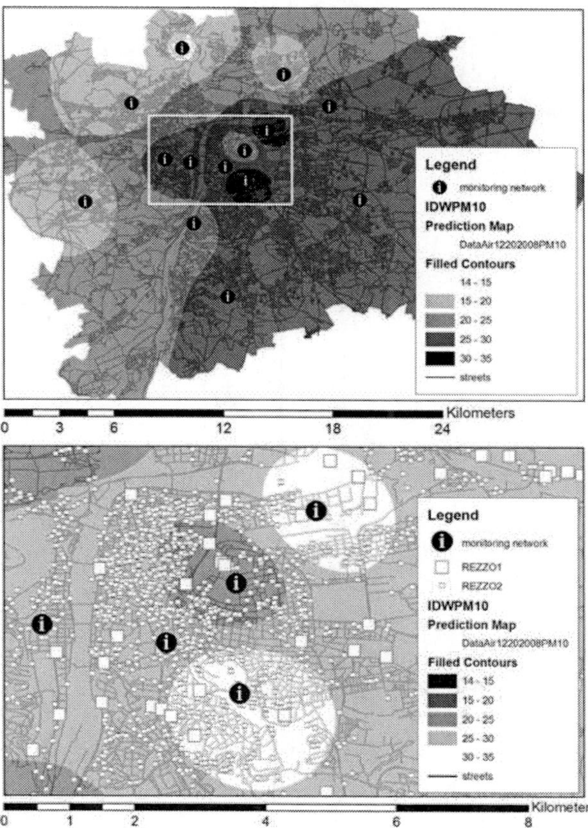

Figure 3. The prediction map of PM_{10} concentrations in the city Prague and its central part

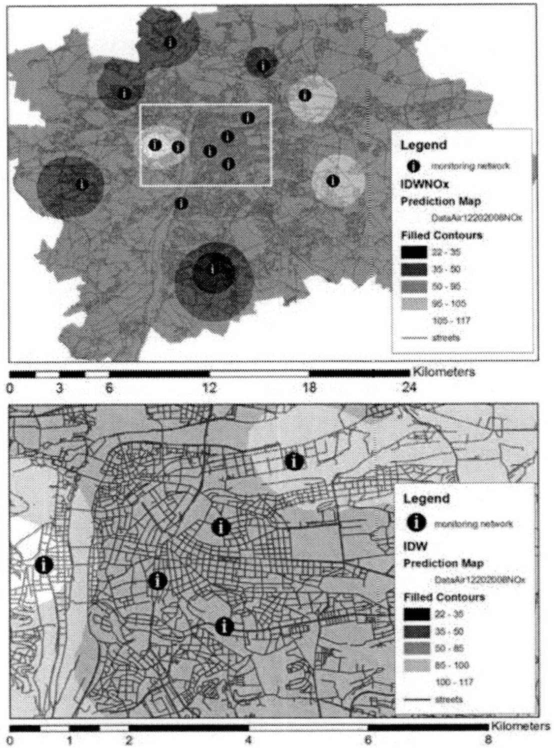

Figure 4. The prediction map of NO_x concentrations in the city Prague and its central part

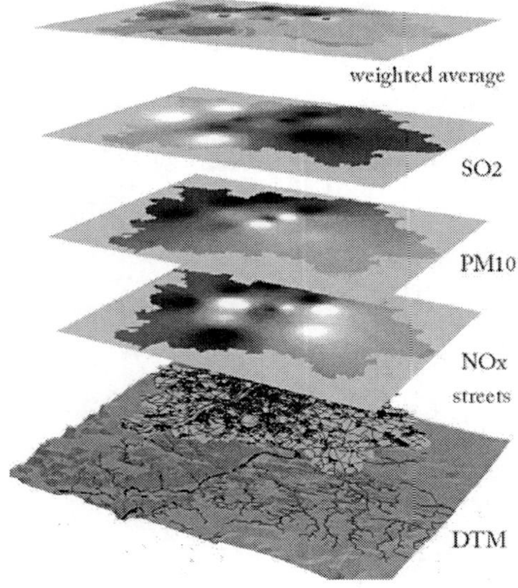

Figure 5. Mapping of suitable locations for residential building based on estimating the environmental effects: the light sites in the average weighted map layer indicate higher levels of mixed air pollution derived from spatial weighting matrixes for prediction of the final rating, the bottom layers represent the DTM with the draped surface water network and the street network

7. Conclusion

The presented research shows various possibilities of using the GIS and its associated systems in the framework of estimating the environmental effects on residential property value. Considering to the complexity of this research, the attached case study is mainly focused on environmental compartments linked to the air pollution and partially to the surface water pollution. The spatial analysis demonstrates the weighting matrixes for prediction of the final rating, which can be extended to the whole range of environmental compartments and other phenomena. In order to provide the described complex analysis, a wide range of spatio-temporal data is needed to create a GIS project and to start database tasks. But, a great part of the digital data can be accessible from the U.S. EPA, landscape and urban planning departments, or local environmental agencies. Also, it is expected that the future GIS development will bring a better access to environmental data and insights to the research focused on building and environment.

Acknowledgments

The presented research was carried out in the framework of the project supported by Academy of Sciences in the Czech Republic (AVCR 1ET400760405). The spatio-temporal data and terrain measurements have been processed by ArcGIS, ERDAS Imagine in the GIS Laboratory at the Faculty of Natural Science, Charles University in Prague. The spatio-temporal data were captured via the Internet and from the annual reports with information on the Prague environment published by the city hall.

References

[1] Cromley, E. K. & McLafferty S. L. (2002). *GIS and Public Health*. New York: The Guilford Press.

[2] Maguire, D. J., Batty, M., & Goodchild, M. F. (2005). *GIS, Spatial Analysis and Modeling*. Redlands, California: ESRI Press.

[3] Rigaux, P., Scholl, M. & Voisard, A. (2002). Spatial *Databases with Application to GIS*. San Francisco, California: Morgan Kaufmann Publishers.

[4] Mitchell, A. (2005). The ESRI Guide to GIS Analysis. Volume 2: Spatial Measurements & Statistics. Redlands, California: ESRI Press.

[5] Fedra, K. (1996). Distributed Models & Embedded GIS: Integration Strategies and Case Studies. In M. F. Goodchild, L. T. Steyaert, B. O. Parks (Ed.), *GIS and Environmental Modeling: Progress and Research Issues* (pp. 413-417). Fort Collins: GIS World, Inc.

[6] Tomlinson, R. F. (1987). Current and potential uses of geographical information systems: The North American experience. *International Journal of Geographic Information Systems, 1(3),* 203-218.

[7] Longley, P. A., Goodchild, M. F., Maguire, D. J. & Rhind, D. W. (2001). *Geographic Information Systems and Science*. New York: John Wiley & Sons Ltd.

[8] Johnston, C. A. (1998). Geographic Information Systems in Ecology. London: Blackwell Science Ltd.
[9] Zeiler, M. (1999). *Modeling Our World*. Redlands, California: ESRI Press.
[10] MacDonald, A. (2001). *Building a Geodatabase*. Redlands, California: ESRI Press.
[11] Zeiler, M. & Arctur, D. (2004). *Desiginig Geodatabases*. California: ESRI Press.
[12] Campbell, J. B. (2002). *Introduction to Remote Sensing*. London: Taylor&Francis.
[13] Li, Z., Zhu, Q. & Gold, C. (2005). *Digital Terrain Modeling: Principles and Methodology*. New York: CRC Press.
[14] Landgrebe, D. A. (2003). *Signal Theory Methods in Multispectral Remote Sensing*. New Jersey: John Wiley and Sons.
[15] Richards, J. A. & Jia, X. (2006). *Remote Sensing Digital Image Analysis*. Berlin: Springer.
[16] Stacey, S. & Pouncey, R. (1997). *Erdas Field Guide*. Atlanta: Erdas, Inc.
[17] Mower, B. (1998). A multiple source approach to acute human health risk assessment. *Waste Management 18*, 377-384.
[18] Bachman, W., Sarasua, W., Hallmark, S. & Guensler, R. (2000). Modeling regional mobile source emissions in a geographic information system framework. *Transportation Research Part C 8*, 205-229.
[19] El-Fadel, M., Shazbak, S., Baaj, M. H. & Saliby, E. (2002). Parametric sensitivity analysis of noise impact of multihighways in urban areas. *Environmental Impact Assessment Review 22*, 145-162.
[20] Yanosky, J. D., Paciorek, C. J., Schwartz, J., Laden, F., Puett, R. & Suh H.H. (2008) Spatio-temporal modeling of chronic PM_{10} exposure for the Nurses' Health Study. *Atmospheric Environment 42*, 4047-4062.
[21] Madsen, C., Carlsen, K. C., Hoek, G., Oftedal, B., Nafstad, P., Meliefste, K., Jacobsen, R., Nystad, W., Carlsen, K. H. & Brunekreef, B. (2007). Modeling the intra-urban variability of outdoor traffic pollution in Oslo, Norway-A GA^2LEN project. Atmospheric *Environment 41*, 7500-7511.
[22] Grivas, G., Chaloulakou, A. & Kassomenos, P. (2008). An overview of the PM10 pollution problem, in the Metropolitan Area of Athens, Greece. Assessment of controlling factors and potential impacts of long range transport. *Science of the Total Environment 389*, 165-177.
[23] Vardoulakis, S., Gonzalez-Flesca, N., Fisher, B. E. A. & Pericleous, K. (2002). Spatial variability of air pollution in the vicinity of a permanent monitoring station in central Paris. *Atmospheric Environment 39*, 2725-2736.
[24] Niemczynowicz, J. (1999). Urban hydrology and water management-present and future challenges. *Urban Water 1*, 1-14.
[25] Matejicek, L. (2006). Modeling of Water Pollution in River Basins with GIS. In A.R.Burk (Ed.), *Focus on Ecology Research* (pp. 247-270). New York: Nova Science Publishers.
[26] Leao, S., Bishop, I. & Evans, D. (2001). Assessing the demand of solid waste disposal in urban region by urban dynamics modelling in a GIS environment. *Resources, Conservation and Recycling 33*, 289-313.
[27] Azócar, G., Romero, H., Sanhueza, R., Vega, C., Aguayo, M. & Muñoz, M.D. (2007). Urbanization patterns and their impacts on social restructuring of urban space in

Chilean mid-cities: The case of Los Angeles, Central Chile. *Land Use Policy 24*, 199-211.

[28] Koomen, E., Dekkers, J. & van Dijk T. (2008). Open-space preservation in the Netherlands: Planning, practice and prospects. *Land Use Policy 25*, 361-377.

[29] Culshaw, M. G., Nathanail, C. P., Leeks, G. J. L., Alker, S., Bridge, D., Duffy, T., Fowler, D., Packman, J. P., Swetnam, R., Wadsworth, R. & Wyatt, B. (2006). The role of web-based environmental information in urban planning-the environmental system for planners. *Science of the Total Environment 360*, 233-245.

[30] Pullar, D. & Tidey E. (2001). Coupling 3D visualisation to qualitative assessment of built environment designs. *Landscape and Urban Planning 55*, 29-40.

[31] Matejicek, L., Engst, P. & Janour, Z. (2006). A GIS-based approach to spatio-temporal analysis of environmental pollution in urban areas: A case study of Prague's environment extended by LIDAR data. *Ecological Modelling 199*, 261-277.

[32] Matejicek, L., Janour, Z. & Strizik, M. (2008). Spatial Modeling of Air Pollution Based on Traffic Emissions in Urban Areas. In S. E. Paterson, L. K. Allan, (Ed.), *Road Traffic: Safety, Modeling, and Impacts* (in print). New York: Nova Science Publishers.

[33] Matejicek, L. (2008). Spatial Modeling and Optimization of Municipal Solid Waste Collection in Urban Regions. In A. S. Martin, J. D. Suarez (Ed.), *Solid Waste: New Research* (in print). New York: Nova Science Publishers.

In: Buildings and the Environment
Editors: Jonas Nemecek and Patrik Schulz
ISBN: 978-1-60876-128-9

Chapter 10

FUNGAL CONTAMINATION IN HOSPITAL BUILDINGS

Isabelle Fournel[c]***, Marc Sautour***[a]***, Frédéric Dalle***[ab]***,***
Coralie L'Ollivier[a]***, Denis Caillot***[d]***, Ingrid Lafon***[d]***, Nathalie Sixt***[e]***,***
Céline Calinon[a]***, Gérard Couillaud***[f]***, LS Aho Glélé***[c]
and Alain Bonnin[ab]

[a]Parasitology and Mycology Laboratory, CHU, Hôpital du Bocage, BP 77908, 21079 Dijon Cedex, France
[b]Interactions Mucosa-Transmissible Agents Laboratory, IFR Santé-STIC, Université de Bourgogne, Faculté de Médecine, BP 77908, 21079 Dijon Cedex, France
[c]Hospital Hygiene and Epidemiology Unit, Hôpital du Bocage, BP77908, 21079 Dijon Cedex, France
[d]Clinical Haematology Unit, Hôpital d'Enfants, BP77908, 21079 Dijon Cedex, France
[e]Environmental Microbiology, Hôpital du Bocage, BP 77908, 21079 Dijon Cedex, France
[f]Paediatric Onco-haematology Unit, Hôpital d'Enfants, BP 77908, 21079 Dijon Cedex, France

ABSTRACT

Molds are spore-forming filamentous microfungi that are saprophytic in the environment. Spores inhaled by humans result in life-threatening infections in immunocompromised patients. This chapter is focused on hospital buildings, where a large concentration of immunocompromised patients is present. The major environmental risk factor occurs during hospital construction. Indeed, demolition renovation or construction work has an impact on the environmental fungal load, and there is a well-recognised relationship between outbreaks of invasive filamentous fungal infections in immunocompromised patients and construction work.

The main environmental risk is due to airborne contamination that is dependent both on outdoor and indoor flora in the hospital. In our experience, most indoor contamination comes from micromycetes, which have a particular ability to adapt to the indoor environment. Thus, *Penicillium* spp. and *Aspergillus* spp. are the most frequent molds indoors. We have also shown that the basidiomycete *Bjerkandera adusta* can find a suitable ecological niche within hospitals, confirming that hospitals have their own flora.

Factors such as moisture, ventilation, temperature, or organic matter present in building materials probably participate in the selection of indoor-adapted fungi. Finally, the significance of fungi, such as *Fusarium* spp., occasionally described in hospital water networks, is unclear.

Seasonal variations in the distribution of species have also been described in our study and others. Indoors, *Aspergillus* was found to be particularly abundant in winter, *Alternaria* and *Penicillium* in autumn, and *Cladosporium* in spring or summer. Outdoors, *Penicillium* spp. and *Aspergillus* spp. were more dominant in winter.

Reducing airborne contamination in hospital buildings is thus an important health concern. High efficiency particulate air filtration with or without laminar air flow ventilation has been shown to reduce airborne fungal contamination. Recently, cheaper alternative devices like mobile filtration have been studied, and we have shown that Plasmair™ units reduce indoor airborne contamination both during periods with and without construction work. We and others also found that airborne contamination was not increased if construction work was performed after implementing protective measures such as keeping windows and doors closed, and reinforcing biocleaning.

Finally, although there is no agreement on a threshold of spore concentration, we believe that environmental surveillance must be established both in clinical units and in mycology laboratories in order to avoid false positive results.

In conclusion, in each hospital a multidisciplinary approach involving biologists, clinicians, hygienists, epidemiologists and environmental engineers is necessary to implement a prevention strategy adapted to local conditions.

Fungal Contamination in Hospital Buildings

Background

Molds are spore-forming filamentous microfungi that are saprophytic and ubiquitous in the environment. Most of these fungi rarely behave as pathogens in a normal host [1]. However, they can trigger life-threatening infections in immunocompromised patients, particularly in those with haematologic malignancies [2]. *Aspergillus* species are the main causative agents of opportunistic mold infections, but other molds such as *Fusarium* spp., *Scedosporium* spp. and mucorales are increasingly involved [3,4].

Molds produce resistant conidia that can travel long distances after airborne dispersion [5]. Humans thus permanently inhale *Aspergillus* and other conidia. In immunocompetent individuals, conidia are eliminated by the innate immune system, whereas in patients with altered immunity, mycelial growth may result in tissue invasion [5,6]. Inhalation being the main portal of entry, invasive aspergillosis (IA) primarily infects the human respiratory tract, prior to possible dissemination. In severely immunocompromised patients, colonization of the respiratory tract by *Aspergillus* is associated with an increased risk of IA. The average incidence rates of IA range from 4 to 26% in allogenic bone marrow transplant patients, from 5 to 25 % for acute leukaemia (AL), 3 to 26% in lung transplant patients and 1 to 15% in heart transplant patients. Invasive aspergillosis is associated with 74 to 92% mortality [5-7].

With increasingly aggressive treatments, fungi have thus emerged as a leading cause of hospital-acquired infections [1], and fungal contamination within hospital buildings has become a major concern for clinicians in charge of severely immunocompromised patients.

Indoor hospital contamination with fungal conidia is a complex and multiparametric phenomenon (figure 1). Contamination of air-handling systems, as well as indoor and outdoor construction work, which liberates large amounts of *Aspergillus* spores, have been identified as the infection sources of multiple outbreaks of nosocomial aspergillosis [8-10] (Table 1). Moreover, molds present in the inside environment of buildings may establish sporulating microcolonies in microenvironments with appropriate building materials, moisture and temperature, and such microcolonies may release fungal spores [11-12]. Based on moisture demand, *Aspergillus* spp. are among the early colonizers [12].

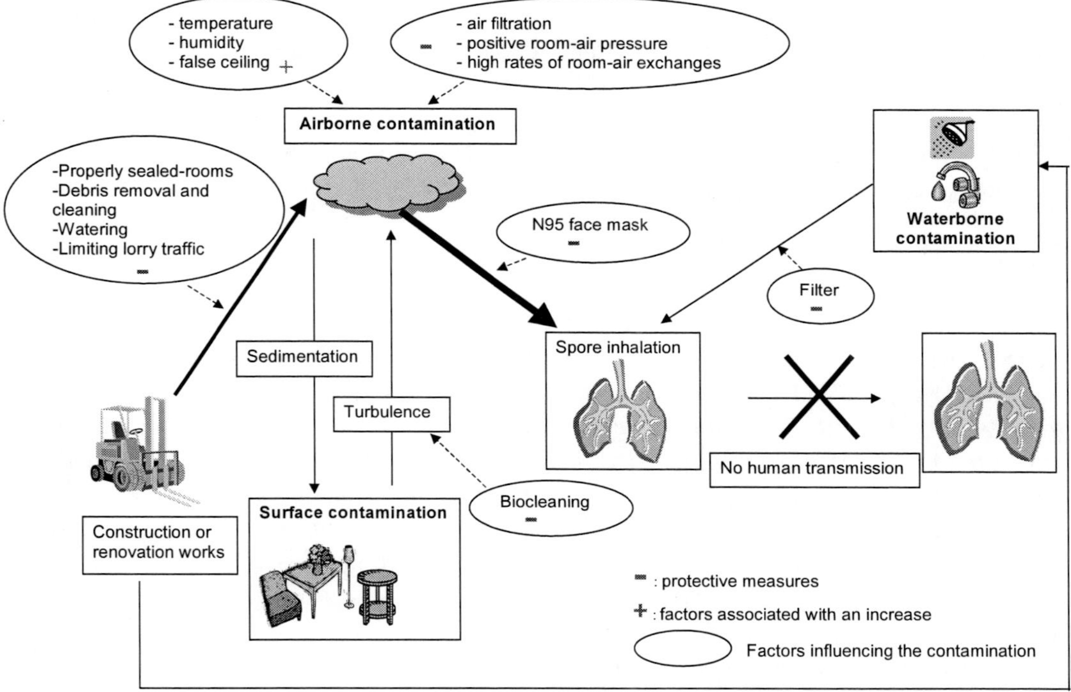

Figure 1. Pathways of hospital contamination with fungal spores

Table 1. Distribution of sources of nosocomial aspergillosis outbreaks (n=53) according to Vonberg [10]

Sources	Number of outbreaks	%
Construction or renovation works	26	49
Contaminated or defective air supply system	9	17
Other sources	6	11
Unknown	12	23
Total	53	100

1. Airborne contamination

1.1. Epidemiology of airborne fungi

Generally speaking, the most common genera of airborne fungi are *Cladosporium*, *Penicillium*, *Alternaria* and *Aspergillus,* both indoors and outdoors [13]. On the Dijon University Hospital campus, *Cladosporium* was found to be the predominant genus cultured from air samples collected outdoors, followed by *Penicillium spp*, non sporulated fungi and *Alternaria* [13]. Most of indoor contamination came from micromycetes which have a particular ability to adapt to indoor environment such as *Penicillium* spp. and *Aspergillus* spp. that were the most frequent pathogenic fungi identified [12-14]. Indeed, ecological niches in the hospital environment seem to exist where fungi are able to survive. Interestingly, in this respect, we showed that *Bjekandera adusta*, a basidiomycete identified by molecular tools and rarely recovered from outdoor samples, could find a suitable niche within Dijon Hospital; our findings confirm that hospitals have their own flora [13].

However, indoor fungal occurrence depends not only on factors such as moisture, ventilation, temperature, or organic matter present in building materials, but also to a certain extent on outdoor parameters [13].Seasonal variations in the distribution and prevalence of species have been described in other studies. This distribution was also studied in Dijon University Hospital. Indoors, the genus *Aspergillus* was found to be particularly abundant in winter, *Alternaria* and *Penicillium* in autumn, and *Cladosporium* in spring or summer [13]. Outdoors, the lowest proportion of *Cladosporium* were recorded in winter, *Penicillium* and *Aspergilus* were more dominant in winter, the lowest proportion of *Alternaria* was found in spring and *Bjekandera* was only detected in summer [13]. This suggests that the presence of such fungi inside the hospital at some periods of the year is probably due to the presence of numerous spores outside during this period.

1.2. Reducing airborne contamination

As fungal infections represent a serious health hazard, reducing airborne contamination is a major concern in hospitals, particularly in tertiary care institutions treating immunocompromised patients at highest risk. Since the hazard of developing an invasive mold infection is linked to inhalation of conidia, it is cumulative, and even concentrations of airborne Aspergillus spores below 1 cfu/m^3 have been shown to cause infection in high-risk patients [10]. In fact, there is no agreement on a threshold of spore concentration at which a significant risk of fungal infection occurs [10,15]. A set of data indicates that the risk increases with the level of contamination. A 50% rate of invasive pulmonary aspergillosis was reported in acute leukaemia patients when *Aspergillus* spore counts were 15 cfu/ m^3 [16]. A decrease in the incidence of IA was observed when *Aspergillus* density dropped from 1-2 to less than 0.2 cfu/m^3 [17]. No cases of IA were reported when the conidia counts was below 0.009 cfu/m^3 [18]. Alberti et al. [19] showed that peaks of air-contamination by *Aspergillus* spores greater than 2 cfu/m^3 were correlated with an increased risk of IA. Consequently the presence of immunocompromised patients in clinical units requires protective measures to be taken. To achieve a minimal concentration of fungal spores , the Centers of Disease Control and Prevention recommend to maintain: a) high-efficiency particulate air (HEPA) filtration;

b) directed room airflow; c) positive room-air pressure relative to the corridor; d) properly sealed rooms; e) high rates of room-air changes [20].

High efficiency particulate air filtration (HEPA), with or without laminar (or unidirectional) air flow ventilation, has been shown to reduce airborne fungal contamination. HEPA filters remove a wide range of particles from the air, with filters rated to trap 99.97% of airborne particles 0.3μ size. However, some have suggested that even with HEPA air filtration systems, *Aspergillus spp* could not be fully eliminated [21,22], particularly when vacuum cleaners were contaminated [21] and when the environmental *Aspergillus* contamination increased dramatically, which might saturate the capacity of HEPA filtration [22]. A filter must be installed in terminal position to avoid airborne contamination. Contamination of air-handling system due to condensation [9] has been reported in operating theatres. This underlines the importance of additional protective measures, of environmental surveillance and of a well-conducted maintenance of filters. Such an expensive equipment requires important installation works. As a consequence, the cost of constructing and maintaining such facilities restricts their use to locations housing patients at high risk [23,24]. This filtration must be associated to a positive-pressure ventilation [25].

Cheaper alternative mobile devices have been developed more recently. Among them, a new technology developed for the Mir Space Station has been adapted to create several mobile air-treatment units aimed at processing contaminated air (Immunair™, Plasmair™, Airinspace, Montigny le Bretonneux, France) [26]. Plasmair™ is a European Community marked mobile-air decontamination unit and FDA-cleared medical device. The system itself is not based on filtration. Airborne organisms are destroyed through a 3-step process based on exposure to high electric fields that deform and alter their membrane or cell wall, subsequent bombardment with positive and negative ions that destroy the internal structure, and a final electrostatic nano filter [23,24]. In our center, it was shown that Plasmair™ reduced indoor airborne contamination both during periods with works and without works [23, 24] (Table 2). In two others French hospital, Immunair™ was also associated with a significant decrease in airborne fungal contamination levels [27].

Table 2. Positivity and mean fungal load of air and surface samples collected in adult haematology unit in rooms equipped with Plasmair™ treatment or with no air treatment, before and during the period of hospital construction in Dijon, France [23,24]

	Before work				During work			
	Air		Surface		Air		Surface	
	Positive samples (%)	Mean fungal load (cfu/m^3)	Positive samples (%)	Mean fungal load (cfu/m^3)	Positive samples (%)	Mean fungal load (cfu/m^3)	Positive samples (%)	Mean fungal load (cfu/m^3)
Air treatment system								
Plasmair ™ treatment	59	4.9	19	2.5	67	4.3	27	1.8
No air treatment	100	10.4	50	2.2	100	14.3	54	1.9
p-value	0.007	0.02	0.036	NS	<0.0001	0.005	0.007	NS

NS: non significant

Mobile air-treatment units might also be useful in clinical departments where construction activities are planned, and as an alternative to protective isolation in departments without HEPA filters. Other mobile filtration devices are available. For instance, Medic CleanAir® is an air filtration system that consists of particle filters placed in the air-intake ports, while the exhaust ports are provided with a HEPA filter. The filter cartridge includes three filtration devices: a pre-filter, an activated carbon filter and HEPA filter. In a study performed in a neonatal intensive care unit, Mahieu et al showed that these devices were associated with a significant decrease in *Aspergillus* spp contamination in the high care unit [28]. Photoclean Quartz system® is another mobile device based on ultraviolet photocatalysis, a process dependent on the strong oxidizing power of titane oxide under UV irradiation, ultimately resulting in the decomposition of organic compound [29]. However, a thorough evaluation of this system in the specific setting of clinical units has not been performed yet.

Particular attention should be paid to ventilation and air-conditioning systems both in terms of microbiological contamination [9] and monitoring of pressures [30,31] when designing hospital buildings and during their maintenance. In France, a specific norm applies to these process [32].

1.3. Impact of demolition, renovation or construction works

Demolition, renovation or construction works have already been reported to have an impact on the environmental fungal load, and there is a well recognised relationship between outbreaks of invasive filamentous fungal infections in immunocompromised patients and the proximity of construction works [14,17,33-35].

The strong association between building works and increased environmental *Aspergillus* contamination must result in taking measures to minimise the exposure of immunocompromised hosts to dust in order to limit the risk of fungal contamination. Thus, the National Disease Surveillance Center recommends different prevention measures that rely on dust control, cleaning, infection control team, ventilation of construction area, debris removal and cleaning, and traffic control [36]. Interestingly, Cooper et al recently demonstrated that airborne contamination was not increased when works were performed after implementing these protective measures [14].

Debris should be removed every day and transported from the site in sealed containers. The construction site should be watered and sealed off from hospital personnel, as well as from visitors and patients. Windows and doors must be closed. Biocleaning, which consists of removing all microbial surface contamination physically and chemically, must be reinforced. However, Lee et al. [37] found *Aspergillus* contamination to be higher one hour after room cleaning than before room cleaning, which suggested a possible correlation between reentrained bioaerosols after cleaning and the risk of nosocomial invasive aspergillosis. The authors suggested that admittance of immunocompromised patients with hematologic malignancy to rooms within 1 hour after terminal cleaning may expose them to higher bioaerosol levels of *Aspergillus species*. The limitation of this study is that sampling was not performed in the same rooms before cleaning and after cleaning, that is why these results should be confirmed by further studies.

Patients must also be reminded to wear a high efficiency face mask (N95 face mask) to protect themselves whenever they are exposed to outdoor dust.

1.4. Environmental surveillance

An environmental surveillance must be set up in clinical units in order to take corrective measures, such as reinforcing biocleaning, reminding the staff to keep windows and doors closed. Although most authors recommend that *Aspergillus* air counts should be fewer than 5 colony forming units /m^3 in protected environment [38], there is no agreement on the threshold of spore concentration at which a significant risk of invasive fungal infection occurs; such a threshold still needs to be defined.

Controling the environment is also critical in mycology laboratories so as to avoid false positive results. Indeed, the identification of filamentous fungi is based on culturing fungi in clinical mycology laboratories. However, molds present on laboratory surfaces may cause contamination of cultures. Thus, there have been reports of pseudo oubreaks of fungal infections due to laboratory contamination [39-41] whose consequences may be serious, as a suspected mold infection will induce aggressive diagnostic procedures, as well as prescription of expensive and poorly tolerated systematic antifungal drugs [29]. It may be beneficial to use Shewhart control charts to monitor fungal contamination in air and on surfaces of mycology laboratories. These graphical tools aim at identifiying an early contamination (Figure 2) [29].

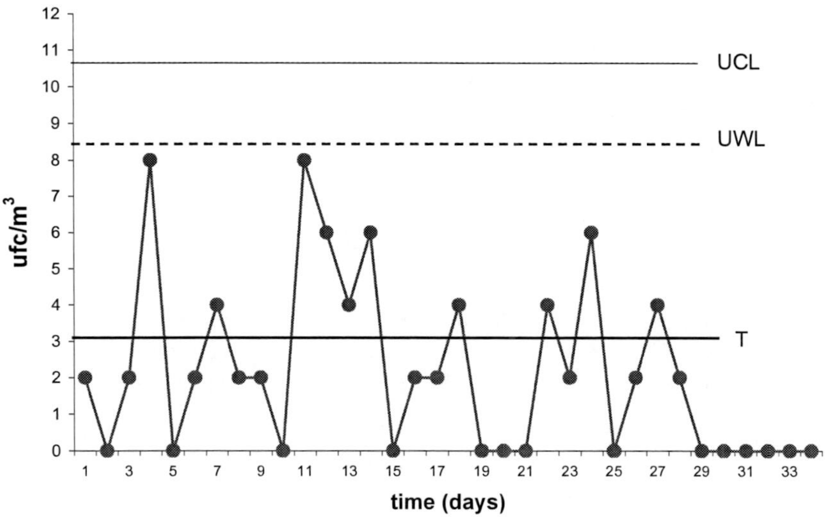

Figure 2. Example of Shewhart chart in a mycology laboratory (Hospital of Dijon).

T = target; UWL = upper warning limit; UCL = upper control limit.

2. Waterborne contamination

Although the environmental risk in a hospital setting is mainly due to airborne contamination, fungi may also be recovered from water samples [1,42,43] (Table 3). For instance, fungi such as *Fusarium* spp., *Acremonium* spp., *Penicillium* spp., *Peacilomyces* spp.

were recovered from tap water in Greek hospitals [42]. Another Greek study performed in a haemodialysis unit [1], showed that 81% of water samples contained filamentous fungi, the most abundant being *Penicillium* spp. and *Aspergilus* spp. In Norway [44], filamentous fungi were recovered from 94% of water samples, and *Aspergillus* spp was found as the most frequent genus.

Table 3. Main studies describing waterborne contamination with micromycetes

Author, Year	Country	Molds positive rate	Sample size	Most prevailing species
Arvanitidou, 1999 [43]	Greece	95% (community water samples) 76% (hospital water samples)	126	*Penicillium, Aspergillus*
Anaissie, 2001 [48]	USA (Texas)	57%	283	Study dealing only on *Fusarium*
Warris, 2001 [44]	Norway	94%	168	*Aspergillus, Paecilomyces, Trichoderma, Penicillinium*
Panagopoulou, 2002 [42]	Greece	12%	186	*Fusarium, Acremonium, Penicillium, Paecilomyces, Blastomyces*
Anaissie, 2003 [46]	USA (Arkansas)	70%	416	*Aspergillus, Paecilomyces, Alternaria*

The risk of water contamination by *Legionnella* spp. and possibly *Aspergillus* may be increased by the interruption of main water supply [45]. The remedial actions suggested are flushing the system, chlorination, or pasteurisation of the water supply [45]. It has been hypothesised that aerosols containing fungal spores are released and inhaled by the patient during showering [46], as sampling performed before and after showering resulted in a significant increase of conidial counts in the air [47]. In another study [44], there was an increase of conidial counts in the air during the day after multiple showers but not in case of a single shower.

Fusarium has also been recovered from hospital water distribution systems [48] and from showerheads [49]. Warris [49] suggested that the type of water source had a significant impact on the recovery of molds from hospital water, and closed water supply systems were associated with a lower fungal contamination. Moreover, current water purification procedures such as chlorination might not be effective against fungal spore elimination [49].The hospital water system contaminated with pathogenic *Fusarium* species may be the source of nosocomial fusariosis [48]. At Dijon University Hospital, *Fusarium* has also been identified in the hospital water system (unpublished data). The question of water filtration was raised, particularly in clinical units with immunocompromised patients, as clinical significance of positive water samples remains a subject of debate. Waterborne contamination varies according to season [44]. Moreover, great spatial variability of *Fusarium* in water has been described in Greece. *Fusarium* was recovered in Heraklion [42] but no fungi were found in Thessalonika or Athens [50]. At Dijon Hospital, *Fusarium* contamination was observed only in a few rooms of some clinical units, which suggested that some specific areas of the hospital water distribution system might be a reservoir for *Fusarium* species.

Conclusion

Invasive fungal infections are a major concern in clinical units with immunocompromised patients. Airborne dissemination is the main vector, but the role of water is still to be discussed. In conclusion, prevention relies on environmental surveillance and protective measures. A multidisciplinary approach is required, including biologists to identify fungi for environmental and clinical diagnosis, clinicians for medical evaluation of high-risk patients, hygienists to make protective measures widely known, and technical staff to provide information on construction progress.

Acknowledgements

We thank Michel Fournel for his editing help.

References

[1] Arvanitidou, M; Spaia, S; Velegraki, A; Pazarloglou, M; Kanetidis, D; Pangidis, P; Askepidis, N; Katsinas, C; Vayonas, G; Katsouyannopoulos, V. High level of recovery of fungi from water and dialysate in haemodialysis units. *J Hosp Infect,* 2000, 45, 225-230.

[2] Warris, A; Gaustad, P; Meis, JF; Voss, A; Verweij, PE; Abrahamsen, TG. Recovery of filamentous fungi from water in a paediatric bone marrow transplantation unit. *J Hosp Infect,* 2001, 47, 143-148.

[3] Faure, O; Fricker-Hidalgo, H; Lebeau, B; Mallaret, MR; Ambroise-Thomas, P; Grillot, R. Eight-year surveillance of environmental fungal contamination in hospital operating rooms and haematological units. *J Hosp Infect,* 2002, 50, 155-160.

[4] Marr, KA; Carter, RA; Crippa, F; Wald, A; Corey, L. Epidemiology and outcome of mould infections in hematopoietic stem cell transplant recipients. *Clin Infect Dis*, 2002, 34, 909-917

[5] Denning, DW. Invasive aspergillosis.*Clin Infect Dis,* 1998, 26, 781-805.

[6] Singh, N; Paterson, DL. Aspergillus infections in transplant recipients. *Clin Microbiol Rev,* 2005, 18, 44-69.

[7] Latge, JP. Aspergillus fumigatus and aspergillosis. *Clin Microbiol Rev,* 1999, 12, 310-350.

[8] Cheng, SM; Streifel, AJ. Infection control considerations during construction activities: land excavation and demolition. *Am J Infect Control,* 2001, 29, 321-328.

[9] Lutz, BD; Jin, J; Rinaldi, MG; Wickes, BL; Huycke, MM. Outbreak of invasive Aspergillus infection in surgical patients, associated with a contaminated air-handling system. *Clin Infect Dis*, 2003, 37, 786-793.

[10] Vonberg, RP; Gastmeier, P. Nosocomial aspergillosis in outbreak settings. *J Hosp Infect,* 2006, 63, 246-254.

[11] Krause, M; Geer, W; Swenson, L; Fallah, P; Robbins, C. Controlled study of mold growth and cleaning procedure on treated and untreated wet gypsum wallboard in an indoor environment. *J Occup Environ Hyg*, 2006, 3, 435-441.

[12] Pasanen, AL; Kalliokoski, P; Jantunen, M. Recent studies on fungal growth on building materials. Amsterdam: Elsevier; 1994.

[13] Sautour, M; Sixt, N; Dalle, F; L'Ollivier, C; Fourquenet, V; Calinon, C; Paul, K; Valvin, S; Maurel, A; Aho, S; Couillault, G; Cachia, C; Vagner, O; Cuisenier, B; Caillot, D; Bonnin, A.. Profiles and seasonal distribution of airborne fungi in indoor and outdoor environments at a French hospital. *Sci Total Environ*, 2009,407,3766-71.

[14] Cooper, EE; O'Reilly, MA; Guest, DI; Dharmage, SC. Influence of building construction work on Aspergillus infection in a hospital setting. *Infect Control Hosp Epidemiol,* 2003, 24, 472-476.

[15] Perdelli, F; Sartini, M ; Spagnolo, AM ; Dallera, M ; Lombardi, R ; Cristina, ML. A problem of hospital hygiene: the presence of aspergilli in hospital wards with different air-conditioning features. *Am J Infect Control,* 2006, 34, 264-268.

[16] Oren, I; Haddad, N; Finkelstein, R; Rowe, JM. Invasive pulmonary aspergillosis in neutropenic patients during hospital construction: before and after chemoprophylaxis and institution of HEPA filters. *Am J Hematol,* 2001, 66, 257-262.

[17] Arnow, PM; Andersen, RL; Mainous, PD; Smith, EJ. Pulmonary aspergilosis during hospital renovation. *Am Rev Respir Dis, 1978*, 118, 49-53.

[18] Sherertz, RJ; Belani, A; Kramer, BS; Elfenbein, GJ; Weiner, RS; Sullivan, ML; Thomas, RG; Samsa, GP. Impact of air filtration on nosocomial Aspergillus infections. Unique risk of bone marrow transplant recipients. *Am J Med*, 1987, 83, 709-718.

[19] Alberti, C; Bouakline, A; Ribaud, P; Lacroix, C; Rousselot, P; Leblanc, T; Derouin, F; Aspergillus Study Group. Relationship between environmental fungal contamination and the incidence of invasive aspergillosis in haematology patients. *J Hosp Infect*, 2001, 48, 198-206.

[20] Centers for Disease Control and Prevention Guidelines for prevention of nosocomial pneumonia. *MMWR Recomm Rep*, 1997, 46,1-79.

[21] Anderson, K; Morris, G; Kennedy, H; Croall, J; Michie, J; Richardson, MD; Gibson, B. Aspergillosis in immunocompromised paediatric patients: associations with building hygiene, design and indoor air. *Thorax,* 1996, 51, 256-261.

[22] Cornet, M; Levy, V; Fleury, L; Lortholary, J; Barquins, S; Coureul, MH; Deliere, E; Zittoun, R; Brücker, G; Bouvet, A. Efficacy of prevention by high-efficiency particulate air filtration or laminar airflow against Aspergillus airborne contamination during hospital renovation. *Infect Control Hosp Epidemiol,* 1999, 20, 508-513.

[23] Sautour, M; Sixt, N; Dalle, F; L'Ollivier, C; Calinon, C; Fourquenet, V; Thibaut, C; Jury, H; Lafon, I; Aho, S; Couillault, G; Vagner, O; Cuisenier, B; Besancenot, JP; Caillot, D; Bonnin, A. Prospective survey of indoor fungal contamination in hospital during a period of building construction.*J Hosp Infect*, 2007, 67, 367-373.

[24] Sixt, N; Dalle, F; Lafon, I; Aho, S; Couillault, G; Valot, S; Calinon, C; Danaire, V; Vagner, O; Cuisenier, B; Sautour, M; Besancenot, JP; L'Ollivier, C; Caillot, D; Bonnin, A. Reduced fungal contamination of the indoor environment with the Plasmair system (Airinspace).*J Hosp Infect*, 2007, 65, 156-162.

[25] Humphreys, H. Positive-pressure isolation and the prevention of invasive aspergillosis. What is the evidence ? *J Hosp Infect,* 2004, 56, 93-100.

[26] Challier, S; Poirot, JL; Le Guinche, I; Malbernard, M; Laudinet, N; Bergeron, V, Une nouvelle unité mobile de décontamination de l'air pour applications médicales, *Tech Hosp,* 2004, 684, 62–67.

[27] Poirot, JL; Gangneux, JP; Fischer, A; Malbernard, M; Challier, S; Laudinet, N; Bergeron, V. Evaluation of a new mobile system for protecting immune-suppressed patients against airborne contamination. *Am J Infect Control*, 2007, 35, 460-466.

[28] Mahieu, LM; De Dooy, JJ; Van Laer, FA; Jansens, H; Ieven, MM. A prospective study on factors influencing Aspergillus spore load in the air during renovation works in a neonatal intensive care unit. *J Hosp Infect,* 2000, 45, 191-197.

[29] Sautour, M; Dalle, F; Olivieri, C; L'Ollivier, C; Enderlin, E; Salome, E; Chovelon, I; Vagner, O; Sixt, N; Fricker-Pap, V; Aho, S; Fontaneau, O; Cachia, C; Bonnin A. A prospective survey of air and surface fungal contamination in a medical mycology laboratory at a tertiary care university hospital. *Am J Infect Control* 2009, 37, 189-94.

[30] Rhame, FS. Prevention of nosocomial aspergillosis. *J Hosp Infect*, 1991, 18, 466-72.

[31] Thio, CL; Smith, D; Merz, WG; Streifel, AJ; Bova, G; Gay, L; Miller, CB; Perl TM. Refinements of environmental assessment during an outbreak investigation of invasive aspergillosis in a leukemia and bone marrow transplant unit *Infect Control Hosp Epidemiol* 2000, 21, 18-23.

[32] Association Française de normalisation. Norme AFNOR NF S90-351. Établissement de santé - Salles propres et environnements maîtrisés apparentés - Exigences relatives pour la maîtrise de la contamination aéroportée. Saint-Denis La Plaine; 2003.

[33] Dewhurst, AG; Cooper, MJ; Khan, SM; Pallett, AP; Dathan, JRE. Invasive aspergillosis in immunosupressed patients: potential hazard of building work. *BMJ,* 1990, 301, 802-804.

[34] Opal, SM; Asp, AA; Cannady, PB; Morse, PL; Burton, LJ; Hammer, PG. Efficacy of infection control measures during a nosocomial outbreak of disseminated aspergillosis associated with hospital construction. *J Infect Dis,* 1986, 153, 634-637.

[35] Perraud, M; Piens, MA; Nicoloyannis, N; Girard, P; Sepetjan, M; Garin, JP. Invasive nosocomial pulmonary aspergillosis: risk factors and hospital building works. *Epidemiol Infect,* 1987, 99, 407-412.

[36] Sub-Committee of the Scientific Advisory Committee of the National Disease Surveillance Centre. National guidelines for the prevention of nosocomial invasive aspergillosis during construction/renovation activities. National Disease Surveillance Centre, Dublin, Ireland; 2002.

[37] Lee, LD; Berkheiser, M; Jiang, Y; Hackett, B; Hachem, RY; Chemaly, RF; Raad, II. *Infect Control Hosp Epidemiol,* 2007, 28, 1066-70.

[38] Morris, G; Kokki, MH; Andersen, K; Richardson, MD. Sampling of Aspergillus spores in air. *J Hosp Infect,* 2000, 44, 81-92.

[39] Weems, JJ Jr; Andremont, A; Davis, BJ; Tancrede, CH; Guiguet, M; Padhye, AA; Squinazi, F; Martone, WJ. Pseudoepidemic of aspergillosis after development of pulmonary infiltrates in a group of bone marrow transplant patients. *J Clin Microbiol* , 1987, 25, 1459-62.

[40] Hruszkewycz, V; Ruben, B; Hypes, CM; Bostic, GD; Staszkiewicz, J; Band, JD. A cluster of pseudofungemia associated with hospital renovation adjacent to the microbiology laboratory. *Infect Control Hosp Epidemiol,* 1992,13, 147-50.

[41] Laurel, VL; Meier, PA; Astorga, A; Dolan, D; Brockett, R; Rinaldi, MG. Pseudoepidemic of Aspergillus niger infections traced to specimen contamination in the microbiology laboratory. *J Clin Microbiol,* 1999, 37, 1612-1616.

[42] Panagopoulou, P; Filioti, J; Petrikkos, G; Giakouppi, P; Anatoliotaki, M; Farmaki, E; Kanta, A; Apostolakou, H; Avlami, A; Samonis, G; Roilides, E. Environmental surveillance of filamentous fungi in three tertiary care hospitals in Greece. *J Hosp Infect*, 2002, 52, 185-191.

[43] Arvanitidiou, M; Kanellou, K; Constantinides, TC; Katsouyannopoulos, V. The occurrence of fungi in hospital and community potable waters. *Lett Appl Microbiol*, 1999, 29, 81-84

[44] Warris, A; Gaustad, P; Meis, JF; Voss, A; Verweij, PE; Abrahamsen, TG. Recovery of filamentous fungi from water in a paediatric bone marrow transplantation unit . *J Hosp Infect*, 2001, 47, 143-148.

[45] 45 The Newcastle upon tyne hospitals NHS foundation trust. Infection control committee. Prevention of aspergillosis during building work policy. United Kingdom; 2007.

[46] Anaissie, EJ; Stratton, SL; Dignani, MC; Lee, CK; Summerbell, RC; Rex, JH; Monson, TP; Walsh, TJ. Pathogenic molds (including Aspergillus species) in hospital water distribution systems: a 3-year prospective study and clinical implications for patients with hematologic malignancies. *Blood,* 2003, 101, 2542-2546.

[47] Anaissie, EJ. Emerging Fungal Infections: Don't Drink the Water, ICAAC, San Diego;1998.

[48] Anaissie, EJ; Kuchar, RT; Rex, JH; Francesconi, A; Kasai, M; Müller, FM; Lozano-Chiu, M; Summerbell, RC; Dignani, MC; Chanock, SJ; Walsh, TJ. Fusariosis associated with pathogenic fusarium species colonization of a hospital water system: a new paradigm for the epidemiology of opportunistic mold infections. *Clin Infect Dis*, 2001, 33, 1871-1878.

[49] Warris, A; Voss, A; Abrahamsen, TG; Verweij, PE. Contamination of hospital water with Aspergillus fumigatus and other molds.*Clin Infect Dis,* 2002, 34, 1159-1160.

[50] Panagopoulou, P; Filioti, J; Farmaki, E; Maloukou, A; Roilides, E. Filamentous fungi in a tertiary care hospital: environmental surveillance and susceptibility to antifungal drugs. *Infect Control Hosp Epidemiol,* 2007, 28, 60-67.

In: Buildings and the Environment
Editors: Jonas Nemecek and Patrik Schulz
ISBN: 978-1-60876-128-9

Chapter 11

BUILDING FORM AND AIR POLLUTION IN COMPACT CITY: A REVIEW FOR THE MACAO PENINSULA

***U Wa Tang*[1*]*, Ni Sheng*[2] *and Zhishi Wang*[1]**
[1]Department of Civil and Environmental Engineering,
University of Macau, Macao, China
[2]General Studies, Macau University of Science and Technology, Macao, China

ABSTRACT

This chapter reviews selected monitoring and modelling studies co-conducted by the authors in the past decade. These studies were either awarded by influential professional organizations (URISA, CPGIS, HKSTS) or granted by national foundation (NSFC) etc. As the study area is the Macao Peninsula, a highly compact and the wealthiest urban area in Asia, the study approaches and results can be used as a base for further discussions of the building form and environment in compact urban areas. The findings will also be very relevant to urban planners as they prepare for higher population densities in cities throughout the world.

1. INTRODUCTION

The importance of building form on sustainable development has been recognized in recent years. However, the impacts of building form on traffic-induced air pollution are still not well understood. Since 1995, the authors have co-conducted a number of research studies on the traffic-induced air pollution in the Macao Peninsula. The building forms in the Macao Peninsula are highly compact due to the high population density (38,000 inhabitants/km^2) and mixed-use development (i.e., a mixture of residential, commercial, industrial, or other land uses in a building or set of buildings).

The chapter reviews selected monitoring and modelling studies which were either awarded by influential professional organizations or granted by national foundation. The

---

*

national foundation which supported our studies is the National Natural Science Foundation of China and the organizations which recognised our studies as the best annual papers include The Urban and Regional Information Systems Association in the United States (2008), The International Association of Chinese Professionals in Geographic Information Sciences in the United States (2008) and The Hong Kong Society for Transportation Studies in Hong Kong (2007). In particular, as the study area (the Macao Peninsula) is the urban core of a highly compact and the wealthiest city (Macao) in Asia, the findings will be very relevant to urban planners as they prepare for higher population densities in cities throughout the world.

2. The Study Area

Macao is a place of yesterday, today and tomorrow. In 1557, the Portuguese explorers came to Macao, making it the first place in China open to the West and one of the most important trading stations in Asia. After the rapid development in Hong Kong, Macao became a seedy sideshow to its neighbouring metropolitan. Pursuant to an agreement signed by China and Portugal on 13 April 1987, Macao reunified to China as one of the two Special Administrative Regions (SARs) of the nation on 20 December 1999 (another one is Hong Kong). Under the "one country, two systems" formula, Macao enjoys a high degree of autonomy and particularly, the blessing from Beijing. Since then, Macao has experienced a tremendous success of economic development and thus been in the spotlight of the mass media again. In 2007, Macao became the largest gambling hub in the world and the richest place in Asia (in terms of per capita gross domestic product). The economy of Macao is based largely on tourism and gambling. By 2002, the tourist-gambling sector has accounted for over a quarter of GDP; this tiny Macao has received more than 8 million visitors each year. The city is so tiny and lack of natural resources that it depends on China for most of its food, fresh water, and energy imports.

The rapid expansion of tourist industry of Macao is changing the city landscape to bring about rapid urban development, which is in turn exerting environmental stresses on the air, water and land environments of Macao. The unique combination of population density and economic activity no doubt poses a formidable challenge to the government and citizens of Macao in terms of managing adverse environmental impacts but also serves as a special model for urban development and management in an environmentally friendly way that other rapidly growing cities will face in future.

As the first place in China open to the West in the 16th century, Macao has just been awarded United Nations World Cultural Heritage status for the preservation of its dramatic mixing of eastern and western cultural relic sites. Therefore, preservation of air quality is particularly important for the tourism industry in the Macao Peninsula. Since 1995, the research team in Faculty of Science and Technology, University of Macau has been making a series of project researches focussed on vehicular exhaustion emission pollution in urban areas of Macao. These research projects were jointly sponsored by University of Macau, Macao Foundation and National Natural Science Foundation of China. The Chinese entities include Research Centre for Environmental Science, Beijing University; Department of Environmental Science and Engineering, Tsinghua University; and Guangzhou Research Institute of Goechemistry, Chinese Science Academy.

This paper reviews results of innovative or representative monitoring and modelling studies in the Macao Peninsula, which includes (1) real-world vehicle emission factors by mobile laboratory; (2) building-based air quality modelling; (3) aerodynamic tests for wind speed profile on a road; (4) field measurements of vertical concentration profiles of particles at roadside; (5) field measurements of particle concentrations on leeward/windward sides of the road; (6) spatial distributions of polycyclic aromatic hydrocarbons (PAHs) in dustfall.

3. Real-World Emission Factors by Mobile Laboratory

Vehicle emission factors are essential input parameters in air quality modelling. Laboratory tests and real-world tests are two major approaches to measuring vehicle emissions. The types of laboratory tests include chassis dynamometer tests and idle tests. However, the operating conditions of the vehicle in the laboratory tests may not fully represent the complex real-world conditions. Moreover, many high-emitting vehicles would not volunteer for government emission testing, so emission inventories may be underestimated. The types of real-world tests include remote sensing tests and on-board emission measurements. A remote sensing test typically uses a stationary analyzer to measure emissions of moving real-world vehicles. However, it can capture only a snapshot (0.5 sec) emission from each moving vehicle, which cannot reflect continuous emissions of individual vehicles in different operating modes. For the on-board emission measurements, an on-street vehicle carries an analyzer to measure emissions from the vehicle itself. However, similar to the laboratory tests, only a few vehicles are available for measurement because cooperation from the vehicle's owners is necessary.

Based on the measured on-road emissions by Pirjola et al. (2004), Yli-Tuomi et al. (2005) calculated the overall fuel-based emission factors of the fleet for size-resolved particle numbers, CO, NO and NO_x. Nevertheless, due to the lack of vehicle samples and the low time resolution, the fleet emission factors were calculated and results only reflected general emission characteristics for all vehicles being followed.

Recently, the authors successfully obtained the emission factors for each category of vehicle (motorcycle, passenger car, taxi, truck and bus) under traffic congestion conditions at the urban hot spots in the Macao Peninsula (Tang et al., accepted; Tang and Wang, 2006). In comparison with the traditional emission tests such as chassis dynamometer and idle-emission tests, vehicle chase tests reflect real-world situations and more vehicle samples are available because it is not necessary to obtain the cooperation of the vehicle owners.

To measure the CO, HC, NO, CO_2 and O_2 concentrations from the preceding vehicles, a five-gas analyser licensed by the Physikalisch-Technische Bundesanstalt (the national metrology institute) in Germany (Model DiCom 4000 from AVL, Austria) was assembled in a sedan mobile laboratory with the inlet mounted on the sedan's front bumper. The detection ranges (ppmv) were as follows: CO 0–100,000; CO_2 0–200,000; NO 0–4,000; O_2 40,000–220,000; and HC 0–200,000. With irradiation (IR) measurement, the response time for CO, CO_2 and HC is ~100 ms. With electrochemical measurement, the response time for O_2 and NO may take more than a second. A laptop computer logged data from the five-gas analyser via a standard RS232 serial port every two seconds and simultaneously recorded the driver's forward view captured by video camera.

In urban hot spots, the local background concentrations of traffic air pollutants are expected to be high, and the measurements from vehicle-following experiments may not represent the emissions of a single vehicle. Therefore, the lowest fifth percentile of values over 1 min was selected to estimate the background concentration (Bukowiecki et al., 2002; Pirjola et al., 2004). This low-percentile method aimed to avoid the influence of signal noise that could be induced by use of the minima method. The selection of 1-min duration was based on the estimation that the average time behind a single vehicle in urban traffic conditions was ~15–45 s.

After elimination of the background concentrations by the 1-min fifth percentile method, the CO, HC, and NO emission factors in 2-s intervals were calculated by the fuel-based method (Singer and Harley, 2000). In the experiment, as the effective plume path length and amount of plume measured depend on turbulence and wind, it is only possible to determine molar ratios of CO/CO_2, HC/CO_2, and NO/CO_2. These molar ratios are constants for a given exhaust plume. By using the fuel-based method, the amount of pollutants per kilogram of fuel burned can be expressed in terms of the molar ratios. For the purpose of roadside air quality modelling, fuel-based emission factors (grams per kilogram of fuel) were converted to distance-based emission factors in grams per kilometre based on the fuel consumption data.

A total of 178 vehicles were individually followed and labelled and the measurements were re-sampled according to the licence plates, vehicle category, vehicle model and registration year. The CO, HC and NO emission factors for each vehicle category were obtained and compared with those estimated from the traditional MOBILE5 emission model (Hao et al., 2003). It can be seen that the CO and HC emissions of petrol passenger cars obtained from the vehicle chase tests are close to those from MOBILE5. The other emission factors from MOBILE5, however, are significantly higher than those from the vehicle chase tests. To evaluate the emission factors obtained from the vehicle chase tests and MOBILE5, these emission factors were input into an operational air quality model (OSPM) to predict the roadside air quality and results were compared with measured values. Evaluation results showed that the emission factors obtained from the vehicle-following measurement techniques are more realistic than those from MOBILE5 in Macao.

4. Building-Based Air Quality Modelling

The method and accuracy of data capture dominate the spatial distribution of urban air pollution. Due to limited budget, installation space, and labor resource, permanent or temporary air pollution monitoring sites are very scattered. Air quality assessment of a city based on scattered monitoring sites may be incorrect because non-homogeneous distribution of air quality is neglected.

A number of model systems have been developed to estimate urban air quality at unsampled sites. Typical model systems include Atmospheric Dispersion Modelling System (ADMS-urban), the Air Quality Information System (AirQUIS), the Environmental Management (EnviMan), and the Urban Simulation Model (Urban SIM), which show that the input/output spatial data are commonly stored in regular grids with resolutions of 1-2 km, regardless of the complexity of building form. Nevertheless, the scale or spatial resolution is still an important research topic.

Recently, the author has developed a model system which can estimate air quality in front of individual buildings along both sides of the road (Tang and Wang, 2007). Compared with the grid-based approach with spatial resolutions of 1-2 km, the present building-based approach can predict the complex spatial variation of traffic emission, urban geometry, dispersion and air pollution. The present prototype air pollution model system integrates the operational air pollution model OSPM (Hertel and Berkowicz, 1989), digital maps of the road network, building layout and topographic information, and an urban landscape model.

The air quality in front of a building on the Macao Peninsula is determined by use of the OSPM model. The OSPM model is a practical street pollution model, developed by the Department of Atmospheric Environment, National Environmental Research Institute, Denmark (Hertel and Berkowicz, 1989). Concentrations of exhaust gases are calculated by a combination of a simple plume model for the direct contribution and a box model for the recirculating part of the pollutants in the street. The traffic emission field is treated as a number of infinitesimal line sources aligned perpendicular to the wind direction at the street level. The plume expression for a line source is integrated along the path defined by the street-level wind. In the box model, it is assumed that the canyon vortex has the shape of a trapeze and the ventilation of the recirculation zone takes place through the edges of the trapeze.

The OSPM model is integrated with ArcView GIS by a prototype urban landscape model programmed in the Avenue language. When the urban landscape model is executed, a target building is selected from a cadastral map. The 2D street configurations around the building are extracted from the cadastral map. To obtain the building height relative to the ground surface, the elevation of the ground surface provided by the terrain map is subtracted from the building elevation. Subsequently, the road segment in front of the target building is selected from a road network map to collect the traffic volume, traffic composition and traffic speed. The extracted street configurations and traffic conditions are input into the OSPM model to simulate the air pollution in front of the target building. The inputs and modelling results are stored in the attribute fields of a point feature created in a street environment digital map based on the modelling position. When all the buildings in the cadastral map are selected and manipulated, the urban landscape model stops the execution.

Applying the developed model system, a preliminary study of the influences of existing building forms on the Macao Peninsula on vehicle transport and street environment is performed. Four building forms in historic and modern areas are considered. Results show that compared with modern building forms, historic building forms with a high percentage (> 48.7%) of narrow roads (less than 10 m wide), complex road network and high density of intersections limit the traffic and parking capacities and therefore lead to lower passenger car and motorcycle volumes. In particular, the greater street canyon effects in historic building forms lead to higher CO concentrations. The present model system has potential for realistically modelling the complex distributions of air pollution in compact cities, which is characterized by complex micro-environmental conditions created by unique building forms.

5. Aerodynamic Tests

The presence of street canyons is a key feature of the highly compact urban forms in the Macao Peninsula. As a historical city which is awarded United Nations World Cultural Heritage status, historical narrow street networks have been successfully preserved and this results in the existence of street canyons in historical urban areas. With one of the highest population density in the world, the Macao Peninsula contains numerous high-rise residences and office buildings, which also result in street canyons in modern urban areas.

To study wind field patterns in Macao with typical narrow street morphology, which may lead to adverse dispersions of traffic emissions, a busy street namely Rua do Campos with an aspect ratio (H/W) of 20 m: 10 m was selected for aerodynamic tests. The tests were conducted in the College of Environmental Sciences and Atmospheric Simulation & Pollution Control State Key Joint Laboratory, Peking University with three cases according to the wind directions assignment (northerly, easterly and southerly winds) (Liu, et al., 2000; Wang, 2002).

For the experimental set up, a scale model representing Rua de Campo and its surrounding urban area of 0.36 km^2, i.e., 600 m x 600 m, was made of plastic foam at a scale of 1:200. The average wind speed was assigned to be 2.3 m/s and it reduced to 1.8 m/s in the free flow section of the testing tunnel. Two rows of point sources and two moving belts were centred at ground level of the selected street canyon to simulate the emission and movement of two-lane traffic, respectively. The emitted pollutant concentration was assigned to be 1 litre/min. Typical inflow directions and speeds were generated in the wind tunnel. Monitoring probes were arranged to measure the wind speed, wind direction, and dimensionless concentration of the tracer gas at different elevations representing three different heights of 2 m, 30 m, and 100 m, respectively. The model area was squared with 10,000 net points and the mass consistent model is used for wind field simulation in the whole study area from the data obtained in the tunnel test.

The resulting wind fields of the northerly, easterly and southerly winds at a common wind speed are shown in Figures 1 a-c, respectively. It was found from the Figure 1a that both the wind speed and direction within the street did not change too much even in the presence of high buildings at upstream. Only at the eastern side of the street, the wind speed and direction changed to some extent. It is because of the complex terrain at that side (buildings and hills). As for the case of easterly winds (see Figure 1b), the wind speed within the street dramatically decreased and wind direction changed randomly. It is because of the influence of high buildings and complex terrain in the eastern part of the tested district. Such testing results may indicate that the vehicular emission pollution will be serious in the middle of the street when the main wind blows from the east. In the case of southerly winds, the wind speed within the street was remarkably reduced due to the influence of the upstream high buildings. It was also found that in the southern end of the street, the wind direction and speed both varied randomly, which indicates that if the main wind comes from the south, the actual pollution in the street may be quite serious.

The results by the tunnel test and the mass consistent model also showed that when the easterly and northerly winds blew over Macao with low speeds, the wind speed with Rua do Campo would be even lower with scattering directions so that vehicle emission could not easily be dispersed. When the northerly wind blew, the situation became a little better because of the less variation of heights of the buildings in the north side. Liu et al. (2000) concluded that Rua de Campo had typical wind flow and dispersion characteristics in street canyons (i.e., high turbulence, low wind speed, and poor dispersion conditions), which resulted in high concentrations of pollution at the ground level.

6. Field Measurements of Vertical Concentration Profiles

To investigate the vertical concentration profiles of traffic-induced air pollution in the Macao Peninsula, field measurements of particulate matters < 1 μm [PM_1], < 2.5 μm [$PM_{2.5}$], and <10 μm [PM_{10}] have been co-conducted by the author and the research team from Department of Environmental Science and Engineering, Tsinghua University in a local street Avenidia de Horta e Costa (Wu et al., 2002). Six particle analyzers (DustTrak model 8520 from TSI), namely D1 to D6, were installed in a vertical profile (2m, 8m, 19m, 30m, 59m, and 79m) on a tall building. Concentrations of PM_1, $PM_{2.5}$, and PM_{10} were measured independently in three consecutive days.

The vertical concentration profiles show that there is a significant decrease in PM_{10}, $PM_{2.5}$ and PM_1 as the height above the ground increases. For the PM_{10} samples, the concentration at the height of 8m decays to 74% of the maximum occurring at 2m above the ground. At the height of 79 m, the concentration of PM_{10} decreases to about 60% of the maximum. The $PM_{2.5}$ samples also show a significant decreasing trend of concentrations as the height increases, although the attenuation does not follow the same pattern as PM_{10}. At the height of 8 m, the daytime mean concentration decays to only 87% of the maximum at the ground level, while the concentrations of $PM_{2.5}$ decreased significantly until the height reached 19m. Then, the decrease of $PM_{2.5}$ also slows down as the height increases. At the height of 79m, the concentration of $PM_{2.5}$ decays to 62% of the maximum. The PM_1 samples exhibit a decay of concentrations that is less significant than the PM_{10} and $PM_{2.5}$'s as the height increases. At the height of 8m, the daytime mean PM_1 concentration decays to about 86% of the maximum at the ground level. At the height of 79 m, the total decrease of PM_1 concentration is about 20%.

The vertical profiles suggest that PM_{10}, $PM_{2.5}$ and PM_1 concentrations are affected significantly by traffic sources at ground level, e.g., resuspended street dust for PM_{10}, tailpipe exhaust from motor vehicles for $PM_{2.5}$ and PM_1, etc., which results in a significant decrease in the concentration of PM_{10}, $PM_{2.5}$ and PM_1, as the height above the ground increases. Furthermore, the street is narrow with width less than 15m, surrounded by high buildings with height more than 30m at both sides, which undoubtedly results in significant street canyon. As the aspect ratio (building height to street width) is high, the flow at the bottom of street canyon is more stable than upper part, which makes air pollutant accumulate at the bottom.

7. FIELD MEASUREMENTS OF CONCENTRATIONS ON LEEWARD/WINDWARD SIDES

To investigate the concentrations of PM_{10}, $PM_{2.5}$ and PM_1 on the leeward and windward sides of street canyons, field measurements have also been in two local streets (Wu et al., 2002). Six particle analyzers (DustTrak model 8520 from TSI), namely D1 to D6, were installed at both sides of the two streets. The two streets are oriented in the same direction (NEE-SWW). During the sampling days in Street 1, wind direction was in NE, E, and SE; while N winds blew during the sampling days in Street 2. Both streets are narrow (less than 15 m wide) and surrounded by buildings with height more than 20 m at both sides, which again, undoubtedly results in remarkable street canyon. Special flow circulations will be generated and result in higher air pollutant concentrations on the leeward side than the windward side. It is found that the mean concentrations of PM_{10}, $PM_{2.5}$ and PM_1 on the leeward side of Street 1 and Street 2 are all higher than those on the windward side, which revealed that the effect of street canyon exactly occurred during the sampling periods.

8. SPATIAL DISTRIBUTIONS OF DUSTFALL PAHS

Monitoring studies co-conducted by the author and Guangzhou Research Institute of Goechemistry, Chinese Science Academy have shown the non-homogeneous distribution of air pollutants, i.e., polycyclic aromatic hydrocarbons (PAHs), in dustfall samples on the Macao Peninsula (Qi et al., 2001).

The study was conducted on November 19, 1998. A total of 13 dustfall samples were collected using dustfall cans at eleven points in Macau Peninsula. The sampling sites are described in Figure 5a. The dustfall samples were dried in a desiccator for 48 h, then sieved (80 mesh) to remove large particles and plant debris. Two to five grams of dry sample were wrapped with pre-cleaned filter paper, and stored in a glass jar under –20°C after 3–5 mL of dichloromethane had been added.

Polycyclic aromatic hydrocarbons (PAHs) are widespread environmental pollutants, which are mainly formed in combustion processes of carbonaceous materials at high temperature. These combustion sources include emissions from automobiles, industrial processes, domestic heating systems, waste incineration facilities, tobacco smoke, and several natural sources including forest fires and volcanic activity. In the Macao Peninsula with very few industrial emissions, automobiles are believed to be an important contributor to the roadside PAHs.

It was found that the polycyclic aromatic hydrocarbons (PAHs) levels of dustfall samples at eleven sites varied to a great extent at different sampling points from 2.72 to 24.83 μg/g. We divided the concentrations of the total PAHs into three levels. The highest level (≥12 μg/g) were observed in Samples 4, 8, 3 and 5, the middle level (6–12 μg/g) in Sample 1, 2, 9, 10 and 11, and the lowest level (< 6μg/g) in Samples 6, 7 and 8. The significant variations of PAHs reveal the measurable impact of complex traffic emission intensities and complex wind flow conditions on the highly compact urban form.

9. CONCLUSIONS

This paper reviews results of innovative or representative monitoring and modelling studies in the Macao Peninsula, which includes (1) real-world emission factors by mobile laboratory; (2) building-based air/noise modelling; (3) aerodynamic tests for wind speed profile on a road; (4) field measurements of vertical concentration profiles of particles at roadside; (5) field measurements of particle concentrations on leeward/windward sides of the road; (6) spatial distributions of polycyclic aromatic hydrocarbons (PAHs) in dustfall. It is found that (1) the emission factors obtained from the vehicle-following measurement techniques are more realistic than those from modeling method (MOBILE5); (2) The building-based air quality model system has potential for realistically modelling the complex distributions of air pollution in compact cities, which is characterized by complex micro-environmental conditions created by unique building forms; (3) Aerodynamic tests in wind tunnels and field measurements for vertical and windward/leeward profiles at roadsides are proper approaches to identify "street canyon effect" in various building forms; (4) Due to the complex wind flow conditions (e.g. street canyon effects) and the complex traffic emission intensities, the polycyclic aromatic hydrocarbons (PAHs) in dustfalls are found to be varied to a great extent at different roads.

REFERENCES

Hao, J. M., Hu, J. N., Fu, L. X., Wu, Y. & Wang, L. T. (2003). Vehicle emission characteristics and control in Macao. In: Wang, Z.S. (Ed.), *Scientific Innovation for Environmental Assessment of Coastal Cities*. University of Macau, Macao, pp. 67-130.

Hertel, O. & Berkowicz, R. (1989). Modelling NO_2 Concentrations in a Street Canyon. DMU Luft A-131. National Environmental Research Institute, Roskilde, 31 p.

Liu, B., Kang, L. & Lin G. (2000). Experimental study on the diffusion low of automobile flux at the Street Rua de Campos in Macao, *Acta Scientiae Circumstantiae*, Vol.*20*, Suppl, 27-33.

Pirjola, L, Parviainen, H., Hussein, T., Valli, A., Hämeri, K., Aaalto, P., Virtanen, A., Keskinen, J., Pakkanen, T. A., Mäkelä, T. & Hillamo, R. E. (2004). '"Sniffer" – a novel tool for chasing vehicles and measuring traffic pollutants', *Atmospheric Environment*, Vol. *38*, pp.3625–3635.

Qi, S. H., Yan, J., Zhang, G., Fu, J., Sheng, G., Wang, Z. S., Tong, S. M., Tang, U. W., Min, & Y. S. (2001). Distribution of polycyclic aromatic hydrocarbons in aerosols and dustfall in Macao. *Environmental Monitoring and Assessment 72*, 115–127.

Singer, B. C. & Harley, R. A. (2000). 'A fuel-based inventory of motor vehicle exhaust emissions in the Los Angeles Area during summer 1997', *Atmospheric Environment* Vol. *34*, pp.1783–1795.

Tang, U. W., Sheng N. & Wang Z. S., accepted. Mobile Laboratories for Particle and Gaseous Pollutants. *Traffic-Related Air Pollution, Nova Science Publishers, Inc., New York, the United States.*

Tang, U. W. & Wang, Z. S. (2006). Determining gaseous emission factors and driver's particle exposures during traffic congestion by vehicle-following measurement techniques, *Journal of Air and Waste Management Association*, Vol. *56*, pp. 1532–1539.

Tang, U. W. & Wang, Z. S. (2007). Influences of urban forms on traffic-induced noise and air pollution: results from a modelling system. *Environmental Modelling and Software 22*: 1750–1764.

Wang, Z. S. (2002). Annual Scientific Research Report On Macao Environment and City Development.

Wu, Y., Hao, J. M., Fu, L. X., Wang, Z. S. & Tang, U. W. (2002). Vertical and horizontal profiles of airborne particulate matter near major roads in Macao, China. *Atmospheric Environment 36*, 4907–4918.

Yli-Tuomi, T., Aarnio, P., Pirjola, L., Mäkelä, T., Hillamo, R. & Jantunen, M. (2005), 'Emissions of fine particles, NO_x, and co from on-road vehicles in Finland', *Atmospheric Environment*, Vol. *39*, pp. 6696–6706.

In: Buildings and the Environment
Editors: Jonas Nemecek and Patrik Schulz

ISBN: 978-1-60876-128-9

Chapter 12

USE OF SEMANTIC WEB TECHNOLOGIES IN DESIGN OF INFORMATIONAL RETRIEVAL SYSTEMS

Anatoly Gladun[1)] and Julia Rogushina[2)]
[1)]International Research and Training Centre of Information Technologies and Systems of NASU, Kiev, Ukraine,
[2)]Institute of Software Systems of NASU, Kiev, Ukraine

ABSTRACT

For more relevant informational retrieval and matching of user request with metadata about informational recourses it is necessary to formalize the user knowledge about subject domain of search and handle the semantics of user's request as well as the semantics or informational recourses. We propose to base on Semantic Web technologies and use the ontologies and associated with them thesauri of the appropriate subject domains for representation of domain knowledge. The algorithms of formation and normalization of the multilinguistic thesauri, and also methods of their comparison are given in this work.

Key words: an information resource, ontology, thesaurus, informational retrieval.

INTRODUCTION

During last years the Web becomes one of the main means of the information publication. It is dynamical distributed environment and the information resources (IR), presented in it, are heterogeneous. Now Web 2.0 and Web 3.0 concepts are widely discussed [1]. But the most important questions of intellectualization of the Web are related with the Semantics Web project [2].

The effective retrieval in the Web becomes very laborious and demands intelligent technologies. The quality estimation of information retrieval systems (IRS) is a complex question. The problem concerns with parameters of IRS estimation. A lot of existing techniques analyze such IRS parameters as relevance, completeness, accuracy and their various combinations. Relevance is a thematic correspondence of the information, received as a result of search, to request. The completeness of search is a ratio of the correctly found documents amount to the total relevant documents known to IRS. Accuracy of search is a ratio of correctly found documents amount to the total amount of the documents given by IRS in reply to request.

However it is necessary to take into account, that the formal request to IRS is the user attempt to formalize his/her information need that, unfortunately, not always really reflects this need. It results in degradation of Internet use. Therefore more important such parameter of IRS quality estimation as pertinence – a ratio of amount of the information interesting for user to total amount of the received information. To increase the pertinence of informational retrieval IRS requires information about area of the user interests. This information applies by IRS for choose among accessible resources what are interesting to user and not only formally correspond to request. Such information should be submitted in the form suitable for automatic processing and reuse, and their formation must be automatized.

For formalization of domain knowledge now ontologies and thesauri are widely used. Standards and software tools for their development are created in Semantic Web project.

Semantic Web Technologies

The Semantic Web provides a common framework that allows data to be shared and reused across application, enterprise, and community boundaries. It is a collaborative effort led by W3C with participation from a large number of researchers and industrial partners.

The Semantic Web is the extension of the World Wide Web that enables people to share content beyond the boundaries of applications and web-sites. The Semantic Web has inspired and engaged many people to create innovative semantic technologies and applications. (semanticweb.org).

In the Semantic Web data itself becomes part of the Web and is able to be processed independently of application, platform, or domain. This is in contrast to the World Wide Web as we know it today, which contains virtually boundless information in the form of documents. We can use computers to search for these documents, but they still have to be read and interpreted by humans before any useful information can be extrapolated. Computers can present you with information but can't understand what the information is well enough to display the data that is most relevant in a given circumstance. The Semantic Web, on the other hand, is about having data as well as documents on the Web so that machines can process, transform, assemble, and even act on the data in useful ways.

Ontologies are considered as one of the pillars of the Semantic Web, although they do not have a universally accepted definition.

OWL is a third W3C specification for creating Semantic Web applications. Building upon RDF and RDFS, OWL defines the types of relationships that can be expressed in RDF using an XML vocabulary to indicate the hierarchies and relationships between different resources. In fact, this is the very definition of "ontology" in the context of the Semantic Web: a schema that formally defines the hierarchies and relationships between different resources. Semantic Web ontologies consist of a taxonomy and a set of inference rules from which machines can make logical conclusions.

It's important to note that OWL has three sub languages, each with increasing complexity: OWL Lite, OWL DL, and OWL Full. OWL DL includes OWL Lite, and OWL Full includes OWL DL and OWL Lite. Developers choose which OWL dialect to use based on the level of complexity and level of detail required by their semantic model [3].

Semantic Web agents, which are computer programs capable of interpreting RDF and OWL semantic information, are also required for harnessing the power of the Semantic Web [4].

Semantic Web proposes:

- A global naming scheme (URI);
- A standard syntax for describing data (RDF);
- A standard means for describing of the data properties (RDF-schema);
- A standard means for describing of data relationships (OWL);
- A standard means of Web-service describing (OWL-S);
- A language for queries to knowledge bases (SPARQL);
- The means to support trust and security.

Implementing the Semantic Web requires adding semantic metadata, or data that describes data, to information resources. This will allow machines to effectively process the data based on the semantic information that describes it. When there is enough semantic information associated with data, computers can make inferences about the data, i.e., understand what a data resource is and how it relates to other data.

INFORMATIONAL RESOURCES OF THE WEB

Among IR, potentially accessible by the Web, still prevails the textual information however it's share constantly decreases due to multimedia IR increase. The subject domain that is characterized by these IR can be represented by two ways: 1) analyzing textual information and 2) considering metadata of these IR.

Metadata contains machine-readable information about the document, which can be automatically processed by computer. Now the most perspective and common metadata model is RDF (Resource Description Framework) [5] based on XML [6] created by W3C in Semantic Web project. With the help of RDF one can describe both structure of a site and connected with appropriate domain. RDF describes informational resources in oriented marked graph form - each IR can have properties, which in turn also can be IR or their collections. Most widespread set of elements for metadata specification is Dublin Core Metadata Elements. Metadata can be built in IR or be stored and updated independently of resources.

Web-services. Initially World Wide Web technology was focused on work with static hypertext documents represented in the Internet. But then sites offering to the clients not only the documents, but also service (for example, sites of e-commerce) began to occur. Many such sites use application servers which not only return the document but can process the data entered by the user (queries, completed form etc.) and dynamically generate the documents depending on the parameters, specified by the user. Such dynamic component of the Internet grows much faster then static one and requires application of more complex information technologies. In this connection it is possible to consider a separate class of IR - Web-services.

Web-service is a set of logically connected and program-accessible through the Internet functions [7]. There is the program identified on UR. It's interface can be determined by XML structures. Web-services are based on three basic Web-standard: SOAP (Simple Object Access Protocol) - the protocol for sending of messages by the HTTP and other Internets protocols; WSDL (Web Services Description Language) - language for the description of program interfaces of Web-services [8]; UDDI (Universal Description, Discovery and Integration) - indexing standard of Web-services. For intelligent Web-service definition OWL-S is developed [9]. This language describes the semantics of Web-service and can be used by software agents.

Statement of Problem

For effective search of the information that user needs (textual and multimedia documents, information services etc.) there is necessary to generate the model of user interests domain (for example, as ontology) and use this model when IRS fulfils the user's query. It is very important to use some international modern standards of domain knowledge representation that provide knowledge interoperability.

Thesauri and Ontologies as Means of Domain Knowledge Representation

For the successful informational retrieval it is necessary to present user knowledge about domain of her/his interests in some form suitable for computer processing. The specifications of high-level domain are formed by integration of the domain structures of low-level domains. It is important to achieve an interoperability of domain knowledge representation. Ontological approach is an appropriate tool for solution of this task. Ontology is an agreement about common use of concepts that contains means of representation of subject knowledge and agreements on methods of reasons. It can be considered as the certain description of the views on the world in some specific sphere of interests.

Ontological commitments are the agreements aimed at coordination and consistent use of the common dictionary. The agents (human beings or software agents) that jointly use the dictionary do not feel necessity of common) knowledge base: one agent can know something that don't know the other ones, and the agent that handles the ontology is not required the answers to all questions that can be formulated with the help of the common dictionary. The ontology is knowledge base of a special kind with the semantic information about some domain. It is a set of definitions in some formal language of declarative knowledge fragment focused on joint repeated use by the various users in the applications. Ontology consists of:

1. a set of the terms;
2. a set of relations between terms;

3. a set of rules of their use that limit their meanings in the context of concrete domain [9].

The formal model of domain ontology O is an ordered triple O = < X, R, F >, where X - finite set of subject domain concepts that represents ontology O; R - finite set of the relations between concepts of the given subject domain; F - finite set of interpretation functions of given on concepts and relations of ontology O.

Every domain with the certain subject of research has it's own terminology, original dictionary used for discussion of typical objects and processes of this domain. The library, for example, involves the dictionary relating to the books, references, bibliographies, magazines etc. Thus, pattern of domain is discovered by its dictionary - the set of words that are used in tis domain. Clearly, however, that the specificity of domain is shown not only in the appropriate dictionary. Besides, it is necessary: (i) to provide strict definitions of grammar managing of combining the dictionary terms into the statements, and (ii) to clear logic connections between such statements. Only when this additional information is accessible, it is possible to understand both nature of domain objects and important relations established between them. Ontology - structured representation of this information [10].

Until recently term "thesaurus" was used as a synonym of ontology, however now in IT with the help of the thesauruses frequently describe domain lexicon in a semantic projection, and ontology apply to modeling semantics and pragmatists in a projection to representation language [11,12]. The models either of ontologies or of thesauruses include as the basic concept the terms and connections between these terms.

The term "thesaurus" for the first time was used still in XIII century by B.Datiny as the name of the encyclopedia. In translation from Greek "thesaurus" means treasure, riches. The thesaurus is the complete systematized data set about some field of knowledge allowing the human or the computer to orient in it.

The thesaurus is a dictionary where the descriptors of the certain field of knowledge with ordering of their hierarchical and correlative relations are represented. The descriptors are given in alphabetic order but they are grouped semantically; the search is carried out from concept to a word. Collection of the domain terms with indication of the semantic relations between them is a domain thesaurus. The thesaurus can be considered as a special case of ontology. The formal model of the thesaurus is a pair Th = < T, R >, where T - finite set of the terms; and R - finite set of the relations between these terms[13].

The multilingual thesaurus [14] is a coordinated set of the monolingual thesauruses containing equivalent descriptors on languages-components necessary and sufficient for interlingual exchange, and including means for the indication of their equivalence. At an recognizing of equivalence of descriptors of the various monolingual versions it is necessary to distinguish on different languages-components the following degrees of equivalence of the terms: 1) complete; 2) incomplete; 3) partial; 4) absence of the equivalent term. Incomplete equivalents are the terms, for which the volumes of concepts, expressed by them, are crossed. Partial equivalents are the terms, for which volume of concept expressed by one equivalent, is included into volume of concept expressed by other equivalent. One way to the recognizing of equivalence to a various degree bases on appropriate domain ontology use: every word from the monolingual thesauruses refers to one of the ontology terms that helps to make connection between words of the various thesauruses. If some words from thesaurus refer to one ontology term then they are equivalent. If some words refer to ontological terms being a subclass one another then these words are in relation of incomplete equivalence.

USE OF THESAURUSES FOR IR RETRIEVAL

For taking into account semantics of area of user interests in process of retrieval of IR satisfying his/her informational need o it is necessary (Figure 1)

1. to generate the domain thesaurus corresponding to information needs of the user (by analysis of IR that this user considers relevant to this domain [15];
2. to construct the thesaurus for every IR known to IRS (simple dictionary without stop-words);
3. to compare the thesauruses of IR relevant to user query to IRS with the domain thesaurus and to find those ones that contain the maximum number of words in intersection.

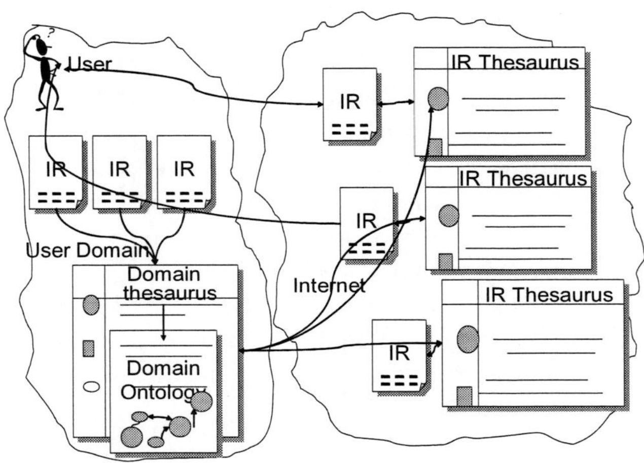

Figure 1. Informational retrieval on base of thesauruses

At thesaurus construction it is necessary to use ontologies of the appropriate areas (with higher level in comparison with user domain to normalize the multilingual thesauruses). Normallization procedure is similar to stemming and provides for integrated processing of words in different morphologic forms and multilingual representations. Normalysed thesaurus contains relation between equivalent terms in different languages. As every thesaurus is constructed from the user point of view (which is reflected in user domain ontology), therefore it`s forming is the user task.

Constructing of Domain Thesaurus

At first user should independently select the set of IR that he/she considers relevant to domain of his/her interests. Every IR is described by not empty set of the textual documents connected with this IR - text of content, metadescriptions, results of indexing etc. The domain thesaurus is formed as a result of the automated analysis of these documents (the user actions of the are reduced to constructing of semantic bunches - by linking of each word of the formed thesaurus with some term of domain ontology. Algorithm of domain thesaurus construction consists from the following steps:

1. Formation of initial set of the textual documents relevant to domain. At the input of algorithm the set A of the textual documents describing chosen IR comes (each of documents from A can have the coefficient of importance and the coefficient of IR relevance ИР that allows to define differently weight of words from these documents for the IR description).
2. Creation of domain information space. For every document from A $a_i \in A, i=\overline{1,n}$ the IR thesaurus $T(a_i)$ - dictionary that contains all words occurred in the document a_i - is constructed . The IR thesaurus is formed as union of the thesauruses a_i: $T_{IR} = \bigcup_{i=1}^{n} T(a_i)$, and domain thesaurus - as association of the IR thesauruses.
3. Clearing of the thesauruses. User should specify dictionary for every $a_i \in A, i=\overline{1,n}$ containing a stop-words (for example, prepositions and conjunctions of language of the document are stop-words for it but prepositions and conjunctions of other language used as examples do not concern to them) $s_{j,} s_j \in Voc$. It is necessary to remove words contained in $s_{j,} s_j \in Voc$ from the thesauruses. Then all service information is rejected (for hypertext, for example, there are marking tags). The cleared thesauruses $T`(a_i), \forall p \in T(a_i) \Rightarrow p \in T`(a_i) \vee p \in s_j, T`(a_i) \cap s_j =$ thus are formed. The cleared thesaurus ИР is under construction as association of the cleared thesauruses $a_i$: $T_{ИР} = \bigcup_{i=1}^{n} T(a_i)\; T`_{ИР} = \bigcup_{i=1}^{n} T`_{ИР}(a_i)$, and cleared domain tesaurus - as association of the IR thesauruses.
4. Linking of thesaurus with domain ontology. To integrate processing of words with equivalent semantics (for example, synonyms, translations of the term on different languages, various kinds of a spelling) the domain thesaurus is associated with some domain ontology (the user can form it himself, use ready ontology or it`s modification).

Each word from the thesaurus it is necessary to link with one of the ontological terms. If the relation is lacking the word is considered as a stop-word or marking element (for example, HTML tag) and should be rejected. $\forall p \in T`(a_i) \exists t = Term(p,O) \in T_O$. The group of the IR thesaurus words terms connected with one ontological term named *the semantic bunch* $R_j, j=\overline{1,n}$ is considered as a single unit. $\forall p \in T`_{ИР} \exists R_j = \{r : r \in T`_{ИР}, Term(p,O) = Term(r,O)\}$. It allows to integrate processing of semantics of the documents written on various languages and, thus, to ensure the multilinguistic analysis of the Internet IR.

5. Extension of ontology. If the IR thesaurus contains words that can't be linked with ontological terms but user considers that these words are significant than it is necessary to add the appropriate terms to domain ontology, specify their connection with other terms of ontology and return to step 4.
6. Construction of the normalized domain thesaurus , i.e. association of all terms of domain ontology that are connected with words from the normalized IR thesaurus (Figure 2):

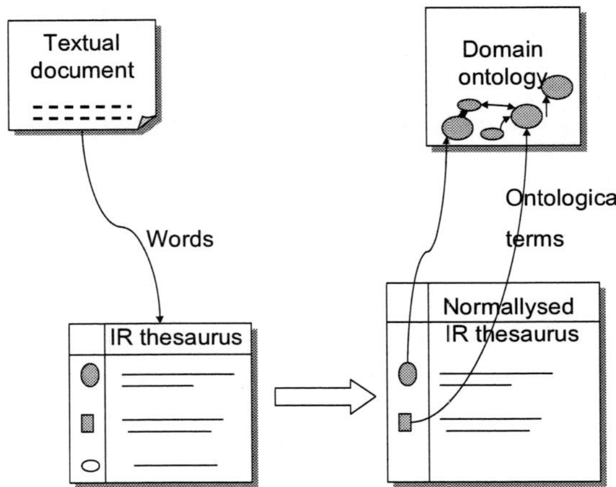

Figure 2. Building of normalized IR thesaurus

The normalized thesaurus is a projection of set of the IR thesaurus words on set of the domain ontology terms . $L_{ИР} = \{t : p \in T`(a_i), i = \overline{1,n}, t = Term(p,O) \in T_O\}$, and normalized domain thesaurus is a union of the normalized IR thesauruses (Figure 3).

BUILDING OF IR THESAURUS

The thesaurus of IR found by IRS as a result of the user query execution is simple a dictionary that does not contain the relations between words (discovery of such connections from the text is rather difficult and in this case is not justified).

The algorithm of the IR thesaurus building consists of the following steps:

1. Formation of the initial IR set U, $U = \{R_j, j = \overline{1,m}\}$.
2. Formation of the IR thesauruses from U. For each IR a thesaurus is formed and cleared.
3. Construction of the normalized IR thesauruses: for normalization the semantical bunches generated by the user during formation of the domain thesaurus are used .

ALGORITHM OF DOMAIN AND IR THESAURUSUS COMPARISON

The normalized IR thesauruses L_{IR} and domain thesaurus L_{domain} are the subsets of the domain ontology terms O chosen by the user: $L_{IR} \subseteq Term(O)$, $L_{domain} \subseteq Term(O)$. If IR description contains more words linked with terms of domain interest for user (that is reflected in the normalized domain thesaurus) then it is possible to suppose that this IR can satisfy informational needs of the user with higher probability than other IR relevant to same formal query. Thus, it is necessary to find IR q satisfied the conditionst $f(q, L_{domain}) = \max f(L_{IR}, L_{domain})$ where the function f is defined as number of elements in crossing of sets L_{IR} and L_{domain}: $f(A,B) = |A \cap B|$. If the various terms of the normalized thesauruses have for the user different importance it is possible to use the appropriate weight coefficients w_j that take into account their importance. In that case the criterion function is $f(A,B)=\sum_{j=1}^{z} y(t_j)$, where the function y is determined for all terms of domain ontology and $y(t_j)=\begin{cases} 0, t_j \notin A \vee t_j \notin B \\ w_j, t_j \in A \wedge t_j \in B \end{cases}$.

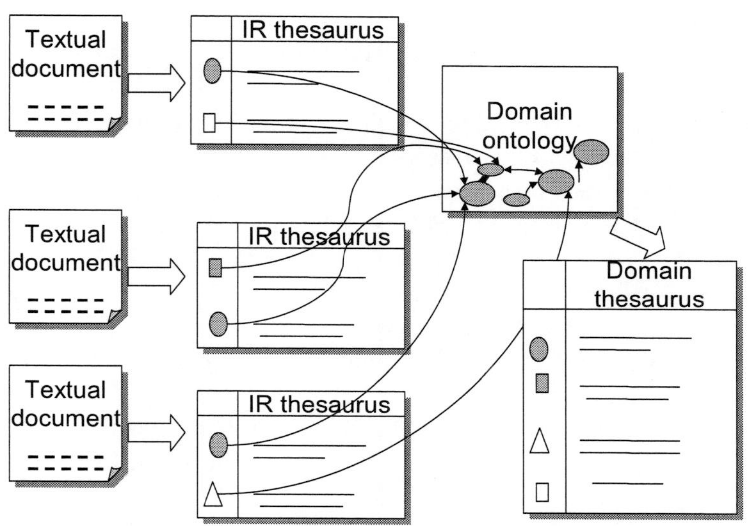

Figure 3. Building of domain thesaurus

Authors' Information

Anatoly Gladun, PhD, since 1997 works as Senior Researcher in International Research and Training Centre of Information Technologies and Systems, National Academy of Sciences and Ministry of Education of Ukraine, 44 Glushkov Pr., Kiev, 03680, Ukraine.

Phone: (+38-044)-4191812, fax (+38-044)-419-18-12, e-mail: glanat@yahoo.com

Graduated from Polytechnic University, Lviv. Post-graduate course in Sankt-Petersburg, Electrotechnical University. A specialist in the field of high-speed intelligent network, Simulation of Telecommunication Systems by Petrinetworks, mobility management, agent-based technologies, intelligence services and web-design. 71 publications.

Julia Rogushina, PhD, since 1997 works as Senior Researcher in Institute of Software Systems, National Academy of Sciences of Ukraine, 44 Glushkov Pr., Kiev, 03680, Ukraine

Phone: (+38-044)-5684698, E-mail: _jjj_@cybergal.com

Graduated from Kyiv State T. Shevchenko University, Department of cybernetics. Post-graduate course in Glushkov Institute of Cybernetics, Kiev, Ukraine. A specialist in the field of knowledge acquisition, data mining, agent-based technologies, ontological analysis.

CONCLUSION

The proposed approach to use of domain ontology for creation and normalization of the IR thesauruses allows to fulfill informational retrieval at a semantic level. The application of thesaurus measure of the information allows to offer to the user only understandable to him/her items of information that provides pertinence of information retrieval .

Semantics Web technologies are used for interoperable knowledge representation.

RFEERENCES

[1] Tim O'Reilly What Is Web 2.0, Available from: http://www.oreillynet.com/pub/a/-oreilly/tim/news/2005/09/30/what-is-web-20.html

[2] Tim Berners-Lee, James Hendler & Ora Lassila. *The Semantic Web*, Available from: http://www.sciam.com/article.cfm?id=the-semantic-web

[3] Lacy, Lee W. OWL: (2005). *Representing Information Using the Web Ontology Language*. Trafford Publishing. Victoria, Canada:. Available from: http://www.lacydatasystems.com/owlbook/

[4] What is the Semantic Web? - Available from: http://www.altova.com/semantic web.html

[5] Resource Description Framework (RDF) Model and Syntax Specification. W3C Proposed Recommendation. - January 1999. Available from: http://www.w3.org/TR/-PR-rdf-syntax.

[6] Extensible Markup Language (XML) 1.0 (Fourth Edition) W3C Recommendation. Available from: http://www.w3.org/TR/REC-xml/

[7] Cowles P. Web Services and the Semantic Web. Available from: – http://www.sys-con.com/webservices/article.cfm?id=419.

[8] D. Martin, M. Burstein, O. Lassila, M. Paolucci, T. *Payne, & McIlraith* S. Describing Web Services using OWL-S and WSDL. – http://www.daml.org/services/owl-s/1.0/owl-s-wsdl.html.

[9] OWL, S. Semantic Markup for Web Services. Available from: http://www.daml.org/-services/owl-s/1.0/owl-s.html.

[10] The differences between a vocabulary, a taxonomy, a thesaurus, an ontology, and a meta-model. Available from: http://www.metamodel.com/article.php?story=-20030115211223271.

[11] Matthews, B. M., Miller, K. & Wilson, M. D. "A Thesaurus Interchange Format in RDF". Available from: http://www.limber.rl.ac.uk/External/SW_conf_thes_paper.htm

[12] IDEF5 Method Report. *Knowledge Based Systems*, Inc.1408 University Drive East College Station, Texas 77840, 1994. – 175 pp.

[13] Narinyani, A. S. The centaur by name TEON: Thesaurus + Ontology. – Available from: http://www.artint.ru/articles/narin/teon.htm.

[14] Gladun, A., Rogushina, J. & Shtonda, V. Ontological Approach to Domain Knowledge Representation for Informational Retrieval in Multiagent Systems // International Jornal "Information *Theories and Applications*", V.13, N.4, 2006. – P.354-362.

[15] Gladun, A. & Rogushina J. Ontologies as a Perspective Direction of Intellectualization of Informational Retrieval in Multiagent Systems of E-commerce // The Proceedings of XI-th Intern. Conf. "Knowledge-Dialogue-Solution", KDS'2005, Varna, Bulgaria, pp.112-120.

In: Buildings and the Environment
Editors: Jonas Nemecek and Patrik Schulz
ISBN: 978-1-60876-128-9

Chapter 13

MEREOLOGICAL ASPECTS OF ONTOLOGICAL ANALYSIS FOR THESAURI CONSTRUCTING

Anatoly Gladun [1*] and Julia Rogushina [2]

[1] International Research and Training Centre of Information Technologies and Systems, NASU, Kiev, Ukraine,
[2] Institute of Software Systems, National Academy of Sciences of Ukraine.

ABSTRACT

One of the significant steps in ontological analysis is the development of relations between the ontology's concepts. All mistakes of this phase result to the great problems on the subsequent steps and in ontology use. Though we propose to use some elements of mereology for more precise definition of relation "part-of" subclasses. This approach is applicable for development of user thesauri that used in different distributed systems (such as informational retrieval systems) as a special case of domain ontology.

Keywords: ontology, thesaurus, mereology, taxonomy, informational retrieval.

INTRODUCTION

Today a lot of IT products use ontologies and thesauri for knowledge representation. It provides the informational processing on semantic level. One of the significant steps in ontological analysis is the development of relations between the ontology's concepts. Mistakes of this phase result to the great problems on the subsequent steps and in ontology use.

Though methodology of construction of ontologies of different levels (global ontologies, ontologies of the appendices, ontologies of subject domains etc.) becomes very urgent. We

* Corresponding Author : *glad@irtc.kiev.ua*

understand that this task cannot be solved automatically but use of some formalisms in ontological analysis has to facilitate the work of domain experts and knowledge engineers.

SEMANTIC WEB AND ONTOLOGIES

The World Wide Web (WWW), also called "the Web", was invented in 1989 by Tim Berners-Lee and his colleagues at CERN and it changed the way people gather and access information [1]. Nowadays, the Web is an ever-growing huge data repository. As a consequence, a major bottleneck has emerged when trying to exploit the information represented in the Web, namely, how to find a specific piece of information we may be interested in.

The purpose of work is to offer methods of the description of ontological relations deal with "the part of" semantic. Elements of the mereological analysis are used for it.

In modern researches in the field of the distributed knowledge management the term "ontology" is used for explicit conceptualization of some subject domain. The ontologies provide the common dictionary of the certain sphere of activity and determine - with various levels of formality - meaning of the terms and relations between them. The knowledge in ontologies is mainly formalized by use of five kinds of components [2]:

- classes,
- relations,
- functions,
- axioms,
- instances.

The formal model of domain ontology is an ordered triple $O = < T, R, F >$, where

- T - final set of the domain terms that is described by ontology O;
- R - final set of the relations between the terms of given domain T;
- F - final set of functions of interpretation given on the terms T and relations R of ontology O.

The set T of the terms of ontology is completely determined by it`s domain but the set R of the relations used in ontologies is more domain-independent.

Nowadays, a lot of ontology representation languages (e.g., DAML-OIL, OWL [3]), free distributed software tools for an ontology design (e.g., Protégé [4], OntoEdit [5], OilEd [6]) and for an ontology analysis (mapping, alignment and merging, e.g., PROMPT [7], Chimaera [8], OntoMerge) are developed.

There are different techniques for ontology constructing: symbolic techniques, statistical techniques and machine learning techniques. The symbolic techniques use lexical-syntactic patterns to obtain semantic relationships. Hearst [9] developed an algorithm to obtain taxonomic relationships from patterns. These techniques are based on heuristic methods which use regular expressions which were previously applied in information acquisition [10]. The statistical techniques are based on statistical data about words concurrency about concepts. The main idea is that the semantic identity of one word is related to its distribution in different contexts, so the meaning of a word is represented by means of the concurrency of words with its frequency of occurrence. The Mo'K tool [11] supports the development of conceptual clustering methods for ontological analysis on base of clustering methods for conceptual hierarchies (taxonomies) supported by knowledge engineers. Finally, the machine learning techniques use learning methods (supervised or not) to obtain ontology components.

Nowadays, there are two main trends in ontology building: (i) manual ontology building, and (ii) (semi-) automatic ontology building (i.e. ontology learning). Ontology population is the process through which a given ontology is populated with instances. Different flaws have been detected on all the existing ontology learning systems. Most of these systems are focused on the extraction of concepts hierarchies [12-13]. Several European projects have been involved in the development of ontology learning systems based on Natural Language Processing (NLP) tools, but these applications suffer from two main drawbacks: they are too language-dependent and they usually need human-expert supervision (i.e. semi-automatic).

The tools of ontology integration help users to find similarity and difference between initial ontologies and create resulting ontology that contains elements of initial ontologies. For this purpose, they automatically determine conformity between the concepts in initial ontologies or provide environment where the user can easily find and determine these conformities. These tools are known as tools of ontology mapping, alignment and merging because of they carry out similar operations.

The relations represent a type of interaction between concepts of the domain. They are formally determined as any subset of products (crossings) of sets n, such as:

R: C1 x C2 x... x Cn.

Examples of the binary relations: "A is a subclass of B " and "A is connected with B". In [14] the most common relations in real domains are discriminated:

- connection of equivalence,
- taxonomical connection,
- structural connection,
- connection of dependence,
- topological connection,
- connection of the reason and consequence,
- functional connection,
- chronological connection,
- connection of similarity,
- conditional connection,
- target connection.

However not all these ontological relations are equal in value. There is possible to allocate a some the most fundamental relations in this set, such as taxonomy and mereology.

The relation is named fundamental if on the basis of this one given relation the formal system allowing to express the basic mathematical concepts can be constructed.

There are four fundamental relations, on the basis of each of which the mathematics can be constructed:

- the relation of belonging (set theory ZF and NF),
- the relation between function, its argument and result (fon Neumann set theory),
- the naming relation (ontology of Lesnevsky),
- the relation "a part of" (mereology) [15].

Taxonomy is a system of classification. A typical example of taxonomy is the hierarchical list (qualifier). The classes in ontology are usually organized in the taxonomy. The taxonomy in which the object can meet more than in one branch, refers to as "poly-hierarchical". The word "taxonomy" comes from two Greek words: "taxis" (order) and "nomos" (agreement). Term is borrowed from philosophy. The taxonomy is the science which studies sharing into the ordered groups or categories. From the ontological point of view, taxonomy is the ontological organization based on the partial ordered relation called «x is A» by means of which the objects are grouped together or more high level is referred to classes.

Only a few of common taxonomic properties exists:

1. asymmetry;
2. transitivity;
3. antireflexiveness.

The good taxonomy should have a resolution threshold that is able to don't group various objects in the same category and to ensure the appropriate number of categories for domain modeling in a controlled way.

The term "taxonomy" has a widespread and does not require an additional definition but the term "mereology" in researches connected with IT is applied more occasionally and that`s why requires an additional decoding.

Mereology (from Greek "part" и "study") - formal theory about parts and concepts connected to them. This term was used by the Polish philosopher Lesnevsky [16] who had analyzed philosophical, logic and mathematical components of the mathematics bases .

The calculus of names of Lesnevsky is constructed as alternative to logic system of Rassel for purposes of natural generalization of traditional logic. In contrast to Rassel who precisely distinguished proper names and general names Lesnevsky considered all names (general, individual and empty), as belonging to the same semantic category. It is necessary to notice, that the Lesnevsky systems in general were stated in the so original form that researchers till now has no standard formulation of a triad of his systems (protothetics, ontology and mereology).

Mereology is the part of this triad of the deductive theories (Lesnevsky considered an ontology only as a system with the unique relation "is_a"). Protothetics is a logic basis of other systems. Protothetics is based on the logic system containing only propositional variables which can be combined by quantifiers and functors of any orders including usual connective of statements. Functors can be present at the form of constants and variables, last also are capable to communicate by quantifiers. In the different versions of protothetics the various axioms can be accepted. Protothetics is a calculus of the statements with quantifiers on propositional variables.

It is more correct to speak not about three systems Lesnevsky but about three section of one system named as "the basis of mathematics" that consists of the appropriate formal theories.

The relation "part - whole" is extremely important because it forms a concept basis of system that is central in modern scientific knowledge.

In mereology the relation "part-of" is expressed through the operator "<". The use of two meanings of such relation is taken into account here:

1. Something that includes concept of equality between instances is named as a characteristic part (" $<<$ ");
2. Something that excludes occurrence of equality is named as a part ("<").

Both aspects can be formalized as follows. Two objects, x and y, are given, and x - part y, then x < y if and only if: 1. $x << y$ or 2. $x = y$, where $x << y$ designates that x is the essential part of y / The relation "part - whole " has the following properties:

1. Skewness : $(x << y) \Rightarrow \neg(y << x)$
2. Transitivity: $(x << y) \wedge (y << z) \Rightarrow x << z$.

The *system* represents structural connection of it's elements. Its base formal characteristic is the property that the elements not are simply included into system, but are included into it as a result of interaction with other elements. As initial the relation M (a, b, c, s) is used. It means that in system s a incorporates with b by means of c, that designate usually as $a \xrightarrow{c} b$ if S is explicit from a context, and c connects only these two elements.

The capacity axiom is applied for mereology : the systems connecting the same objects by the same ways are equal.

Axioms of Mereology

Axiom 1. An empty system $\exists s \forall x, y, c \neg M(x, y, c) = \varnothing$ exists, and such system is unique.

Axiom 2. The system having even one part is an own part of itself : $\forall s(\exists x, y, cM(x, y, c, s) \Rightarrow \exists zM(s, s, z, s))$.

Axiom 3. The other connections of two objects can be more then one but the presence of empty connection denies an opportunity of any other connection:

$$\forall x, y, s(M(x, y, \varnothing, s) \Rightarrow \forall c(M(x, y, c, s \vee M(y, x, c, s) \Rightarrow c = \varnothing).$$

Other axioms of mereology describe relations between system and its elements. Mereology exceeds the bounds of study of the partial relations between elements of general systems. It also is engaged in those objects which parts are relevant to whole. Such objects are identified as instances.

Among the mereological relations it is possible to allocate seven various classes, and in general, the transitivity is not accepted among instances of different classes:

1. Component - object: page - book;
2. Member - collection: tree - wood;
3. Part - weight: piece - bread;
4. Material -object: aluminium - airplane;
5. Property - activity: to see – to read;
6. Stage - process: boiling - preparation of tea;
7. Place - area: Ukraine- Europe.

A lot of researches of the relations "part - whole" are devoted to study of parts, and it is possible also to identify various types of whole agrees to the following properties:

1. Is a part separable from whole ("melody - song" or "coach- train").
2. Are the parts spatial or temporary ("room - apartment" or "winter - year");
3. Does a part play the certain functional role concerning to whole ("engine - automobile").
4. Are the parts indivisible ("atom - molecule").

As an example, mereological approach can be used in creation of semantical definitions of Web-services hat based on OWL-S ontology.

Use of the mereological relations for the description of semantics of web-services

Process of construction of ontology, according to methodology of development of ontological models IDEF5 (Integrated DEFinition Methods) [17], consists of five basic actions:

1. Study and systematization of the entry conditions. This action establishes the basic purposes and contexts of the project of development of ontology, and also allocates roles between the members of the project
2. Collection and accumulation of the data. At this stage a collection and accumulation of initial data that are necessary for ontology construction is carried out.
3. Analysis of the data. This stage consists of the analysis and grouping of the assembled data and is intended for simplification of terminology construction.
4. Initial development of ontology. At this stage the preliminary ontology is formed on the basis of the selected data.
5. Specification and confirmation of ontology - Final stage of process.

For maintenance of ontologies construction process in IDEF5 special schematic language (Schematic Language-SL) exists. There is a graphic language specially intended for a statement of main data of system by the competent experts in considered area in the ontological form. SL allows to build various types of the diagrams and schemes in IDEF5. A designation of a class and designation of a separate element enter to it.

The Web-service can be transformed for the description as follows:

1. Description of Web-service, analysis of its purpose and functions, definition of its application domain.
2. Description of Web-service IOPEs (inputs, outputs, preconditions, and effects), creation of terminology for service domain.

3. Analysis of domain information. On this step it is recommended to apply considered above theoretical approaches and allocate mereological relations that are distinctive for service domain.
4. Initial development of ontology. Formation of OWL-S for service domain. At this stage the preliminary ontology is formed, and then an implementation of the analysed above properties of mereological and taxonomical relations is checked.
5. Analysis of received OWL-S and its suitability for automatic identification of Web-service semantics. Specification and confirmation of ontology

Conclusion

In work the problems connected to the ontological approach to representation of knowledge of Web-services (OWL-S) are considered (examined). However open there are questions as creations of the ontologies adequately reflecting specificity of certain subject domains, and problems connected to comparison and an establishment of conformity between various ontologies. Besides separate complex (difficult) tasks are correct configuration, orchestration and choreography of atomic Web-services, semantically suitable for the decision of the certain task, and also quality check of services.

Authors' Information

Anatoly Gladun, PhD, since 1997 works as Senior Researcher in International Research and Training Centre of Information Technologies and Systems, National Academy of Sciences and Ministry of Education of Ukraine (Glushkov Institute of Cybernetics), 44 Glushkov Pr., Kiev, 03680, Ukraine.

Phone: (+38-044)-4191812, fax (+38-044)-419-18-12, e-mail: glanat@yahoo.com

Graduated from Polytechnic University, Lviv. Post-graduate course in Sankt-Petersburg, Electrotechnical University. A specialist in the field of high-speed intelligent network, Simulation of Telecommunication Systems, mobility management, Semantic Web, Agent-based Technologies, Intelligence Networks and Web-services. 71 publications.

Julia Rogushina, PhD, since 1997 works as Senior Researcher in Institute of Software Systems, National Academy of Sciences of Ukraine, 44 Glushkov Pr., Kiev, 03680, Ukraine

Phone: (+38-044)-5684698, E-mail: _jjj_@cybergal.com

Graduated from Kyiv State T. Shevchenko University, Department of cybernetics. Post-graduate course in Glushkov Institute of Cybernetics, Kiev, Ukraine. A specialist in the field of knowledge acquisition, data mining, agent-based technologies, ontological analysis.

References

[1] Berners-Lee, T., Hendler, J. & Lassila, O. (2001). The Semantic Web. *Scientific American*, May, 2001, 34-43.

[2] Gruber, T. R. A. (1993). translation approach to portable ontology specifications // *Knowledge Acquisition*, V. *5*, – P.199-220.

[3] Horrocks, I., Patel-Schneider, P. & van Harmelen, F. (2003). From SHIQ and RDF to OWL: The making of a Web Ontology Language // Journal of Web Semantics: Science, *Services and Agents on the World Wide Web*, 1 (2003), 7-26.

[4] The Protégé Ontology Editor and Knowledge Acquisition System. Available from: - http://protege.stanford.edu/

[5] OntoEditTM Datasheet. Available from: - http://www.ontoprise.de/customercenter/-index.html

[6] Bechhofer, S., Horrocks I., Goble, C. & Stevens, R. (2001). OilEd: A reason-able ontology editor for the Semantic Web. In Proceedings of the Joint German Austrian Conference on Artificial Intelligence, volume 2174 of Lecture Notes In *Artificial Intelligence, Springer*, - P. 396–408.

[7] Prompt. Available from: - http://protege.stanford.edu/plugins/prompt/prompt.html

[8] The Chimaera Ontology Environment. Available from: - http://ksl.stanford.edu/-software/chimaera/

[9] Hearst, M. A. (1992) Automatic acquisition of hyponyms from large text corpora. In Proceedings of the 14th International Conference on Computational Linguistics. Nantes, France. Maedche, A. and Staab, S. (2001). *Ontology Learning for the Semantic Web. IEEE Intelligent Systems*, *16(2)*, 72-79.

[10] Gladun, A. & Rogushina, J. Formalization of Search Context on Base of Ontologies and Multilinguistic Thesauruses // Computing, v.*6*, issue 3, P.16-22

[11] Bisson, G., Nedellec, C. & Canamero, D. (2000). Designing Clustering Methods for Ontology Building - The Mo'K Workbench. In: S. Staab, A. Maedche, C. Nedellec, P. Wiemer-Hastings (eds.) // Proceedings of the Workshop on Ontology Learning, 14th European Conference on Artificial Intelligence ECAI'00, Berlin, Germany.

[12] Maedche, A. & Staab, S. (2001). Ontology Learning for the Semantic Web // *IEEE Intelligent Systems*, *16(2)*, 72-79.

[13] Shamsfard, M. & Barforoush A. (2004). Learning ontologies from natural language texts //*International Journal of Human-Computer Studies*, *60(1)*, 17-63.

[14] Rogushina, J. & Gladun, A. (2006). Ontological Approach to Search of Web-services in the Distributed Internet Environment //*Magazine Informatics*, №5, Minsk, Belarus.-P.134-146.

[15] Anatoly Gladun, (2007). Julia Rogushina Semantic Search of Internet Information Resources on Base of Ontologies and Multilinguistic Thesauruses // *Information Theories & Applications*, vol.14, №1, -P.48-54.

[16] Stanislaw Lesniewski. About the bases of mathematics Available from: - http://www.philosophy.ru/library/logic/lesnewxi.html (in Russian).

[17] IDEF5 Ontology Description Capture Method. Available from: - http://www.idef.com/-IDEF5.html

INDEX

A

B

C

D

E

F

G

H

I

J

K

L

M

N

O

P

Q

R

S

T

U

V

W

X

Y

Z